Advanced Pure Mathematics

R. G. Meadows and R. Delbourgo

[redacted]

Advanced Pure Mathematics

A Revision Course

R. G. Meadows and R. Delbourgo

Penguin Books

Penguin Books Ltd, Harmondsworth,
Middlesex, England
Penguin Books Inc., 7110 Ambassador Road,
Baltimore, Md 21207, USA
Penguin Books Australia Ltd, Ringwood,
Victoria, Australia

First published 1971

Made and printed in Great Britain by
Hazell Watson & Viney Ltd., Aylesbury, Bucks

Contents

Preface 7

1 Algebra 9

A Algebra: Theory summary 9
B Algebra: Illustrative worked problems 12
C Algebra: Problems 15
D Algebra: Answers and hints 19

2 Trigonometry, Complex Numbers and Hyperbolic Functions 25

A Trigonometry: Theory summary 25
B Trigonometry: Illustrative worked problems 26
C Trigonometry: Problems 27
D Complex numbers: Theory summary 29
E Hyperbolic functions: Theory summary 30
F Complex numbers and hyperbolic functions: Illustrative worked problems 31
G Complex numbers and hyperbolic functions: Problems 32
H Trigonometry: Answers and hints 33
I Complex numbers and hyperbolic functions: Answers and hints 36

3 Further Topics in Algebra 39

A Determinants: Theory summary 39
B Matrices: Theory summary 39
C Systems of linear equations: Theory summary 40
D Determinants and matrices: Illustrative worked problems 41
E Determinants and matrices: Problems 42
F Permutations, combinations and probability: Theory summary 44
G Permutations, combinations and probability: Illustrative worked problems 45
H Permutations, combinations and probability: Problems 46
I Convergence of series: Theory summary 47
J Convergence of series: Illustrative worked problems 47
K Convergence of series: Problems 49
L Determinants and matrices: Answers and hints 50
M Permutations, combinations and probability: Answers and hints 52
N Convergence of series: Answers and hints 52

4 Coordinate Geometry 55

A Coordinate geometry: Theory summary 55
B Coordinate geometry: Illustrative worked problems 61
C Coordinate geometry: Problems 62
D Further properties of triangles: Theory summary 64
E Further properties of triangles: Illustrative worked problems 65
F Further properties of triangles: Problems 66
G Coordinate geometry: Answers and hints 66
H Further properties of triangles: Answers and hints 69

5 Further Geometry 70

A Further geometry: Theory summary 70
B Further geometry: Illustrative worked problems 72
C Further geometry: Problems 73
D Further geometry: Answers and hints 74

6 Vector Analysis 76

A Vector analysis: Theory summary 76
B Vector analysis: Illustrative worked problems 79
C Vector analysis: Problems 81
D Vector analysis: Answers and hints 82

7 Differentiation and Integration 84

A Differentiation: Theory summary 84
B Differentiation: Illustrative worked problems 86
C Differentiation: Problems 88
D Integration: Theory summary 89
E Integration: Illustrative worked problems 91
F Integration: Problems 93
G Mixed integration and differentiation: Problems 95
H Differentiation: Answers and hints 96
I Integration: Answers and hints 98
J Mixed integration and differentiation: Answers and hints 101

8 Further Calculus Topics 104

A Further differentiation: Theory summary 104
B Further differentiation: Illustrative worked problems 104
C Further differentiation: Problems 106
D Further integration: Theory summary 106
E Further integration: Illustrative worked problems 107
F Further integration: Problems 109
G Further applications of calculus: Theory summary 110
H Further applications of calculus: Illustrative worked problems 111
I Further applications of calculus: Problems 112
J Further differentiation: Answers and hints 113
K Further integration: Answers and hints 114
L Further integration and differentiation: Answers and hints 115

9 Differential Equations 119

A Formation and solution of differential equations: Theory summary 119
B Differential equations: Illustrative worked problems 121
C Differential equations: Problems 123
D Differential equations: Answers and hints 125

Index 128

Preface

This volume is intended to provide a revision course in advanced-level mathematics, including special (scholarship) topics, for students who are preparing for examinations in mathematics of GCE or equivalent standard.

The book is divided into nine chapters, which cover most of the syllabus contents of the UK examining boards in pure mathematics. Each topic dealt with in a given chapter follows the same basic structure, being divided into four broad sections:

(a) Summary of the relevant theory, including fundamental results.

(b) Worked problems taken mainly from past papers, which illustrate some of the ways in which the theory is applied.

(c) Questions, again drawn from past papers, which the student is encouraged to work through to make sure that he has grasped the underlying theory.

(d) Answers and hints to the problems in section (c). The comments made here are directed at enabling a student in difficulty to see a means of finding a solution without necessarily asking for help from his teacher.

We would like to emphasize that the book should not be regarded as a first course on the subject but rather as providing a basis for practice and, especially, as an aid to students of varying abilities in their systematic revision of pure mathematics at advanced level. There are ample opportunities within the framework of this book for the less-gifted student as well as the advanced student to consolidate their understanding of basic principles and to gain practice and confidence in the solution of problems of examination standard.

We should like to thank the following examination boards for permission to use their questions:

London University Schools Examination Department	[L]
Oxford Delegacy of Local Examinations	[O]
University of Cambridge Local Examinations Syndicate	[C]
Oxford and Cambridge Schools Examination Board	[O & C]
Associated Examining Board	[AEB]
Joint Matriculation Board	[JMB]
Welsh Joint Education Committee	[W]

The worked problems and all answers and hints supplied to questions are the sole responsibility of the authors.

Chapter One
Algebra

A Algebra: Theory summary

1 *Indices*

$$a^m \times a^n = a^{m+n},$$
$$a^m \div a^n = a^{m-n},$$
$$(a^m)^n = a^{mn},$$
$$a^0 = 1,$$

where a is any positive number.

2 *Logarithms*

If $N = a^x$, then by definition $\log_a N = x$, where a is the base of the logarithm.

Basic properties of logarithms:

(a) $\log_a x + \log_a y = \log_a xy,$

$$\log_a x - \log_a y = \log_a \frac{x}{y},$$

$$\log_a x^n = n \log_a x.$$

(b) For a change of base, use

$$\log_a b = \frac{\log_c b}{\log_c a}$$

and, particularly, $\log_e x = \dfrac{\log_{10} x}{\log_{10} e}$, $\quad \log_{10} e = 0{\cdot}4343.$

(c) *Series expansions.*

$$\log_e(1+x) = x - \frac{x^2}{2} + \frac{x^3}{3} - \ldots - (-1)^n \frac{x^n}{n} + \ldots \quad (-1 < x \leqslant 1),$$

$$\log_e(1-x) = -x - \frac{x^2}{2} - \frac{x^3}{3} - \ldots - \frac{x^n}{n} \ldots \quad (-1 \leqslant x < 1).$$

$$e^x = 1 + x + \frac{x^2}{2!} + \frac{x^3}{3!} + \ldots + \frac{x^n}{n!} + \ldots, \quad \text{for all values of } x,$$

where $e = 1 + 1 + \dfrac{1}{2!} + \dfrac{1}{3!} + \ldots = 2{\cdot}71828\ldots.$

(d) *Application to determine a power-law relationship from experimental data.* For example if the x–y data indicates a law of the form $y = Ax^n$, then

$$\log y = n \log x + \log A,$$

and hence by plotting $\log y$ against $\log x$ we should obtain a straight line of slope n and intercept $\log A$.

3 *Binomial theorem,*

(a) *Binomial series expansions.*

$$(a+x)^n = a^n + {}^nC_1 a^{n-1}x + {}^nC_2 a^{n-2}x^2 + \ldots + {}^nC_r a^{n-r}x^r + \ldots + x^n,$$

where n is a positive integer. Note

$${}^nC_r = \frac{n!}{(n-r)!\,r!}$$

and in modern notation is often denoted as $\dbinom{n}{r}$.

$$(1+x)^n = 1 + nx + \frac{n(n-1)}{2!}x^2 + \frac{n(n-1)(n-2)}{3!}x^3 + \ldots.$$

If n is a positive integer the series is finite, ending in the term in x^n. Otherwise the series is infinite and valid only for $-1 < x < 1$.

(b) *Some relations between the coefficients in the binomial expansion.*

$${}^nC_r = {}^nC_{n-r}.$$
$${}^nC_0 + {}^nC_1 + \ldots + {}^nC_n = 2^n.$$
$${}^nC_0 + {}^nC_2 + {}^nC_4 + \ldots = {}^nC_1 + {}^nC_3 + {}^nC_5 + \ldots = 2^{n-1}.$$

4 *Partial fractions*

(a) The following forms illustrate the resolution of an algebraic fraction into its partial fractions.

(i) $$F(x) = \frac{ax+b}{(x-\alpha)(x-\beta)} = \frac{A}{x-\alpha} + \frac{B}{x-\beta}.$$

(ii) $$F(x) = \frac{ax^2+bx+c}{(x-\alpha)(x^2+dx+e)} = \frac{A}{x-\alpha} + \frac{Bx+C}{x^2+dx+e}.$$

(iii) $$F(x) = \frac{ax^2+bx+c}{(x-\alpha)(x-\beta)^2} = \frac{A}{x-\alpha} + \frac{B}{x-\beta} + \frac{C}{(x-\beta)^2}.$$

(iv) When the degree of the numerator is equal to or greater than the degree of the denominator first divide out.

e.g. $$\frac{3x^4+2x+7}{(x-2)x^2} = 3x + 6 + \frac{12x^2+2x+7}{(x-2)x^2}$$
$$= 3x + 6 + \frac{A}{x-2} + \frac{B}{x} + \frac{C}{x^2}.$$

(b) *Calculation of coefficients in partial fractions.*

(i) By multiplying the identity throughout by the denominators in the l.h.s. and equating the coefficients of powers of x which result on both sides.

e.g. in (a)iv, $12x^2+2x+7 = Ax^2+Bx(x-2)+C(x-2)$

and equating respective coefficients on both sides we obtain

$12 = A+B, \quad 2 = -2B+C, \quad 7 = -2C.$

(ii) 'Cover up' rule. This rule determines, for example, A and B in (a)i, A in (a)ii, and A and C in (a)iii as follows:

In (a)i, ii and iii, $A = \lim_{x\to\alpha} \{F(x)(x-\alpha)\}.$

In (a)i $\quad B = \lim_{x\to\beta} \{F(x)(x-\beta)\}.$

In (a)iii $\quad C = \lim_{x\to\beta} \{F(x)(x-\beta)^2\}.$

The remaining coefficients may be determined by means of method (b)i.

5 *Proof of an identity by mathematical induction*

If we are required to verify the truth of an identity, concerning an integer n, holding for all values of n, then

(a) we assume it to hold for $n = N$ and show that it also holds for $n = N+1$, and

(b) we check that the identity holds for $n = 1$.

If both (a) and (b) are satisfied, the principle of mathematical induction enables us to conclude that the given identity is valid for all values of the integer n.

6 *Finite series: Some standard results*

(a) The sum of the arithmetic series $a+(a+d)+(a+2d)+ \dots$ to n terms is

$$\frac{n}{2}\{2a+(n-1)d\}.$$

(b) The sum of the geometric series $a+ar+ar^2+ \dots$ to n terms is

$$\frac{a(1-r^n)}{1-r}.$$

(c) The sum of the squares of the first n natural numbers is

$$\sum_{r=1}^{n} r^2 = \frac{n}{6}(n+1)(2n+1).$$

(d) The sum of the cubes of the first n natural numbers is

$$\sum_{r=1}^{n} r^3 = \left[\frac{n}{2}(n+1)\right]^2 = \left[\sum_{r=1}^{n} r\right]^2.$$

7 *Finite series: Some standard methods for summing $\sum_{r=1}^{n} u_r$, where u_r is in general a function of r*

(a) Find a sequence v_r such that $u_r = v_{r+1}-v_r$, then

$$\sum_{r=1}^{n} u_r = v_{n+1}-v_1,$$

on cancelling intermediate terms in the summation. Examples: if $u_r = r(r+1)$, use $v_r = \frac{1}{3}r(r+1)(r+2)$; if $u_r = \cos(r\alpha+\beta)$, use

$$v_r = \frac{\sin[(r-\frac{1}{2})\alpha+\beta]}{2\sin\frac{1}{2}\alpha}.$$

(b) Express u_r as a sum of terms $v_r+w_r+ \dots$, where $\sum v_r$, $\sum w_r$, etc. are summable. Example:

$$\sum_{r=1}^{n} r(r+3) = \sum_{r=1}^{n} r^2 + 3\sum_{r=1}^{n} r,$$

then use results 1.A.6(c) and (a) respectively.

(c) If u_r is an algebraic rational function, express u_r as a sum of partial fractions.

E.g. $\quad u_r = \dfrac{2}{r(r+2)} = \dfrac{1}{r} - \dfrac{1}{r+2},$

so then $\quad \displaystyle\sum_{r=1}^{n} u_r = \frac{3}{2} - \frac{1}{n+1} - \frac{1}{n+2}.$

8 *Recurrence relations in sequences*

If $u_1, u_2, \dots, u_n, \dots$ are terms of a sequence and are related by

$u_{r+1} = au_r+b,$

where a and b are constants, and $r = 1, 2, \dots,$

then, if $a = 1, \quad u_r = u_1+(r-1)b,$

but if $a \neq 1, \quad u_r = u_1 a^{r-1} + \dfrac{b(1-a^{r-1})}{1-a}.$

Instead, if the same sequence has the recurrence relation

$u_r + au_{r-1} + bu_{r-2} = 0,$

and if λ, μ are the roots of $u^2+au+b = 0$,

then, if $\lambda = \mu, \quad u_r = \lambda^n(u_1+Ar),$

but if $\lambda \neq \mu, \quad u_r = B\lambda^r + C\mu^r,$

where A, B, C are general constants.

9 *The remainder theorem*

(a) If a polynomial $f(x)$ is divided by $x-a$ the remainder is $f(a)$

(b) If $f(a) = 0$, then $x-a$ is a factor of $f(x)$ and $x = a$ is a root of $f(x) = 0$.

10 *Quadratic and cubic equations*

(a)

(i) The general quadratic equation $ax^2+bx+c = 0$ factorizes into $a(x-\alpha)(x-\beta) = 0$, where, by comparison of coefficients of powers of x,

sum of roots, $\quad \alpha+\beta = -\dfrac{b}{a},$

product of roots, $\quad \alpha\beta = \dfrac{c}{a}.$

(ii) The roots of the above quadratic equation are given by the formula

$$x = \frac{-b \pm \sqrt{(b^2-4ac)}}{2a}.$$

(b) The general cubic equation $ax^3+bx^2+cx+d=0$ factorizes into $a(x-\alpha)(x-\beta)(x-\gamma)=0$, where

$$\alpha+\beta+\gamma=-\frac{b}{a},$$

$$\alpha\beta+\beta\gamma+\gamma\alpha=\frac{c}{a},$$

$$\alpha\beta\gamma=-\frac{d}{a}.$$

11 *General factorization of polynomials*
Any polynomial of degree n,

i.e. $a_0x^n+a_1x^{n-1}+\ldots+a_{n-1}x+a_n \quad (a_0 \neq 0)$

can be uniquely factorized into the form

$a_0(x-\alpha_1)(x-\alpha_2)\ldots(x-\alpha_n)$,

where $\sum_{r=1}^{n}\alpha_r=-\frac{a_1}{a_0} \quad \sum\alpha_r\alpha_s=\frac{a_2}{a_0}, \quad \sum\alpha_r\alpha_s\alpha_t=-\frac{a_3}{a_0},$ etc.

12 *Graphical location and numerical evaluation of the roots of an equation, $f(x)=0$*

(a) If $a<b$ and $f(a)<0<f(b)$ or $f(a)>0>f(b)$, then $f(x)=0$ has at least one root between a and b.

(b) *Descartes' rule of signs.* If $f(x)$ is a polynomial, the equation $f(x)=0$ cannot have more positive roots than there are changes of sign in $f(x)$ or more negative roots than there are changes of sign in $f(-x)$.

(c) *Newton's approximation.* If x_0 is an approximate root of $f(x)=0$, then

$$x_1=x_0-\frac{f(x_0)}{f'(x_0)}$$

is generally a better approximation. This formula may be repeated to improve the approximation, that is, at the nth stage in the iteration,

$$x_{n+1}=x_n-\frac{f(x_n)}{f'(x_n)}.$$

The formula gives convergence to the root provided $|f(x)f''(x)|<\{f'(x)\}^2$ in a range covering the root.

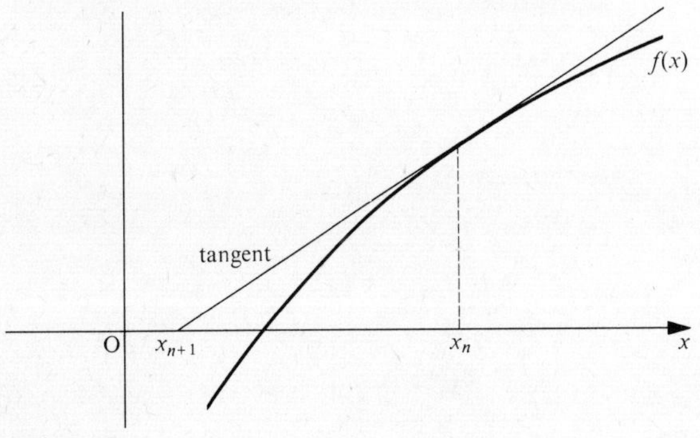

Figure 1

(d) An equation $x-g(x)=0$ can be solved iteratively by using $x_{n+1}=g(x_n) \quad (n=0,1,2,\ldots)$,

provided $|g(x)|<1$ in a range covering the root.

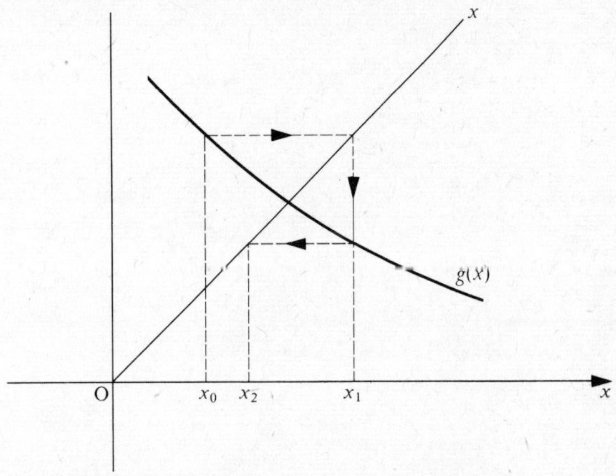

Figure 2

13 *Inequalities*

(a) If $a>0$, then $-a<0$; If $a<b$, then $-a>-b$.

(b) If $a<b$ and $c>0$, then $ca<cb$; but if $c<0$, then $ca>cb$.

(c) If $a<b$ and $b<c$, then $a<c$.

(d) For any two numbers a, b,

$$|a\pm b|\leqslant|a|+|b|$$

and $$|a\pm b|\geqslant\big||a|-|b|\big|.$$

These are referred to as the triangle inequalities.

(e) If a and b are non-negative real numbers,

$\frac{1}{2}(a+b)\geqslant\sqrt{(ab)}$,

that is, the arithmetic mean is always greater than or equal to the geometric mean.

In general, if $a_1, a_2, \ldots, a_n$ are non-negative numbers,

(f) $\frac{1}{n}(a_1+a_2+\ldots+a_n)\geqslant\sqrt[n]{(a_1a_2\ldots a_n)}$.

(i) The inequality $ax+by<c$ defines the region below the line $ax+by=c$ if $b>0$ (see Figure 3a), but above the same line if $b<0$ (Figure 3b).

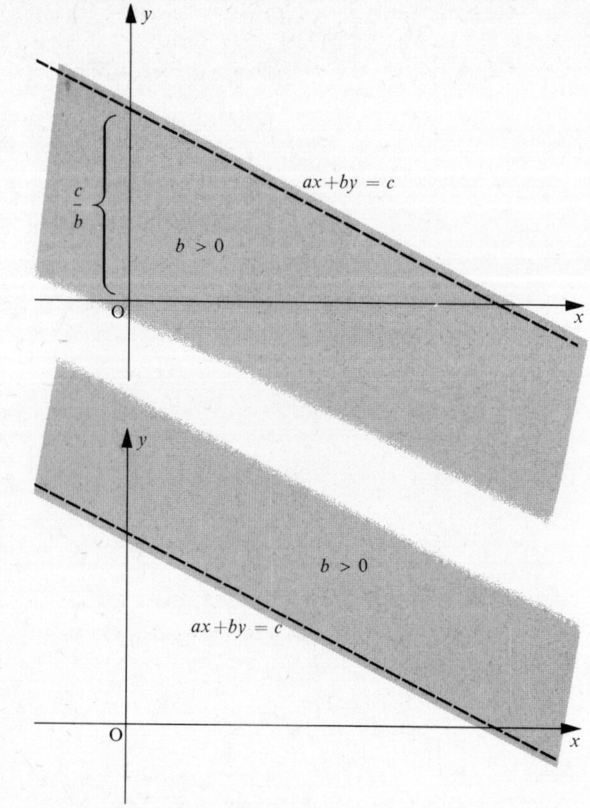

Figure 3

(ii) The inequality $ax+by \geqslant c$ defines the region below and on the line $ax+by = c$ if $b < 0$ but above and on the same line if $b > 0$.

B Algebra: Illustrative worked problems

1

(i) If $\log_x a = 5$ and $\log_x 3a = 9 (x > 0)$, find a and x.

(ii) Resolve

$$\frac{9}{1-x-20x^2}$$

into partial fractions and hence find the coefficient of x^n in the expansion of the expression in ascending powers of x.

(iii) Find the coefficient of x^{2n} in the expansion of

$$\left(\frac{a+x}{a-x}\right)^{\frac{1}{2}}$$

in ascending powers of x for $-a < x < a$. [L, J 1968, P II, Q2]

Solution.

(i) If $\log_x a = 5$, then

$$a = x^5, \qquad \textbf{1.1}$$

and if $\log_x 3a = 9$,

$$3a = x^9. \qquad \textbf{1.2}$$

On dividing equation **1.1** by **1.2** we obtain $x^4 = 3$. Hence $x = 3^{\frac{1}{4}}$ and $a = 3^{\frac{5}{4}}$.

(ii) Let $\dfrac{9}{1-x-20x^2} \equiv \dfrac{9}{(1-5x)(1+4x)} \equiv \dfrac{A}{1-5x}+\dfrac{B}{1+4x}$.

On equating numerators of both sides of the identity, we obtain

$$9 \equiv A(1+4x)+B(1-5x). \qquad \textbf{1.3}$$

Putting $x = \frac{1}{5}$ in identity **1.3** yields $A = 5$, and putting $x = -\frac{1}{4}$ gives $B = 4$.

$$\begin{aligned}\text{Hence}\quad \frac{9}{1-x-20x^2} &= \frac{5}{1-5x}+\frac{4}{1+4x}\\ &= 5(1-5x)^{-1}+4(1+4x)^{-1}\\ &= 5\{1+5x+5^2x^2+\ldots+5^nx^n+\ldots\}+\\ &\quad +4\{1-4x+4^2x^2+\ldots+\\ &\quad +(-1)^n4^nx^n+\ldots\}.\end{aligned}$$

Thus the coefficient of x^n is $5^{n+1}+(-1)^n4^{n+1}$.

$$\begin{aligned}\text{(iii)}\quad \left(\frac{a+x}{a-x}\right)^{\frac{1}{2}} &= \frac{a+x}{(a^2-x^2)^{\frac{1}{2}}}\\ &= \left[1+\frac{x}{a}\right]\left[1-\frac{x^2}{a^2}\right]^{-\frac{1}{2}}\\ &= \left[1+\frac{x}{a}\right]\left[1+\frac{1}{2}\left(\frac{x^2}{a^2}\right)+\frac{\frac{1}{2}\cdot\frac{3}{2}}{2!}\left(\frac{x^2}{a^2}\right)^2+\ldots+\right.\\ &\quad \left.+\frac{\frac{1}{2}\cdot\frac{3}{2}\ldots\frac{1}{2}(2n-1)}{n!}\left(\frac{x^2}{a^2}\right)^n+\ldots\right].\end{aligned}$$

Hence the coefficient of x^{2n} is

$$\frac{\frac{1}{2}\cdot\frac{3}{2}\ldots\frac{1}{2}(2n-1)}{n!}\left[\frac{1}{a^2}\right]^n = \frac{1.3.5\ldots(2n-1)}{n!\,2^na^{2n}}.$$

2

(i) Resolve the expression

$$\frac{x-2}{(x^2+1)(x-1)^2}$$

into its simplest partial fractions.

(ii) If $f(x) \equiv ax^3+bx^2+cx+d$ and $f(0) = -2$, $f(1) = -6, f(-2) = -12$, solve the equation $f(x) = 0$, given that the product of its roots is twice their sum.

[L, S 1967, P II, Q 1]

Solution.

(i) Let $\dfrac{x-2}{(x^2+1)(x-1)^2} = \dfrac{A}{x-1}+\dfrac{B}{(x-1)^2}+\dfrac{Cx+D}{x^2+1}$.

Using the 'cover up' rule,

$$B = \lim_{x\to 1}\frac{x-2}{x^2+1} = -\frac{1}{2},$$

and equating numerators of both sides

$$\begin{aligned}x-2 &= A(x-1)(x^2+1)+B(x^2+1)+(Cx+D)(x-1)^2\\ &= (A+C)x^3+(-A+B-2C+D)x^2+(A+C-2D)x+\\ &\quad +(-A+B+D).\end{aligned}$$

Hence, on comparing corresponding powers of x on both sides,

$$A+C = 0, \qquad -A-\tfrac{1}{2}-2C+D = 0, \qquad A+C-2D = 1,$$
$$-2 = -A-\tfrac{1}{2}+D.$$

Thus $A = -C = 1, \qquad D = -\frac{1}{2}$

and therefore $\dfrac{x-2}{(x^2+1)(x-1)^2} = \dfrac{1}{x-1} - \dfrac{1}{2(x-1)^2} - \dfrac{2x+1}{2(x^2+1)}$.

(ii) Now $f(0) = [ax^3+bx^2+cx+d]_{x=0} = d = -2;$

$f(1) = a+b+c-2 = -6$

i.e. $a+b+c = -4;$ **1.4**

$f(-2) = -8a+4b-2c-2 = -12,$

i.e. $-8a+4b-2c = -10.$ **1.5**

The product of the roots is $-\dfrac{d}{a}$. and their sum is $-\dfrac{b}{a}$;

hence $-\dfrac{d}{a} = -\dfrac{2b}{a}, \qquad b = \frac{1}{2}d = -1.$

On solving equations **1.4** and **1.5** for a and c, we have $a = 2$ and $c = -5$.

Thus $f(x) = 2x^3-x^2-5x-2.$

We now use the remainder theorem to find the factors of $f(x)$ and hence the roots of $f(x) = 0$.

$f(-1) = -2-1+5-2 = 0,$

$x = -1$ is a root;

$f(2) = 16-4-10-2 = 0,$

$x = 2$ is a root;

and if γ is the third root,

$-1+2+\gamma = -\dfrac{b}{a} = \dfrac{1}{2},$

$\gamma = -\frac{1}{2}.$

3

An arithmetic progression $a_1, a_2, \ldots, a_n$ is given, where all the terms are positive. Prove that

(i) $\dfrac{1}{a_1a_2}+\dfrac{1}{a_2a_3}+\ldots+\dfrac{1}{a_{n-1}a_n} = \dfrac{n-1}{a_1a_n},$

(ii) $\dfrac{1}{\sqrt{a_1}+\sqrt{a_2}}+\dfrac{1}{\sqrt{a_2}+\sqrt{a_3}}+\ldots+\dfrac{1}{\sqrt{a_{n-1}}+\sqrt{a_n}} = \dfrac{n-1}{\sqrt{a_1}+\sqrt{a_n}}.$

[O & C, S 1967 M & HM I, Q 2]

Solution.

Let $a_n - a_{n-1} = \ldots = a_3-a_2 = a_2-a_1 = x.$

x is the 'common difference' for the arithmetic progression of positive terms.

(i) $\dfrac{1}{a_ra_{r+1}} = \left(\dfrac{1}{a_r}-\dfrac{1}{a_{r+1}}\right)\dfrac{1}{x}, \quad$ since $a_{r+1}-a_r = x.$

Hence

$$\frac{1}{a_1a_2}+\frac{1}{a_2a_3}+\ldots+\frac{1}{a_{n-1}a_n}$$
$$= \frac{1}{x}\left[\left(\frac{1}{a_1}-\frac{1}{a_2}\right)+\left(\frac{1}{a_2}-\frac{1}{a_3}\right)+\ldots+\left(\frac{1}{a_{n-1}}-\frac{1}{a_n}\right)\right]$$
$$= \frac{1}{x}\left(\frac{1}{a_1}-\frac{1}{a_n}\right) = \frac{a_n-a_1}{xa_1a_n} = \frac{(n-1)x}{x(a_1a_n)} = \frac{n-1}{a_1a_n}.$$

(ii) $\dfrac{1}{\sqrt{a_{r-1}}+\sqrt{a_r}} = \dfrac{\sqrt{a_{r-1}}-\sqrt{a_r}}{a_{r-1}-a_r} = \dfrac{\sqrt{a_{r-1}}-\sqrt{a_r}}{-x}.$

Hence

$$\frac{1}{\sqrt{a_1}+\sqrt{a_2}}+\frac{1}{\sqrt{a_2}+\sqrt{a_3}}+\ldots+\frac{1}{\sqrt{a_{n-1}}+\sqrt{a_n}}$$
$$= -\frac{1}{x}\{(\sqrt{a_1}-\sqrt{a_2})+(\sqrt{a_2}-\sqrt{a_3})+\ldots+(\sqrt{a_{n-1}}-\sqrt{a_n})\}$$
$$= -\frac{1}{x}\{\sqrt{a_1}-\sqrt{a_n}\} = \frac{a_1-a_n}{-x(\sqrt{a_1}+\sqrt{a_n})} = \frac{-(n-1)x}{-x(\sqrt{a_1}+\sqrt{a_n})}$$
$$= \frac{n-1}{\sqrt{a_1}+\sqrt{a_n}}.$$

4

(i) If α, β, γ are the roots of the equation $x^3+ax+b = 0$, prove that

$$\alpha^6+\beta^6+\gamma^6 = \tfrac{1}{3}(\alpha^3+\beta^3+\gamma^3)^2+\tfrac{1}{2}(\alpha^2+\beta^2+\gamma^2)(\alpha^4+\beta^4+\gamma^4).$$

(ii) Find the sum of the first n terms of the series

$1^3+3^3+5^3+\ldots+(2n-1)^3+\ldots.$

(iii) Find the sum to infinity of the series

$1+\dfrac{2^3}{2!}+\dfrac{3^3}{3!}+\ldots+\dfrac{n^3}{n!}+\ldots.$ [L, S 1967, PS, Q 1]

Solution.

(i) As $(x-\alpha)(x-\beta)(x-\gamma)$

$= x^3-(\alpha+\beta+\gamma)x^2+(\alpha\beta+\beta\gamma+\gamma\alpha)x-\alpha\beta\gamma$

$\equiv x^3+ax+b,$

we have $\alpha+\beta+\gamma = 0, \qquad \alpha\beta+\beta\gamma+\gamma\alpha = a, \qquad \alpha\beta\gamma = -b.$

Now as $x^3+ax+b = 0,$ **1.6**

$x^6+ax^4+bx^3 = 0,$ **1.7**

and on substituting $x = \alpha, \beta, \gamma$ successively in equation **1.7** and adding the three individual equations thereby obtained, we have

$(\alpha^6+\beta^6+\gamma^6)+a(\alpha^4+\beta^4+\gamma^4)+b(\alpha^3+\beta^3+\gamma^3) = 0.$ **1.8**

Likewise on repeating the same procedure with **1.6**,

$(\alpha^3+\beta^3+\gamma^3)+a(\alpha+\beta+\gamma)+3b = 0$

and as $\alpha+\beta+\gamma = 0,$

$b = -\frac{1}{3}(\alpha^3+\beta^3+\gamma^3).$ **1.9**

Also $a = \alpha\beta+\beta\gamma+\gamma\alpha = \frac{1}{2}\{(\alpha+\beta+\gamma)^2-(\alpha^2+\beta^2+\gamma^2)\}$

$= -\frac{1}{2}(\alpha^2+\beta^2+\gamma^2).$ **1.10**

Hence on substituting in equation **1.8** for a and b using equations **1.9** and **1.10** we obtain

$$\alpha^6+\beta^6+\gamma^6 = \tfrac{1}{3}(\alpha^3+\beta^3+\gamma^3)^2+\tfrac{1}{2}(\alpha^2+\beta^2+\gamma^2)(\alpha^4+\beta^4+\gamma^4).$$

(ii) Remembering that $\sum_1^n r^3 = \frac{1}{4}n^2(n+1)^2$ we have

$1^3+2^3+3^3+\ldots+(2n-1)^3+(2n)^3 = \sum_1^{2n} r^3$

$= \frac{1}{4}(2n)^2(2n+1)^2,$ **1.11**

$2^3+4^3+\ldots+(2n)^3 = 2^3(1+2^3+\ldots+n^3)$

$= 8\sum_1^n r^3 = 2n^2(n+1)^2.$ **1.12**

Hence, subtracting equation **1.12** from equation **1.11**, we obtain

$1^3+3^3+5^3+\ldots+(2n-1)^3 = \frac{1}{4}(2n)^2(2n+1)^2-2n^2(n+1)^2$

$= n^2(2n^2-1).$

(iii) $1+\frac{2^3}{2!}+\frac{3^3}{3!}+\ldots+\frac{n^3}{n!}+\ldots$

$$= \sum_1^{\infty} \frac{n^3}{n!}$$

$$= \sum \frac{n^2}{(n-1)!}$$

$$= \sum \frac{(n-1)(n-2)+3(n-1)+1}{(n-1)!}$$

$$= \sum \left[\frac{1}{(n-3)!}+\frac{3}{(n-2)!}+\frac{1}{(n-1)!}\right]$$

$$= e+3e+e = 5e.$$

5

A sequence is defined by the relation

$$u_{n+1} = \tfrac{1}{2}(1+u_n^2) \quad (n \geqslant 1).$$

If u_n tends to a finite limit as n tends to infinity, find this limit.

If $u_n \neq 1$, prove that $u_{n+1} > u_n$. (You should make sure that your proof covers the cases $u_n > 1$ and $u_n < 1$.)

Hence, or otherwise, determine the behaviour of u_n as n tends to infinity when (i) $0 \leqslant u_1 < 1$, (ii) $u_1 > 1$.

[O & C, S 1967, M & H M V (PS), Q 9]

Solution. The sequence $u_1, u_2, \ldots, u_n, \ldots$ has

$$u_{n+1} = \tfrac{1}{2}(1+u_n^2) \quad (n \geqslant 1).$$

If u_n tends to a limit l, that is, $u_n \to l$, then $u_{n+1} \to l$ also, and the limit l must satisfy

$$l = \tfrac{1}{2}(1+l^2),$$

i.e. $l^2-2l+1 = 0$

$$l = 1 \quad \text{(double root)}.$$

Now $u_{n+1}-u_n = \tfrac{1}{2}(1+u_n^2)-u_n$

$= \tfrac{1}{2}(u_n^2-2u_n+1)$

$= \tfrac{1}{2}(u_n-1)^2 > 0$ if $u_n > 1$ or if $u_n < 1$,

but $u_{n+1}-u_n = 0$ if $u_n = 1$ (i.e. $u_{n+1} = u_{n+2} = \ldots = 1$).

Hence $u_{n+1} > u_n$ unless $u_n = 1$.

(i) When $0 \leqslant u_1 < 1$, from above,

$$u_1 < u_2 < u_3 < \ldots.$$

Also, if $u_n < 1$, $u_{n+1} < \tfrac{1}{2}(1+1) = 1$,

i.e. $u_{n+1} < 1$, so that all the terms of the sequence, which is increasing, are less than one. Hence the sequence tends to a limit, and this limit is one by the first part above.

(ii) When $u_1 > 1$, again

$$u_1 < u_2 < u_3 < \ldots,$$

so $u_n > 1$ and $\lim u_n$, if it exists, must be greater than u_1, which is greater than one. This is impossible from the above, hence u_n increases without limit.

6 Show that if a_0 is an approximate root of the equation $f(x) = 0$, then

$$a_0-\frac{f(a_0)}{f'(a_0)}$$

is a better approximation.

If $f(x) = x^2-b$ and a_0 is an approximate solution of the equation $f(x) = 0$, show that a closer approximation is

$$\frac{a_0^2+b}{2a_0}.$$

Show that a second approximation is

$$\frac{a_0^4+6a_0^2b+b^2}{4a_0(a_0^2+b)}.$$

Deduce that $\sqrt{27}$ is very nearly $\frac{1351}{260}$.

[L, S1967, FM VI, Q 7]

Solution. Let the exact root of $f(x) = 0$ be $x = a_0+h$,

then $f(a_0+h) = 0$,

but $f(a_0+h) \simeq f(a_0)+f'(a_0)h$

(neglecting terms in $h^2, h^3, \ldots$ etc.),

hence $f(a_0)+f'(a_0)h \simeq 0$ or $h \simeq -\frac{f(a_0)}{f'(a_0)}$.

Thus a better approximation for the root is

$$a_0-\frac{f(a_0)}{f'(a_0)}. \quad \textbf{1.13}$$

In this case $f(a_0) = a_0^2-b$ and $f'(x) = 2x$, therefore $f'(a_0) = 2a_0$ and using expression **1.13** a closer approximation than a_0 is

$$a_0-\frac{a_0^2-b}{2a_0} = \frac{a_0^2+b}{2a_0}.$$

A second approximation may be obtained by re-applying expression **1.13** with a_0 replaced by $(a_0^2+b)/2a_0$. This second approximation is

$$\frac{a_0^2+b}{2a_0}-\frac{\{(a_0^2+b)/2a_0\}-b}{2(a_0^2+b)/2a_0} = \frac{a_0^4+6a_0^2b+b^2}{4a_0(a_0^2+b)} \quad \textbf{1.14}$$

on simplifying.

To find $\sqrt{27}$, we should solve $x^2-27 = 0$. Thus on comparing the latter with the equation $x^2-b = 0$, we have $b = 27$ and choosing the approximate root $a_0 = 5$ we can use formula **1.14** directly,

i.e. $\sqrt{27} \simeq \frac{5^4+6.5^2.27+27^2}{20(5^2+27)} = \frac{1351}{260}$.

7 It is believed that the variables x, y, z are connected by a relationship of the form $z = Ax^my^n$, where A, m, n are constants. Show graphically that the values given in the following tables support this belief and find values for A, m, n.

$x = 0{\cdot}45$	y	175	603	1130	1850
	z	65·5	139	204	276

$x = 5{\cdot}6$	y	175	603	1130	1850
	z	21 600	45 960	67 420	91 070

[L, S1967, PS, Q 3]

Solution. If the variables are connected by $z = Ax^m y^n$, then

$\log z = n \log y + (\log A + m \log x)$.

Thus if the graph of $\log_{10} z$ against $\log_{10} y$ is plotted we should obtain:

(a) for $x = 0{\cdot}45$, a straight line of slope n and intercept

$c_1 = \log_{10} A + m \log_{10} 0{\cdot}45$; **1.15**

(b) for $x = 5{\cdot}6$, a straight line of slope n and intercept

$c_2 = \log_{10} A + m \log_{10} 5{\cdot}6$. **1.16**

It is seen from Figures 4(a) and 4(b) that straight-line graphs of equal slopes are indeed obtained, tending to support the validity of $z = Ax^m y^n$ for the data provided.

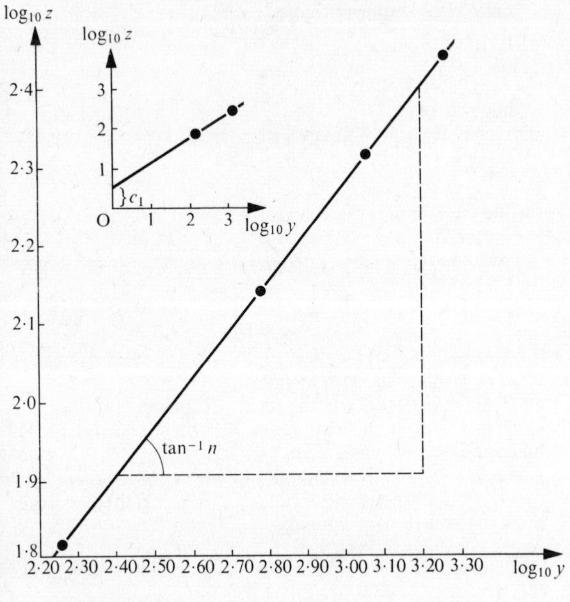

Figure 4(a) $\text{Log}_{10} z$ against $\log_{10} y$ for $x = 0{\cdot}45$

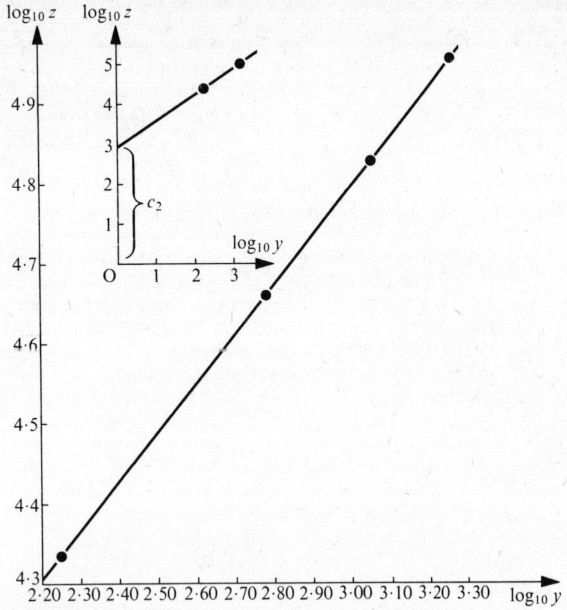

Figure 4(b) $\text{Log}_{10} z$ against $\log_{10} y$ for $x = 5{\cdot}6$

From graph (a), $$n = \frac{2{\cdot}4 - 1{\cdot}91}{3{\cdot}2 - 2{\cdot}4} \simeq 0{\cdot}61$$

$$= \frac{2{\cdot}4 - c_1}{3{\cdot}2},$$

$$c_1 \simeq 0{\cdot}45.$$

From graph (b), $$n = \frac{4{\cdot}918 - 4{\cdot}428}{3{\cdot}2 - 2{\cdot}4} \simeq 0{\cdot}61$$

$$= \frac{4{\cdot}918 - c_2}{3{\cdot}2},$$

$$c_2 \simeq 2{\cdot}97.$$

Solving for m and $\log_{10} A$ using the above values and equations **1.15** and **1.16**, we obtain

$$m = \frac{c_2 - c_1}{\log_{10} 5{\cdot}6 - \log_{10} 0{\cdot}45} \simeq 2{\cdot}30.$$

$$\log_{10} A = \frac{c_1 \log_{10} 5{\cdot}6 - c_2 \log_{10} 0{\cdot}45}{\log_{10} 5{\cdot}6 - \log_{10} 0{\cdot}45} \simeq 1{\cdot}246,$$

$$A \simeq 17{\cdot}6.$$

C Algebra: Problems

Answers and hints will be found on pp. 19–24.

1

(i) If the equation $2x^2 - qx + r = 0$ has roots $\alpha + 1, \beta + 2$, where α, β are the real roots of the equation $x^2 - bx + c = 0$ and $\alpha \geqslant \beta$, find q, r in terms of b, c.

In the case $\alpha = \beta$, show that $q^2 = 4(2r+1)$.

(ii) If a and b are both positive and unequal, and

$\log_a b + \log_b a^2 = 3$,

find b in terms of a. [L, J 1968, PI, Q1]

2

(i) The coefficient of x^2 in the binomial expansion of $(1+nx)^{\frac{5}{2}}$ is three times the coefficient of x^2 in the expansion of $(1+\frac{5}{2}x)^n$. Find the ratio of the coefficients of x^3 in the two expansions.

(ii) If $a > 1$, expand $a-\sqrt{(a^2-1)}$ in powers of $1/a$ as far as the term in $1/a^5$. [L, J 1968, P I, Q 2]

3

(i) The first term of a geometric progression is 8. The sum of its first ten terms is one-eighth of the sum of the reciprocals of these terms. Show that the sum of the first seven terms of the original geometric progression is the same as the sum of the reciprocals of the seven terms.

(ii) When x^3+ax^2+bx+c is divided by $x-1$ there is a remainder 2, when divided by $x+2$ the remainder is -1 and when divided by $x-2$ the remainder is 15. Find the remainder when the divisor is $x+1$ and hence find the three linear factors of the cubic expression. [L, J 1968, P II, Q 1]

4

(i) Prove that $\log_b a \times \log_a b = 1$.

Show that, if n is a positive integer,

$$\log_e(n+1)-\log_e n = 2\left[\frac{1}{2n+1}+\frac{1}{3(2n+1)^3}+\ldots\right].$$

Given that $\log_{10} e = 0{\cdot}43429$, find $\log_{10} 11$ to five places of decimals.

(ii) Sum to infinity the series

$$\frac{3}{2!}+\frac{7}{3!}+\ldots+\frac{n^2-n+1}{n!}+\ldots.$$ [L, J 1968, P II, Q 3]

5

(i) Find the sum of the finite series $\sum_{r=1}^{n} U_r$, where

$$U_r = \frac{r}{(2r-1)(2r+1)(2r+3)},$$

and deduce the sum to infinity.

(ii) Evaluate $\lim_{x\to 0} \frac{\cos^2 3x-\cos^2 4x}{1-\cos 5x}$.

(iii) Find the sum of the finite series $\sum_{r=1}^{n} \sin rx$. [L, S 1968, FM VI, Q 2]

6 Establish Newton's formula for obtaining a closer approximation to a real root of the equation $f(x) = 0$.

Use this method to find, correct to three significant figures, the positive root of the equation $4\cos x-2x-1 = 0$. [L, J 1968, FM VI, Q 4]

7 If $a > 0$, prove that the quadratic expression ax^2+bx+c is positive for all real values of x when $b^2 < 4ac$.

Hence find the range of values of p for which the quadratic function of x

$$f(x) \equiv 4x^2+4px-(3p^2+4p-3)$$

is positive for all real values of x.

Illustrate your result by making sketch graphs of $y = f(x)$ for each of the cases $p = 0$ and $p = 1$. [L, S 1967, P I, Q 1]

8

(i) Find an expression for $\sum_{n=0}^{\infty} ar^n$, where a and r are constants, stating the condition under which your result is valid.

Hence express $0{\cdot}3\dot{2}\dot{1}$ as a rational fraction in its lowest terms.

(ii) Show that the sum of the squares of the first n positive integers is $\frac{1}{6}n(n+1)(2n+1)$.

Find the sum to n terms of the series

$$1\times 5+2\times 6+3\times 7+\ldots.$$ [L, S 1967, P II, Q 2]

9 The following readings were obtained in an experiment:

x	10	20	30	40	50
y	1·21	3·98	7·97	13·08	19·22

It is believed that x and y satisfy a relationship of the form $y = Ax^n$, where A and n are constants. By drawing a suitable graph, show that these readings verify the relationship.

From your graph find approximate values of A and n.

[L, J 1967, P I, Q 7]

10

(i) Show that $x = 1$ is a root of the equation

$$6x^5-x^4-43x^3+43x^2+x-6 = 0$$

and solve the equation.

(ii) Find the sum of the series

$$\frac{4}{3}+\frac{9}{8}+\frac{16}{15}+\ldots+\frac{n^2}{n^2-1}.$$ [L, J 1967, P II, Q 1]

11 Expand in ascending powers of x as far as x^6

(a) $\frac{e^x+1}{2e^{\frac{1}{2}x}}$,

(b) $\frac{1}{x^2}\log_e(1-x^2)$,

(c) $\frac{1}{2x}\log_e\left[\frac{1+x}{1-x}\right]$,

stating in each case the range of values of x for which the expansion is valid.

Find the sum to infinity of the series

$$1+\left[\frac{1}{2}+\frac{1}{3}\right]\frac{1}{2^2}+\ldots+\left[\frac{1}{2n}+\frac{1}{2n+1}\right]\frac{1}{2^{2n}}+\ldots.$$

[L, J 1967, P II, Q 2]

12 If the expansions in ascending powers of x of the functions $e^{ax/(1+bx)}$ and $(1+x)^n$, where n is a given non-zero number, have the same coefficients as far as the term in x^2 inclusive, find a and b in terms of n.

If a and b take these values, show that the expansions cannot have the same coefficients of x^3, but that they have the same coefficients of x^4 if $n = \frac{3}{2}$. [O, S 1968, PM I, Q 2]

13 If

$$x^2+y^2-2(x+y)-23 = 0$$

and

$$xy-6 = 0,$$

show that $(x+y)^2-2(x+y)-35 = 0$,

and find all the solutions of the given equations.

[O, S 1967, PM I, Q 1]

14 Prove by induction, or otherwise, that the sum of the cubes of the first n natural numbers is $\frac{1}{4}n^2(n+1)^2$.

Find the sum of the cubes of the first n even numbers and the sum of the cubes of the first n odd numbers.

[O, S 1967, PM II, Q 1]

15 If the equation

$$ax^3+bx^2+cx+d = 0$$

has a pair of reciprocal roots, α and $1/\alpha$, prove that

$$a^2-d^2 = ac-bd.$$

Verify that this condition is satisfied for the equation

$$6x^3+11x^2-24x-9 = 0,$$

and solve the equation. [O, S 1967, PM II, Q 2]

16 Find the positive root of the equation $x = 2\cos x$, correct to two places of decimals, showing that your solution has this degree of accuracy. [O, S 1967, P II, Q 6]

17 Prove, by mathematical induction or otherwise, that

$$n.1^3+(n-1).2^3+(n-2).3^3+\ldots+1.n^3$$
$$=\tfrac{1}{60}n(n+1)(n+2)(3n^2+6n+1).$$ [O, S 1967, PS, Q 1]

18 Prove that

$$3xyz \equiv x^3+y^3+z^3-(x+y+z)(x^2+y^2+z^2-xy-yz-zx).$$

Given that $x+y+z = -1$,
$x^2+y^2+z^2 = 5$
and $x^3+y^3+z^3 = -7$,

find the value of $xy+yz+zx$ and of xyz. Deduce that x, y and z are the three roots of the equation $t^3+t^2-2t = 0$, and hence solve the given equations in x, y and z. [O, S 1966, P I, Q 1]

19

(i) Prove that if a, b and c are positive then

$$9abc \leqslant (a+b+c)(ab+bc+ca).$$

(ii) Prove that, if x and y are positive,

$$\log_x y = \frac{1}{\log_y x}.$$

(iii) Deduce from (i) and (ii) that, if p, q, r and n are each greater than 1, and if s be written for pqr, then

$$\log_s n \leqslant \tfrac{1}{9}(\log_p n+\log_q n+\log_r n).$$ [O, S 1966, P I, Q 2]

20 Prove that, if p and $n-2$ are positive integers, then

$$(1-x)^n\{1+2x+3x^2+\ldots+(p+1)x^p\}$$
$$\equiv (1-x)^{n-2}\{1-(p+2)x^{p+1}+(p+1)x^{p+2}\}.$$

Deduce that, if $p \leqslant n-2$,

$$\binom{n}{p}-2\binom{n}{p-1}+\ldots+(-1)^p(p+1)\binom{n}{0} = \binom{n-2}{p},$$

where $\binom{n}{p}$ denotes the coefficient of x^p in the expansion of $(1+x)^n$ in ascending powers of x. [O, S 1966, P I, Q 4]

21 Write down the expansion of e^x in ascending powers of x.
Find the values of constants a, b and c such that

$$(r+3)(r+1) \equiv a(r+2)(r+1)+b(r+2)+c.$$

Hence or otherwise show that

$$\frac{2-3x^2}{2x^2}+\frac{e^x}{x^2}(x^2+x-1) = \sum_{r=1}^{\infty}\frac{(r+3)x^r}{r!(r+2)}.$$ [O, S 1966, P I, Q 5]

22 Prove that, if $z = \dfrac{2x}{x^2+1}$ and $x > 1$,

then $$\log\frac{x^2+1}{x^2-1} = \frac{1}{2}z^2+\frac{1}{4}z^4+\frac{1}{6}z^6+\ldots$$

and $$\log\frac{x+1}{x-1} = z+\frac{1}{3}z^3+\frac{1}{5}z^5+\ldots.$$

Show that $\log 5 = 2.\tfrac{3}{5}+\tfrac{1}{2}(\tfrac{3}{5})^2+\tfrac{2}{3}.(\tfrac{3}{5})^3+\tfrac{1}{4}(\tfrac{3}{5})^4+\ldots.$

[O, S 1966, P III Sp., Q 1]

23

(i) Show that if $\log a+\log c = 2\log b$ then a, b, c are in geometric progression.
Show that if $\log x+\log z = 3\log y$ then x, y^2, yz are in geometric progression.

(ii) Show that $x-y-z$ is a factor of the expression

$$x^3+y^3+z^3-yz(y+z)-zx(z+x)-xy(x+y)+2xyz.$$

Without further working write down two other factors of this expression. [C, S 1967, M I, Q 1]

24 Show that as x varies the maximum value of the function $(a-x)(x-b)$ is $\tfrac{1}{4}(a-b)^2$.
Illustrate this result by a sketch-graph of the function when $0 < b < a$.
If $0 < b < a$ find for the equation $(a-x)(x-b) = k$,

(i) the value of k (in terms of a and b) for which the equation has equal roots,

(ii) the range of values of k for which the equation has roots whose values lie between b and a,

(iii) the range of values of k for which both roots of the equation are positive. [C, S 1967, M I, Q 3]

25 Expand the functions

$$(1+x)^p \quad\text{and}\quad \frac{1+ax}{1+bx}$$

in ascending powers of x as far as the terms in x^3.
If these expansions are identical as far as the terms in x^2, express a and b in terms of p.
By taking $p = \tfrac{2}{3}$, $x = \tfrac{1}{8}$ show that an approximation to the cube root of 81 is $\tfrac{212}{49}$.
Find, in terms of p and x, the difference between the terms in x^3 of the two expansions, and evaluate this difference, correct to one significant figure, when $p = \tfrac{2}{3}$, $x = \tfrac{1}{8}$.
[C, S 1967, M I, Q 5]

26 If x, y and z are any three real numbers, prove that

$$k(x^2+y^2+z^2)-2(yz+zx+xy)$$

can never be negative if $k \geqslant 2$.
Find conditions for the expression to vanish, distinguishing between the cases $k > 2$ and $k = 2$.
A closed rectangular box has total external surface area A, and the sum of the lengths of its twelve edges is p. The diagonal (i.e. the straight line joining the intersection of three faces to the intersection of the remaining three faces) is of length d.

Prove that $48d^2 > p^2 > 24A$

unless the box is cubical. [C, S 1967, O S, Q 1]

27 It is known that two of the four roots of the equation

$$x^4-2x^3-18x^2+px+45 = 0$$

are equal and opposite (α and $-\alpha$). Show that the other two are of the form $1+\beta$ and $1-\beta$. Find the possible values of α and β, and the two possible values of p. [C, S 1966, PM 3, Q 2]

28

Prove, by induction or otherwise, that

$$1-x+\frac{x(x-1)}{2!}-\frac{x(x-1)(x-2)}{3!}+\ldots+(-1)^n\frac{x(x-1)\ldots(x-n+1)}{n!} = (-1)^n\frac{(x-1)(x-2)\ldots(x-n)}{n!}.$$

Hence find the roots of the equation

$$1-x+\frac{x(x-1)}{2!}-\frac{x(x-1)(x-2)}{3!}+\ldots+(-1)^n\frac{x(x-1)\ldots(x-n+1)}{n!} = 0.$$

Show how the binomial theorem could be used to show that these roots are solutions of the equation.

[O & C, S 1967, M & HM I, Q 1]

29

A sequence satisfies the relation

$$u_n = u_{n-1}+u_{n-2} \quad (n \geqslant 2).$$

Prove that

(i) $u_n^2-u_{n-1}u_{n+1} = (-1)^{n-1}(u_1^2-u_0u_2)$,

(ii) $u_1u_2+u_2u_3+\ldots+u_{2n}u_{2n+1} = u_{2n+1}^2-u_1^2 \quad (n \geqslant 1)$.

[O & C, S 1967, M & HM I, Q 3]

30

Given that $k > 0, a > 0$, prove by considering the minimum value of the function $x^{-k}+(a-x)^{-k}$, that

$$\frac{1}{x^k}+\frac{1}{(a-x)^k} \geqslant \frac{2^{k+1}}{a^k} \quad \text{when } 0 < x < a.$$

Deduce from this that, if $x > 0, y > 0$,

$$\frac{1}{x^k}+\frac{1}{y^k} \geqslant \frac{2^{k+1}}{(x+y)^k}.$$

[O & C, S 1967, M & HM II, Q 1]

31

State Newton's method of approximation to a root of the equation $f(x) = 0$.

Apply the method to the case $f(x) = x^3-18x+2$, starting with $x = 4$ as the first approximation; prove that the second approximation is $x = 4{\cdot}2$.

Prove that there is a root of the equation

$$x^3-18x+2 = 0$$

between $x = 4$ and $x = 4{\cdot}2$.

[O & C, S 1967, M & HM II, Q 2]

32 If $2u_nu_{n+1}-2u_n+1 = 0\,(n \geqslant 0)$, find one set of complex numbers $a, b, c\,(c \neq 1)$ such that

$$\frac{u_{n+1}-a}{u_{n+1}-b} = c\frac{u_n-a}{u_n-b}.$$

Hence show that $u_{n+4} = u_n$.

[O & C, S 1967, M & HM V (P Sp), Q 2]

33

By considering the expression

$$(a_1x+b_1)^2+(a_2x+b_2)^2+\ldots+(a_nx+b_n)^2,$$

or otherwise, prove that

$$\sqrt{(a_1^2+a_2^2+\ldots+a_n^2)}+\sqrt{(b_1^2+b_2^2+\ldots+b_n^2)} \geqslant \sqrt{\{(a_1+b_1)^2+(a_2+b_2)^2+\ldots+(a_n+b_n)^2\}}.$$

Under what conditions does the equality sign apply?

[O & C, S 1967, M & HM V (P Sp), Q 10]

34

(a) If one of the roots of the equation $x^2+ax+b = 0$ is four times the other, and $b \neq 0$, prove that $4a^2 = 25b$.

(b) Write down the series expansions of $\log_e(1+x)$ and $\log_e(1-x)$ and deduce that for $|x| < 1$,

$$\log_e(1-x^2) = -\left[x^2+\frac{x^4}{2}+\frac{x^6}{3}+\ldots\right].$$

Deduce the sum of the series

$$\tfrac{1}{2}(\tfrac{1}{2})^2+\tfrac{1}{4}(\tfrac{1}{2})^4+\tfrac{1}{6}(\tfrac{1}{2})^6+\ldots.$$

[A E B, S 1968, P & A I, Q 1]

35

(a) The solution of the equation $x^4+x^3+x^2+x = 5$ is known to be $x = 1+h$ where h is small.

Neglecting powers of h above the first and using the binomial theorem show that the solution of the equation is $x = 1{\cdot}1$ approximately.

(b) A square $A_1B_1C_1D_1$ is of side $2a$. The midpoints of the sides are joined to form a second square $A_2B_2C_2D_2$, the midpoints of the sides of this square are joined to form a third square $A_3B_3C_3D_3$ and so on. Prove that the lengths of the sides $A_1B_1, A_2B_2, A_3B_3, \ldots$ form a geometric progression and determine the length of the side A_nB_n of the nth square.

Show that the sum of the areas of the first six squares is $\frac{63}{32}$ times as large as the area of the first square $A_1B_1C_1D_1$.

[A E B, S 1968, P & A II, Q 2]

36

(a) Find the sum to infinity of the binomial series

$$1+\tfrac{1}{3}\times\tfrac{1}{5}-\tfrac{1}{3}\times\tfrac{2}{3}\times\tfrac{1}{5}\times\tfrac{1}{10}+\tfrac{1}{3}\times\tfrac{2}{3}\times\tfrac{5}{3}\times\tfrac{1}{5}\times\tfrac{1}{10}\times\tfrac{1}{15}-\ldots.$$

(b) Write down the first three terms in the expansion of $\log_e(1+x)$ in ascending powers of x; state the coefficient of x^n and the range of values of x for which the expansion is valid.

Express $\dfrac{1}{n(n+1)}$ in partial fractions.

Hence find the sum to infinity of the series

$$\frac{1}{1\times 2}-\frac{1}{2\times 3}+\frac{1}{3\times 4}-\frac{1}{4\times 5}+\ldots$$

in terms of $\log_e 2$.

Hence or otherwise find the sum to infinity of the series

$$\frac{1}{1\times 2\times 3}+\frac{1}{3\times 4\times 5}+\frac{1}{5\times 6\times 7}+\ldots.$$

[JMB, S 1967, P I, Q 12]

37

(a) Each of two progressions, one arithmetic and the other geometric, has a for its first term and l for its nth term. The sth term is b in the arithmetic progression and c in the geometric progression. Express $b-a$ in terms of $l-a$, s and n, and express c/a in terms of l/a, s and n. Hence show that

$$\left(\frac{c}{a}\right)^{l-a} = \left(\frac{l}{a}\right)^{b-a}.$$

(b) Express $\dfrac{1}{r(r+1)(r+2)}$

in partial fractions. Hence or otherwise show that the sum

$$\frac{1}{1\times 2\times 3}+\frac{1}{2\times 3\times 4}+\ldots+\frac{1}{n(n+1)(n+2)}$$

is $\dfrac{n(n+3)}{4(n+1)(n+2)}$. [JMB, S 1966, P II, Q I]

38

(a) By writing the expression $x^2+5y^2+14z^2-16yz-4zx+2xy$ in the form $(ax+by+cz)^2+(dy+ez)^2+(fz)^2$, show that its value is never negative for real values of x, y, z.

Find the integer values of x, y, z for which the expression has the value 2.

(b) If p and q are positive, prove that

$p^p q^q \geqslant p^q q^p$. [JMB, S 1966, P S, Q 3]

39

Find $S(n) = \sum_{r=1}^{n} r^2$

in terms of n. Hence obtain a formula for

$T(n) = \sum pq$,

where p and q take all integer values satisfying $1 \leqslant p < q \leqslant n$.

Evaluate $\lim_{n\to\infty} \dfrac{T(n+1)-T(n)}{S(n)}$. [W, S 1969, P I, Q 2]

40

Given the first term, the nth term and the sum of n terms of an arithmetic progression, find the number of terms and the common difference of the progression.

A roll of adhesive tape is wound around a cylinder of diameter 83 mm. The external diameter of the complete roll is 110 mm. If the length of the tape is 65·84 m, find its approximate thickness. [W, S 1968, P I, Q 1]

41

(a) Express $\dfrac{\sqrt{5}+\sqrt{3}}{\sqrt{6}+\sqrt{2}}$

as a sum of square roots of rational numbers. Hence or otherwise evaluate this quantity as accurately as your tables permit.

(b) If $x = \log_b a$, $y = \log_c b$, $z = \log_a c$, and if $abc \neq 0$, prove that

$(z-1)\log_e a+(x-1)\log_e b+(y-1)\log_e c = 0.$

[W, S 1968, P I, Q 5]

42

Write down the relations between the roots and the coefficients of a cubic equation.

Solve the equation $x^3-6x^2+11x-6 = 0$, given that its roots are in arithmetic progression. [W, S 1968, P I, Q 7]

D Algebra: Answers and hints

1

(i) Sum of roots, $\alpha+1+\beta+2 = \alpha+\beta+3 = \dfrac{q}{2}$.

Product of roots, $(\alpha+1)(\beta+2) = \alpha\beta+2\alpha+\beta+2 = \dfrac{r}{2}$.

But, since α, β are roots of $x^2-bx+c = 0$,

$\alpha+\beta = b, \quad \alpha\beta = c, \quad \alpha = \tfrac{1}{2}\{b+\sqrt{(b^2-4c)}\}$

(since $\alpha \geqslant \beta$).

Therefore $q = 2(\alpha+\beta+3) = 2(b+3)$,

$r = 2\{\alpha\beta+(\alpha+\beta)+\alpha+2\} = 2c+3b+\sqrt{(b^2-4c)}+4.$

If $\alpha = \beta$, $b^2 = 4c$ and then $r = \tfrac{1}{2}b^2+3b+4$, hence

$8r = 4b^2+24b+32$

$= \{2(b+3)\}^2-4 = q^2-4,$

$q^2 = 4(2r+1).$

(ii) $\log_a b = \dfrac{1}{\log_b a}$.

Thus $\log_a b+\log_b a^2 = \dfrac{1}{\log_b a}+2\log_b a = 3,$

or $2(\log_b a)^2-3\log_b a+1 = (2\log_b a-1)(\log_b a-1) = 0.$

Hence $\log_b a = \tfrac{1}{2}$,

i.e. $a = b^{\frac{1}{2}}$ or $b = a^2$.

($\log_b a = 1$ would give $a = b$ and is excluded.)

2

(i) $(1+nx)^{\frac{5}{2}} = 1+\dfrac{5}{2}nx+\dfrac{5}{2}\cdot\dfrac{3}{2}\cdot\dfrac{1}{2!}n^2x^2+\dfrac{5}{2}\cdot\dfrac{3}{2}\cdot\dfrac{1}{2}\cdot\dfrac{1}{3!}n^3x^3+\ldots;$

$$\left(1+\frac{5}{2}x\right)^n = 1+n\frac{5}{2}x+\frac{n(n-1)}{2!}\frac{25}{4}x^2+\frac{n(n-1)(n-2)}{3!}\frac{125}{8}x^3+\ldots.$$

The coefficient of x^2 in the first expression is three times that in the second,

$\tfrac{15}{8}n^2 = 3\times n(n-1)\tfrac{25}{8}$,

yielding $n = \tfrac{5}{4}$, and hence ratio of coefficients of x^3 is -1.

(ii) $$a-\sqrt{(a^2-1)} = a\left[1-\sqrt{\left(1-\frac{1}{a^2}\right)}\right]$$

$$= a\left[1-\left\{1-\frac{1}{2}\cdot\frac{1}{a^2}+\frac{1}{2}\cdot\left(-\frac{1}{2}\right)\cdot\frac{1}{2!}\left(-\frac{1}{a^2}\right)^2+\frac{1}{2}\cdot\left(-\frac{1}{2}\right)\cdot\left(-\frac{3}{2}\right)\cdot\frac{1}{3!}\left(\frac{1}{a^2}\right)^3+\ldots\right\}\right]$$

$$= \frac{1}{2}\cdot\frac{1}{a}+\frac{1}{8}\cdot\frac{1}{a^3}+\frac{1}{16}\cdot\frac{1}{a^5}+\ldots.$$

3

(i) The given geometric progression is

$8(1+r+r^2+\ldots)$ **1.17**

and the reciprocal series is

$\dfrac{1}{8}\left[1+\dfrac{1}{r}+\dfrac{1}{r^2}+\ldots\right]$. **1.18**

The sum of the first ten terms of series **1.17** is

$$\frac{8(1-r^{10})}{1-r} \quad \textbf{1.19}$$

and the sum of the first ten terms of series **1.18** is

$$\frac{\frac{1}{8}(1-1/r^{10})}{1-1/r}. \quad \textbf{1.20}$$

Using sum **1.19** $= \frac{1}{8}\times$ sum **1.20** gives $r^9 = \frac{1}{512}, r = \frac{1}{2}$, giving, for the sum of the first seven terms of series **1.17**,

$$8+4+2+1+\tfrac{1}{2}+\tfrac{1}{4}+\tfrac{1}{8} = \tfrac{1}{8}+\tfrac{1}{4}+\tfrac{1}{2}+1+2+4+8,$$

the sum of the first seven terms of series **1.18**.

(ii) When x^3+ax^2+bx+c is divided by

$x-1$: remainder $\quad 2 = 1^3+a.1^2+b.1+c;$ **1.21**

$x+2$: remainder $\quad -1 = -8+4a-2b+c;$ **1.22**

$x-2$: remainder $\quad 15 = 8+4a+2b+c.$ **1.23**

Solving equations **1.21–23** gives $a = 2, b = 0, c = -1$ and the cubic is x^3+2x^2-1. The remainder when divided by $x+1$ is $-1+2-1 = 0$ so that $x+1$ is a factor.

Finally $\quad x^3+2x^2-1 = (x+1)(x^2+x-1)$

$$= (x+1)(x+\tfrac{1}{2}-\tfrac{1}{2}\sqrt{5})(x+\tfrac{1}{2}+\tfrac{1}{2}\sqrt{5}).$$

4

(i) Let $x = \log_b a$, then $b^x = a$ and $x \log_a b = \log_a a = 1$, hence $\log_b a \times \log_a b = 1$.

Use $\quad \log(1+x) = x-\frac{1}{2}x^2+\frac{1}{3}x^3\ldots,$

$\quad \log(1-x) = -x-\frac{1}{2}x^2-\frac{1}{3}x^3\ldots,$

hence $\quad \log\left[\dfrac{1+x}{1-x}\right] = 2(x+\frac{1}{3}x^3+\ldots).$

The required result follows by putting $\quad x = \dfrac{1}{2n+1}$.

By substituting $n = 10$ in the given series, we have

$$\log_e \frac{11}{10} = 2\left[\frac{1}{21}+\frac{1}{3}\cdot\frac{1}{21^3}+\ldots\right] \approx 0{\cdot}09531,$$

but $\quad \log_{10}\frac{11}{10} = \log_{10} e \log_e \frac{11}{10} \approx 0{\cdot}04139.$

Therefore $\quad \log_{10} 11 = 1{\cdot}04139.$

(ii) $$\frac{3}{2!}+\frac{7}{3!}+\ldots = \sum_2^\infty \frac{n^2-n+1}{n!}$$

$$= \sum_2^\infty \left[\frac{1}{(n-2)!}+\frac{1}{n!}\right]$$

$$= e+e-2 = 2(e-1).$$

5

(i) Resolve u_r into partial fractions, then

$$\sum_1^n u_r = \sum_1^n \frac{1}{16}\left[\frac{1}{2r-1}+\frac{2}{2r+1}-\frac{3}{2r+3}\right] = \frac{n(n+1)}{2(2n+1)(2n+3)}.$$

$$\sum_1^\infty u_r = \lim_{n\to\infty} \frac{n^2}{8n^2} = \frac{1}{8}.$$

(ii) $$\lim_{x\to 0} \frac{\cos^2 3x-\cos^2 4x}{1-\cos 5x}$$

$$= \lim_{x\to 0} \frac{(1-9x^2/2!-\ldots)-(1-16x^2/2!-\ldots)}{1-25x^2/2!-\ldots} = \frac{14}{25}.$$

(iii) Use $\quad \sin rx = v_{r+1}-v_r,$

where $\quad v_r = \dfrac{-\cos(r-\frac{1}{2})x}{2\sin\frac{1}{2}x},$

then $\quad \displaystyle\sum_1^n \sin rx = v_{n+1}-v_1 = \frac{1}{2\sin\frac{1}{2}x}\{\cos\tfrac{1}{2}x-\cos(n+\tfrac{1}{2})x\}.$

6

Write equation as $f(x) = \cos x-\frac{1}{2}x-\frac{1}{4} = 0$. An approximate positive root found by trial is $\frac{1}{4}\pi$. A better approximation using Newton's formula is

$$x_1 = \frac{1}{4}\pi-\frac{f(\frac{1}{4}\pi)}{f'(\frac{1}{4}\pi)} = 0{\cdot}8388$$

and a second time gives $x \simeq 0{\cdot}8378$, hence $x \simeq 0{\cdot}838$.

7 $\quad -\frac{3}{2} < p < \frac{1}{2}$.

8

(i) $$\sum_{n=0}^\infty ar^n = a+ar+ar^2+\ldots$$

$$= \lim_{n\to\infty} \frac{a(1-r^n)}{1-r} = \frac{a}{1-r}, \quad \text{provided } |r| < 1.$$

$$0{\cdot}3\dot{2}\dot{1} = 0{\cdot}3212121\ldots$$

$$= 0{\cdot}3+0{\cdot}021+0{\cdot}00021+\ldots$$

$$= 0{\cdot}3+\frac{0{\cdot}021}{1-0{\cdot}01} = \frac{53}{165}.$$

(ii) $$1\times 5+2\times 6+3\times 7+\ldots = \sum_1^n r(r+4)$$

$$= \sum_1^n r^2+4\sum_1^n r$$

$$= \tfrac{1}{6}n(n+1)(2n+1)+4\frac{n}{2}(n+1)$$

$$= \tfrac{1}{6}n(n+1)(2n+13).$$

9

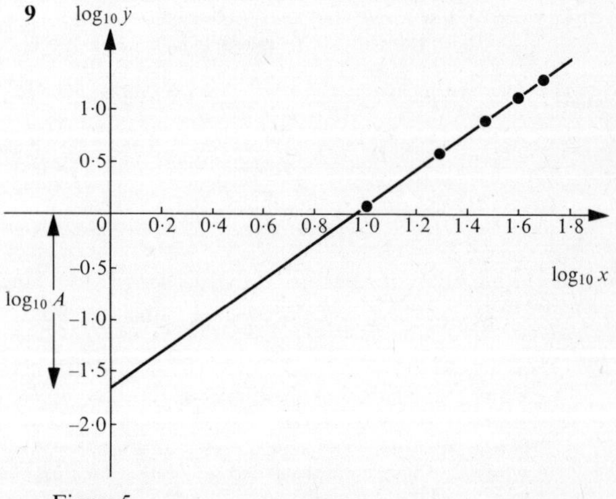

Figure 5

Plot the graph of $\log_{10} y = n \log_{10} x + \log_{10} A$ as shown in Figure 5.

Intercept $\log_{10} A = -1{\cdot}62 = \bar{2}{\cdot}38$.

Thus $A = 0{\cdot}024$.

Slope $n = 1{\cdot}72$.

10

(i) Let $f(x) = 6x^5 - x^4 - 43x^3 + 43x^2 + x - 6$,

then $f(1) = 0$ and $x = 1$ is a root of $f(x) = 0$ (remainder theorem).

Also $f(2) = f(-3) = 0$, hence $x = 2, -3$ are roots. Further, note that

$$f\left(\frac{1}{x}\right) = -x^5 f(x),$$

and so $\frac{1}{2}, -\frac{1}{3}$ are also roots. The solutions of $f(x) = 0$ are $1, 2, -3, \frac{1}{2}, -\frac{1}{3}$.

(ii) $$\sum_{r=2}^{n} \frac{r^2}{r^2-1} = \sum_{2}^{n}\left[1+\frac{1}{r^2-1}\right]$$

$$= \sum_{2}^{n}\left[1+\frac{1}{2}\left(\frac{1}{r-1}-\frac{1}{r+1}\right)\right]$$

$$= (n-1)+\frac{1}{2}+\frac{1}{4}-\frac{1}{2n}-\frac{1}{2(n+1)}$$

$$= \frac{4n^3+3n^2-5n-2}{4n(n+1)}.$$

11

(a) $$\frac{e^x+1}{2e^{\frac{1}{2}x}} = \frac{1}{2}(e^{\frac{1}{2}x}+e^{-\frac{1}{2}x})$$

$$= 1+\frac{(\frac{1}{2}x)^2}{2!}+\frac{(\frac{1}{2}x)^4}{4!}+\frac{(\frac{1}{2}x)^6}{6!}+\ldots \quad \text{for } -\infty < x < +\infty.$$

(b) $$\frac{1}{x^2}\log_e(1-x^2) = \frac{1}{x^2}\left[-x^2-\frac{x^4}{2}-\frac{x^6}{3}-\frac{x^8}{4}-\ldots\right]$$

$$= -1-\frac{x^2}{2}-\frac{x^4}{3}-\frac{x^6}{6}-\ldots \quad \text{for } -1 < x < 1.$$

(c) $$\frac{1}{2x}\log_e\left[\frac{1+x}{1-x}\right] = \frac{1}{2x}\{\log_e(1+x)-\log_e(1-x)\}$$

$$= 1+\frac{x^2}{3}+\frac{x^4}{5}+\frac{x^6}{7}+\ldots \quad \text{for } -1 < x < 1.$$

$$1+\left[\frac{1}{2}+\frac{1}{3}\right]\frac{1}{2^2}+\left[\frac{1}{4}+\frac{1}{5}\right]\frac{1}{2^4}+\ldots$$

$$= 1+\tfrac{1}{3}(\tfrac{1}{2})^2+\tfrac{1}{5}(\tfrac{1}{2})^4+\ldots+$$

$$+\tfrac{1}{2}(\tfrac{1}{2})^2[1+\tfrac{1}{2}(\tfrac{1}{2})^2+\tfrac{1}{3}(\tfrac{1}{2})^4+\ldots]$$

$$= \log_e 3+\tfrac{1}{8}[-4\log_e \tfrac{3}{4}]$$

$$= \tfrac{1}{2}\log_e 3+\log_e 2.$$

12 $a = n, \quad b = \frac{1}{2}$.

Note the exponential may be expanded as follows:

$$e^{ax/(1+bx)} = 1+ax(1+bx)^{-1}+\frac{1}{2!}a^2x^2(1+bx)^{-2}+$$

$$+\frac{1}{3!}a^3x^3(1+bx)^{-3}+\ldots$$

$$= 1+ax(1-bx+b^2x^2-b^3x^3+\ldots)+$$

$$+\tfrac{1}{2}a^2x^2(1-2bx+3b^2x^2\ldots)+\ldots$$

$$= 1+ax+(-ab+\tfrac{1}{2}a^2)x^2+$$

$$+(ab^2-a^2b+\tfrac{1}{6}a^3)x^3+\ldots.$$

13

$$(x+y)^2-2(x+y) = x^2+y^2-2(x+y)+2xy$$

$$= 23+12,$$

hence $(x+y)^2-2(x+y)-35 = 0$. **1.24**

To effect solution substitute $z = x+y$ in equation **1.24** and solve,

i.e. $z^2-2z-35 = (z-7)(z+5) = 0$.

Hence $x+y = 7$ or -5 and, using $y = 6/x$, the solutions follow:

$x = 6, \ y = 1; \qquad x = 1, \ y = 6;$

$x = -2, \ y = -3; \qquad x = -3, \ y = -2.$

14 $$2^3+4^3+\ldots+(2n)^3 = 2^3(1^3+2^3+\ldots+n^3)$$

$$= 2^3\tfrac{1}{4}n^2(n+1)^2 = 2n^2(n+1)^2.$$

$$1^3+3^3+\ldots+(2n-1)^3 = \sum_{1}^{2n} r^3-\{2^3+4^3+\ldots+(2n)^3\}$$

$$= n^2(2n^2-1).$$

15 Let the third root be β, then use

$$\alpha+\frac{1}{\alpha}+\beta = -\frac{b}{a}, \qquad \alpha\frac{1}{\alpha}+\alpha\beta+\beta\frac{1}{\alpha} = \frac{c}{a}, \qquad \alpha\frac{1}{\alpha}\beta = -\frac{d}{a}$$

to prove $a^2-d^2 = ac-bd$.

Roots of the given equation are $1\frac{1}{2}, -3, -\frac{1}{3}$.

16 Let $f(x) = \frac{1}{2}x-\cos x, \quad f'(x) = \frac{1}{2}+\sin x$.
For an approximate solution try $x = \frac{1}{3}\pi$; $f(\frac{1}{3}\pi) = 0{\cdot}024$, $f'(\frac{1}{3}\pi) = 1{\cdot}366$. Then, using Newton's formula, positive root is approximately

$$\frac{\pi}{3}-\frac{f(\frac{1}{3}\pi)}{f'(\frac{1}{3}\pi)} = 1{\cdot}03 \text{ radians}.$$

17 Given series is

$$S_n = \sum_{r=1}^{n} (n-r+1)r^3.$$

It may easily be shown that

$$S_{n+1} = S_n + \sum_{1}^{n+1} r^3 = S_n + \tfrac{1}{4}(n+1)^2(n+2)^2. \qquad \mathbf{1.25}$$

Check that the r.h.s. holds for $n = 1$; assume that it holds for S_n and find S_{n+1} using equation **1.25**, which gives $\frac{1}{60}(n+1)(n+2)(n+3)\{3(n+1)^2+6(n+1)+1\}$,

that is, if S_n is correct, then r.h. expression also holds for S_{n+1}.

18 Expand r.h.s. using identity and given equations,

$$xy+yz+zx = \tfrac{1}{2}\{(x+y+z)^2-(x^2+y^2+z^2)\} = -2,$$
$$xyz = \tfrac{1}{3}[-7-1(5+2)] = 0.$$
$$\begin{aligned}(t-x)(t-y)(t-z) &= t^3-(x+y+z)t^2+(xy+yz+zx)t-xyz\\ &= t^3+t^2-2t\\ &= t(t+2)(t-1), \quad \text{hence roots are } 0,\ -2,\ 1.\end{aligned}$$

19

(i) Use $(a+b+c)(ab+bc+ca)-9abc$
$$= a(b-c)^2+b(c-a)^2+c(a-b)^2.$$

(ii) Let $y = x^a$, hence $x = y^{1/a}$, and by taking logarithms the result follows.

(iii) Note $\log_p n+\log_q n+\log_r n = \dfrac{1}{\log_n p}+\dfrac{1}{\log_n q}+\dfrac{1}{\log_n r}$

and $\log_{pqr} n = \dfrac{1}{\log_n pqr} = \dfrac{1}{\log_n p+\log_n q+\log_n r}$.

20 $(1-x)^n\{1+2x+3x^2+\ldots(p+1)x^p\}$
$$\begin{aligned}&\equiv (1-x)^{n-2}[(1-x)^2\{1+2x+\ldots+(p+1)x^p\}]\\ &\equiv (1-x)^{n-2}\{1-(p+2)x^{p+1}+(p+1)x^{p+2}\}\end{aligned}$$
on multiplying out.

Second part requires comparison of coefficient of x^p on l.h.s. and r.h.s. of above identity.

21 Substitute respectively $r = -2, -1, 0$ in identity, then

$c = -1, \quad b = 1, \quad a = 1.$

$$\begin{aligned}\text{r.h.s.} &= \sum_{1}^{\infty}\frac{(r+3)x^r}{r!(r+2)} = \sum_{1}^{\infty}\frac{(r+3)(r+1)x^r}{(r+2)!}\\ &= \sum_{1}^{\infty}\left[\frac{x^r}{r!}+\frac{x^r}{(r+1)!}-\frac{x^r}{(r+2)!}\right]\\ &= (e^x-1)+\left(\frac{1}{x}e^x-1-x\right)-\frac{1}{x^2}\left(e^x-1-x-\frac{x^2}{2!}\right)\\ &= \text{l.h.s.}\end{aligned}$$

22 Use $\{(1+z)(1-z)\}^{-\frac{1}{2}} = \dfrac{x^2+1}{x^2-1}$, where $z < 1$,

then
$$\begin{aligned}\log\left[\frac{x^2+1}{x^2-1}\right] &= -\frac{1}{2}\{\log(1+z)+\log(1-z)\}\\ &= \tfrac{1}{2}z^2+\tfrac{1}{4}z^4+\tfrac{1}{6}z^6+\ldots.\end{aligned}$$

Similarly use $\left[\dfrac{1+z}{1-z}\right]^{\frac{1}{2}} = \dfrac{x+1}{x-1}$.

For second part substitute $x = 3$, $z = \frac{3}{5}$ in both of given logarithmic series and add.

23

(i) Take exponentials of both sides (i.e. antilogs), obtain $ac = b^2$. Similarly, $xz = y^3$; show this implies

$$\frac{y^2}{x} = \frac{z}{y} = \frac{zy}{y^2}, \quad \text{whence result.}$$

(ii) Show that expression vanishes when x is replaced by $y+z$. Two other factors are $x-y+z$ and $x+y-z$.

24 The maximum value of the quadratic expression occurs at $x = \frac{1}{2}(a+b)$. The given equation has equal roots if $k = \frac{1}{4}(a-b)^2$, roots whose values lie between a and b only if $0 < k < \frac{1}{4}(a-b)^2$, and positive roots if $-ab < k < \frac{1}{4}(a-b)^2$.

25 Expand each by the binomial series, and prove that $a = \frac{1}{2}(1+p)$, $b = \frac{1}{2}(1-p)$. The substitution gives

$$\frac{\sqrt[3]{81}}{4} \simeq \frac{53}{49}.$$

The difference between the two expansions is approximately $-\frac{1}{12}p(p^2-1)x^3$ and, with the given values, works out to approximately 0·00006.

26 The expression is greater than or equal to $(x-y)^2+(y-z)^2+(z-x)^2 \geqslant 0$ for $k \geqslant 2$. If $k = 2$, the expression vanishes provided $x = y = z$, but if $k > 2$, it vanishes provided $x = y = z = 0$. Interpret x, y, z to be the lengths of the three edges of the box. The stated inequalities amount to $3(x^2+y^2+z^2) > (x+y+z)^2 > 3(xy+yz+zx)$. The first part of this is the case $k = 2$, and the second part is equivalent to this case also.

27 The form of the roots must be as stated, since the sum of all the roots is two. The equation factorizes into $(x^2-\alpha^2)\{x^2-2x+(1-\beta^2)\} = 0$. Compare the coefficients on expanding, and from the three equations obtained prove $p^2-36p+180 = 0$. Thus $p = 6$ and hence $\alpha = \pm\sqrt{3}$, $\beta = \pm 4$; or $p = 30$ and hence $\alpha = \pm\sqrt{15}$, $\beta = \pm 2$.

28 Use induction on n. The roots of the given equation are

$$x = 1, 2, \ldots, n.$$

For the last part consider the binomial series expansion of $(1-1)^x$, where x is positive.

29

(i) Prove that $u_n^2-u_{n-1}u_{n+1} = (-1)^1(u_{n-1}^2-u_{n-2}u_n)$; repeated use of this gives the final result.

(ii) Prove $u_r u_{r+1}+u_{r+1}u_{r+2} = u_{r+2}^2-u_r^2$. Sum this result for $r = 1, 3, \ldots, 2n-1$.

30 Denoting the function by y, prove that $y' = 0$ when $(a-x)^{-k-1} = x^{-k-1}$ and so when $x = \frac{1}{2}a$. With this value of x, show that $y'' > 0$, and so $y_{\min} = 2^{k+1}/a^k$. The inequality follows. The last part is a restatement, with $a = x+y$.

31 The Newton formula, with $x_0 = 4$, gives

$$x_1 = 4 - \frac{4^3 - 18 \times 4 + 2}{3 \times 4^2 - 18} = 4{\cdot}2.$$

$f(4) = -6 < 0$, and $f(4{\cdot}2) = 0{\cdot}488 > 0$, hence there is a root between these two values of x.

32 Substitute for u_{n+1} into the identity. From the quadratic identity in u_n obtained, deduce that $1-a = c(1-b)$, $b-ac = 0$ and $\frac{1}{2}(c-1) - b(1-a) + ac(1-b) = 0$. From these three, show by elimination that (since $c \neq 1$), $c = -i$ only, and $a = \frac{1}{2}(1+i)$, $b = \frac{1}{2}(1-i)$, or vice versa. With either possibility, we have

$$\frac{u_{n+4}-a}{u_{n+4}-b} = (-i)^4 \frac{u_n - a}{u_n - b}.$$

Now $(-i)^4 = 1$; and by subtracting 1 from both sides of this result, we obtain

$$\frac{b-a}{u_{n+4}-b} = \frac{b-a}{u_n - b},$$

whence $u_{n+4} = u_n$.

33 $$\sum_{r=1}^{n} (a_r x + b_r)^2 = x^2 \sum a^2 + 2x \sum ab + \sum b^2$$

$$\equiv Ax^2 + 2Bx + C, \quad \text{say,}$$

and is non-negative for all x.

Hence $$Ax^2 + 2Bx + C \equiv \frac{1}{A}\{(Ax+B)^2 + (AC - B^2)\} \geqslant 0,$$

and in particular for $x = -B/A$, we obtain $AC \geqslant B^2$,

i.e. $$\sum a^2 \sum b^2 \geqslant \left(\sum ab\right)^2.$$

The stated inequality (on squaring) is equivalent to

$$\sum a^2 + \sum b^2 + 2\sqrt{\left(\sum a^2\right)}\sqrt{\left(\sum b^2\right)} \geqslant \sum (a+b)^2,$$

which can be reduced to

$$\sqrt{\left(\sum a^2 \sum b^2\right)} \geqslant \sum ab.$$

Hence the result. The equality sign applies only if

$$\frac{b_1}{a_1} = \frac{b_2}{a_2} = \ldots = \frac{b_n}{a_n}.$$

34

(a) Let roots be α, 4α, then $(x-\alpha)(x-4\alpha) \equiv x^2 + ax + b$. Hence, equating coefficients, $a = -5\alpha$, $b = 4\alpha^2$,

i.e. $4a^2 = 25b$.

(b) $$\log(1+x) = x - \tfrac{1}{2}x^2 + \tfrac{1}{3}x^3 + \ldots,$$
$$\log(1-x) = -x - \tfrac{1}{2}x^2 - \tfrac{1}{3}x^3 \ldots;$$
$$\log(1+x) + \log(1-x) = \log(1-x^2) = -\left(x^2 + \frac{x^4}{2} + \frac{x^6}{3} + \ldots\right).$$
$$\tfrac{1}{2}(\tfrac{1}{2})^2 + \tfrac{1}{4}(\tfrac{1}{2})^4 + \tfrac{1}{6}(\tfrac{1}{2})^6 + \ldots = \tfrac{1}{2}[(\tfrac{1}{2})^2 + \tfrac{1}{2}(\tfrac{1}{2})^4 + \tfrac{1}{3}(\tfrac{1}{2})^6 + \ldots]$$
$$= -\tfrac{1}{2}\log\{1 - (\tfrac{1}{2})^2\}$$
$$= -\tfrac{1}{2}\log\tfrac{3}{4} = \log 2 - \tfrac{1}{2}\log 3$$

35

(a) $(1+h)^4 + (1+h)^3 + (1+h)^2 + (1+h) - 5 = 0$,

i.e. $1 + 4h + 1 + 3h + 1 + 2h + 1 + h - 5 \simeq 0$

or $10h - 1 \simeq 0$.

Thus $h \simeq 0{\cdot}1$ and the solution is approximately $1{\cdot}1$.

(b) Lengths of sides are $2a$, $\sqrt{2}a$, a, $a/\sqrt{2}$, $\frac{1}{2}a$, $a/2\sqrt{2}$, ...

$$A_n B_n = 2a\left(\frac{1}{\sqrt{2}}\right)^n = 2^{(1-\frac{1}{2}n)}a.$$

Sum of the areas $= 4a^2 + 2a^2 + a^2 + \frac{1}{2}a^2 + \frac{1}{4}a^2 + \frac{1}{8}a^2$
$= \frac{63}{32} \times 4a^2$.

36

(a) Sum $= (1 + \frac{1}{5})^{\frac{1}{3}} = \sqrt[3]{(\frac{6}{5})}$

(b) $$\log_e(1+x) = x - \frac{x^2}{2} + \frac{x^3}{3} - \ldots (-1)^{n+1}\frac{x^n}{n} + \ldots$$

for $-1 < x \leqslant 1$.

$$\frac{1}{n(n+1)} = \frac{1}{n} - \frac{1}{n+1}; \qquad 2\log_e 2 - 1; \qquad \log_e 2 - \tfrac{1}{2}.$$

37

(a) $$b - a = (s-1)\frac{(l-a)}{(n-1)}, \qquad \frac{c}{a} = \left(\frac{l}{a}\right)^{(s-1)/(n-1)}.$$

(b) $$\frac{1}{r(r+1)(r+2)} = \frac{\frac{1}{2}}{r} - \frac{1}{r+1} + \frac{\frac{1}{2}}{r+2}.$$

38

(a) Expression becomes $(x+y-2z)^2 + (2y-3z)^2 + z^2 \geqslant 0$ for all real x, y, z and has the value 2 when

$x, y, z = 0, 2, 1;\quad 0, -1, 1;\quad -1, -2, -1;\quad 1, -2, -1$,

respectively.

39 We have $S(n) = \frac{1}{6}n(n+1)(2n+1)$,

$T(n) = 1(2+\ldots+n)+2(3+\ldots+n)+\ldots+(n-1)n$.

Show $T(n) = \sum_{r=1}^{n-1} \{r.\frac{1}{2}(n-r)(n+r+1)\}$

$$= \frac{1}{2}\left[n^2 \sum_{r=1}^{n-1} r - \sum_{r=1}^{n-1} r^3 + n \sum_{r=1}^{n-1} r - \sum_{r=1}^{n-1} r^2\right].$$

Note that the second term in the brackets is $\{\frac{1}{2}(n-1)n\}^2$. Use the result for $S(n)$ to simplify $T(n)$ to $\frac{1}{24}n(n-1)(3n^2+5n+2)$. The given ratio simplifies to $3(n+1)/(2n+1)$ and tends to 3/2 as $n \to \infty$.

40 If a is the first term, l is the nth term, and S_n is the sum of n terms, the number of terms is $2S_n/(a+l)$, and the common difference is

$$\frac{l^2-a^2}{2S_n-(a+l)}.$$

In the problem, $l = \pi \times 110$ mm, $a = \pi \times 83$ mm, $S_n = 65\,840$ mm. The common difference is twice the thickness. Thus the thickness is 0·2 mm.

41

(a) The expression equals $\sqrt{\frac{15}{8}}+\sqrt{\frac{9}{8}}-\sqrt{\frac{5}{8}}-\sqrt{\frac{3}{8}} = 1{\cdot}027$ approximately.

(b) The l.h.s. gives $z \log_e a + x \log_e b + y \log_e c - \log_e abc$, which equals $\log_e c + \log_e a + \log_e b - \log_e(abc) = 0$.

42 The first part is standard theory. In the second part let the roots be $\alpha-\delta, \alpha, \alpha+\delta$. Use the relations between the roots, and solve the resulting equations for α, δ, showing $\alpha = 2, \delta = 1$. The roots are then 1, 2, 3.

Chapter Two
Trigonometry, Complex Numbers and Hyperbolic Functions

A Trigonometry: Theory summary

1 *Radian measure and small angles*

(a) $\theta \text{ radians} = \dfrac{180\theta^\circ}{\pi}$, $\quad \pi = 3{\cdot}141593$ (to six decimal places).

(b) $\tan\theta > \theta > \sin\theta$, for $0 < \theta < \frac{1}{2}\pi$.

(c) $\lim\limits_{\theta\to 0}\left[\dfrac{\sin\theta}{\theta}\right] = 1$; $\quad \lim\limits_{\theta\to 0}\left[\dfrac{\tan\theta}{\theta}\right] = 1$.

2 *Graphs of* $\sin\theta$, $\cos\theta$, $\tan\theta$

See Figure 6.

3 *Signs of* $\sin\theta$, $\cos\theta$, $\tan\theta$

(a) $0 < \theta < 90^\circ$: all ratios are positive.
(b) $90^\circ < \theta < 180^\circ$: only $\sin\theta$ is positive.
(c) $180^\circ < \theta < 270^\circ$: only $\tan\theta$ is positive.
(d) $270^\circ < \theta < 360^\circ$: only $\cos\theta$ is positive.

4 *Values of ratios for some special angles*

(a) $\sin 0^\circ = 0$, $\quad \cos 0^\circ = 1$, $\quad \tan 0^\circ = 0$;

$\sin 30^\circ = \dfrac{1}{2}$, $\quad \cos 30^\circ = \dfrac{\sqrt{3}}{2}$, $\quad \tan 30^\circ = \dfrac{1}{\sqrt{3}}$;

$\sin 45^\circ = \dfrac{1}{\sqrt{2}}$, $\quad \cos 45^\circ = \dfrac{1}{\sqrt{2}}$, $\quad \tan 45^\circ = 1$;

$\sin 60^\circ = \dfrac{\sqrt{3}}{2}$, $\quad \cos 60^\circ = \dfrac{1}{2}$, $\quad \tan 60^\circ = \sqrt{3}$;

$\sin 90^\circ = 1$, $\quad \cos 90^\circ = 0$, $\quad \tan 90^\circ = \infty$.

(b) $\sin(90^\circ - \theta) = \cos\theta$; $\quad \tan(90^\circ - \theta) = \cot\theta$.
$\sin(180^\circ - \theta) = \sin\theta$; $\quad \cos(180^\circ - \theta) = -\cos\theta$;
$\tan(180^\circ - \theta) = -\tan\theta$.
$\sin(-\theta) = -\sin\theta$; $\quad \cos(-\theta) = \cos\theta$;
$\tan(-\theta) = -\tan\theta$.

5 *Pythagoras theorem identities*

$\cos^2\theta + \sin^2\theta = 1$.
$1 + \tan^2\theta = \sec^2\theta$.
$\cot^2\theta + 1 = \operatorname{cosec}^2\theta$.

Note. $\tan\theta = \dfrac{\sin\theta}{\cos\theta}$, $\quad \sec\theta = \dfrac{1}{\cos\theta}$,

$\operatorname{cosec}\theta = \dfrac{1}{\sin\theta}$, $\quad \cot\theta = \dfrac{1}{\tan\theta}$.

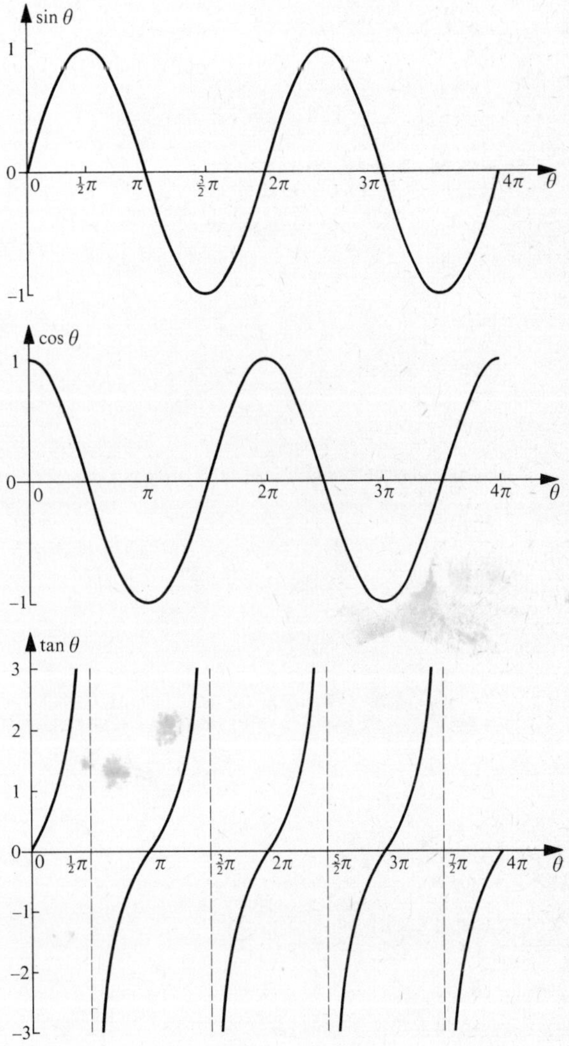

Figure 6

6 *Compound angles*

(a) $\sin(A+B) = \sin A \cos B + \cos A \sin B$.

(b) $\sin(A-B) = \sin A \cos B - \cos A \sin B$.

(c) $\cos(A+B) = \cos A \cos B - \sin A \sin B$.

(d) $\cos(A-B) = \cos A \cos B + \sin A \sin B$.

(e) $\tan(A+B) = \dfrac{\tan A + \tan B}{1 - \tan A \tan B}$.

(f) $\tan(A-B) = \dfrac{\tan A - \tan B}{1 + \tan A \tan B}$.

7 *Sums and differences as products*

(a) $\sin A + \sin B = 2 \sin \frac{1}{2}(A+B) \cos \frac{1}{2}(A-B)$.

(b) $\sin A - \sin B = 2 \cos \frac{1}{2}(A+B) \sin \frac{1}{2}(A-B)$.

(c) $\cos A + \cos B = 2 \cos \frac{1}{2}(A+B) \cos \frac{1}{2}(A-B)$.

(d) $\cos A - \cos B = 2 \sin \frac{1}{2}(A+B) \sin \frac{1}{2}(B-A)$.

8 *Multiple and submultiple angles*

(a) $\sin 2\theta = 2 \sin \theta \cos \theta$.

(b) $\cos 2\theta = \cos^2\theta - \sin^2\theta = 2\cos^2\theta - 1 = 1 - 2\sin^2\theta$.

(c) $\tan 2\theta = \dfrac{2 \tan \theta}{1 - \tan^2\theta}$.

(d) $\sin \theta = \dfrac{2t}{1+t^2}, \qquad \cos \theta = \dfrac{1-t^2}{1+t^2}, \qquad \tan \theta = \dfrac{2t}{1-t^2},$

where $t = \tan \frac{1}{2}\theta$.

(e) $\sin 3\theta = 3 \sin \theta - 4 \sin^3\theta$.

(f) $\cos 3\theta = 4 \cos^3\theta - 3 \cos \theta$.

9 *Maclaurin series expansions for* $\sin \theta$ *and* $\cos \theta$

$$\sin \theta = \theta - \frac{\theta^3}{3!} + \frac{\theta^5}{5!} - \frac{\theta^7}{7!} + \ldots.$$

$$\cos \theta = 1 - \frac{\theta^2}{2!} + \frac{\theta^4}{4!} - \frac{\theta^6}{6!} + \ldots.$$

10 *Inverse functions*

(a) $\theta = \sin^{-1} y$ denotes $y = \sin \theta$, where $-90° \leqslant \theta \leqslant 90°$.
$\theta = \cos^{-1} y$ denotes $y = \cos \theta$, where $0° \leqslant \theta \leqslant 180°$.
$\theta = \tan^{-1} y$ denotes $y = \tan \theta$, where $-90° \leqslant \theta \leqslant 90°$.

(b) $\tan^{-1} m_1 \pm \tan^{-1} m_2 = \tan^{-1} \dfrac{m_1 \pm m_2}{1 \mp m_1 m_2}$.

11 *Solution of equations*

(a) General solution of:

(i) $\sin \theta = A$ is $\theta = n\pi + (-1)^n \sin^{-1} A$.
or $\theta = 2n\pi + \sin^{-1} A$ and $\theta = (2n+1)\pi - \sin^{-1} A$.

(ii) $\cos \theta = A$ is $\theta = 2n\pi \pm \cos^{-1} A$.

(iii) $\tan \theta = A$ is $\theta = n\pi + \tan^{-1} A$.

(b) $a \cos \theta + b \sin \theta = \sqrt{(a^2+b^2)} \cos(\theta - \alpha)$ where $\alpha = \tan^{-1} \dfrac{b}{a}$,

$= \sqrt{(a^2+b^2)} \sin(\theta + \beta)$ where $\beta = \tan^{-1} \dfrac{a}{b}$,

used to solve equations of the type $a \cos \theta + b \sin \theta = c$, giving

$$\theta = 2n\pi \pm \cos^{-1} \frac{c}{\sqrt{(a^2+b^2)}} + \alpha.$$

12 *Triangle formulae*

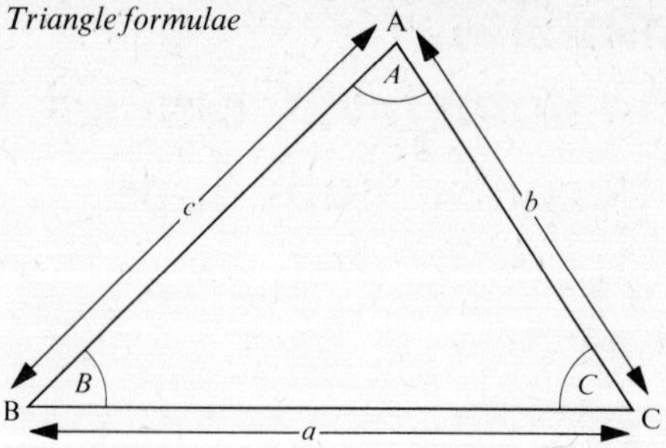

Figure 7

(a) *The sine rule.*

$$\frac{a}{\sin A} = \frac{b}{\sin B} = \frac{c}{\sin C} = 2R,$$ where R is the radius of the circumcircle of $\triangle$ABC.

(b) *The cosine rule.*

(i) $a^2 = b^2 + c^2 - 2bc \cos A$.

(ii) $a = (b+c) \cos \alpha$, where $\alpha = \sin^{-1} \dfrac{2\sqrt{(bc \cos \frac{1}{2}A)}}{(b+c)}$.

(c) $\tan \frac{1}{2}(A-B) = \dfrac{a-b}{a+b} \cot \frac{1}{2}C$.

(d) $a = b \cos C + c \cos B$.

(e) $\sin \frac{1}{2}A = \sqrt{\left[\dfrac{(s-b)(s-c)}{bc}\right]}, \qquad \cos \frac{1}{2}A = \sqrt{\left[\dfrac{s(s-a)}{bc}\right]},$

$\tan \frac{1}{2}A = \sqrt{\left[\dfrac{(s-b)(s-c)}{s(s-a)}\right]},$

where $s = \frac{1}{2}(a+b+c)$.

(f) *Area of triangle.*

(i) $\triangle = \frac{1}{2}bc \sin A = \frac{1}{2}ac \sin B = \frac{1}{2}ab \sin C$.

(ii) $\triangle = \sqrt{\{s(s-a)(s-b)(s-c)\}}$ (Hero's formula).

(iii) $\triangle = \dfrac{abc}{4R} = rs$,

where R is the radius of the circumscribed and r is the radius of inscribed circle of $\triangle$ABC.

B Trigonometry: Illustrative worked problems

1

(i) Solve the equation

$\sin \theta + \sin 3\theta + \sin 5\theta + \sin 7\theta = 0$

for $0° < \theta < 360°$.

(ii) If $3 \tan \theta - \sec \theta = 1$, find the possible values of

$3 \sec \theta + \tan \theta$. [L, J 1968, P II, Q 4]

Solution.

(i) $(\sin \theta + \sin 7\theta) + (\sin 3\theta + \sin 5\theta) = 0$,
i.e. $2 \sin 4\theta \cos 3\theta + 2 \sin 4\theta \cos \theta = 0$.
Hence $2 \sin 4\theta(\cos 3\theta + \cos \theta) = 4 \sin 4\theta \cos 2\theta \cos \theta = 0$
and therefore the general solution is $\theta = \frac{1}{4}n180° \, (n = 0, 1, 2, \ldots)$.
Solutions for $0 < \theta < 360°$ are $\theta = 45°, 90°, 135°, 180°, 225°, 270°, 315°$.

(ii) If $\qquad 3\tan\theta - \sec\theta = 1,$ **2.1**

then $\quad 3\tan\theta - \sqrt{(1+\tan^2\theta)} = 1,$

i.e. $\qquad (3\tan\theta - 1)^2 = 1+\tan^2\theta$

or $\quad 8\tan^2\theta - 6\tan\theta \equiv 2\tan\theta(4\tan\theta - 3) = 0.$

Thus $\tan\theta = 0$ or $\frac{3}{4}$. Now from equation **2.1**,

$$\sec\theta = 3\tan\theta - 1.$$
$$3\sec\theta + \tan\theta = 10\tan\theta - 3.$$

Hence the possible values of $3\sec\theta+\tan\theta$ are $-3(\tan\theta = 0)$ and $4\frac{1}{2}(\tan\theta = \frac{3}{4})$.

2

(i) Solve the simultaneous equations

$$\tan x + \tan y = 1,$$
$$\cot x + \cot y = -1,$$

giving all the values of x between 0° and 180°, together with the corresponding values of y.

(ii) Eliminate ϕ from the equations

$$\tan\phi + \cot\phi = a^3, \qquad \sec\phi - \cos\phi = b^3,$$

where a and b are constants. [L, S 1967, P I, Q 3]

Solution.

(i) Let $\tan x = t_1$ and $\tan y = t_2$,

then $\quad t_1 + t_2 = 1$ **2.2**

and $\quad \dfrac{1}{t_1} + \dfrac{1}{t_2} = -1,$

i.e. $\quad t_1 + t_2 = -t_1 t_2 = 1.$

Hence $\quad t_1 = -\dfrac{1}{t_2}$

and substituting back in equation **2.2** we obtain

$$t_1 - \frac{1}{t_1} = 1 \quad \text{or} \quad t_1^2 - t_1 - 1 = 0.$$

$$t_1 = \tan x = \frac{1\pm\sqrt{5}}{2} = 1{\cdot}618 \quad \text{or} \quad -0{\cdot}618;$$

$$t_2 = \tan y = -0{\cdot}618 \quad \text{or} \quad +1{\cdot}618.$$

Hence $x = 58°\,17'$, $y = 148°\,17'$; or $x = 148°\,17'$, $y = 58°\,17'$.

(ii) $\tan\phi + \cot\phi = \dfrac{\sin\phi}{\cos\phi} + \dfrac{\cos\phi}{\sin\phi} = \dfrac{1}{\sin\phi\cos\phi} = a^3,$ **2.3**

$\sec\phi - \cos\phi = \dfrac{1-\cos^2\phi}{\cos\phi} = \dfrac{\sin^2\phi}{\cos\phi} = b^3.$ **2.4**

Squaring equation **2.3** and multiplying by equation **2.4** gives

$$a^6 b^3 = \frac{1}{\cos^3\phi},$$

thus $\quad \cos\phi = \dfrac{1}{a^2 b}$

and on substituting back in equation **2.4** we obtain

$$a^2 b - \frac{1}{a^2 b} = b^3,$$

i.e. $\quad a^4 b^2 - 1 = a^2 b^4 \quad$ or $\quad a^4 b^2 - a^2 b^4 = 1.$

3

(i) Prove that

$$\tan^{-1}(\tfrac{1}{3}) + \tan^{-1}(\tfrac{1}{17}) = \tfrac{1}{2}\sin^{-1}(\tfrac{20}{29}).$$

(ii) Show that $\theta = 54°$ is a solution of the equation

$$\cos 3\theta + \sin 2\theta = 0,$$

and find the other solutions in the range 0° to 180°. Show that $\sin 54° = (1+\sqrt{5})/4$.

(iii) Prove that, if $\tanh x = \sin\phi$, where $0 < \phi < \frac{1}{2}\pi$,

$$x = \log_e \tan(\tfrac{1}{4}\pi + \tfrac{1}{2}\phi).$$

[O & C, S 1967, M for Sc (with Stats) III S, Q 1]

Solution.

(i) From $\quad \tan^{-1}x + \tan^{-1}y = \tan^{-1}\left[\dfrac{x+y}{1-xy}\right],$

we obtain $\quad \tan^{-1}(\tfrac{1}{3}) + \tan^{-1}(\tfrac{1}{17}) = \tan^{-1}\left[\dfrac{\frac{1}{3}+\frac{1}{17}}{1-\frac{1}{51}}\right]$

$$= \tan^{-1}(\tfrac{2}{5}) = \theta, \quad \text{say}.$$

We need to show that $\theta = \frac{1}{2}\sin^{-1}(\frac{20}{29})$.

Now $\quad \sin 2\theta = \dfrac{2\tan\theta}{1+\tan^2\theta},$

so that $\quad \sin 2\theta = \dfrac{2(\frac{2}{5})}{1+(\frac{2}{5})^2} = \dfrac{\frac{4}{5}}{\frac{29}{25}} = \dfrac{20}{29}.$

Hence $\quad \theta = \frac{1}{2}\sin^{-1}(\frac{20}{29}).$

(ii) $\theta = 54°$ is indeed a solution of $\cos 3\theta + \sin 2\theta = 0$ because $\cos 162° = -\cos 18° = -\sin 72° = -\sin 108°$,

i.e. $\quad \cos 162° + \sin 108° = 0.$

To find all the solutions of $\cos 3\theta + \sin 2\theta = 0$, use the double-angle and triple-angle formulae. We obtain

$$(4\cos^3\theta - 3\cos\theta) + 2\sin\theta\cos\theta = 0,$$
$$(\cos\theta)(4\cos^2\theta + 2\sin\theta - 3) = 0$$
$$\text{or} \quad -(\cos\theta)(4\sin^2\theta - 2\sin\theta - 1) = 0.$$

Hence either $\theta = 90°$, or $\sin\theta = \frac{1}{4}(1+\sqrt{5})$, on solving the quadratic equation and noting that $0 \leqslant \theta \leqslant 180°$.

$\theta = 54°$ is a solution; hence so also is its supplement, 126°,

i.e. $\quad \sin 54° = \frac{1}{4}(1+\sqrt{5}) = \sin 126°.$

(iii) $\dfrac{1+\tanh x}{1-\tanh x} = \dfrac{\cosh x + \sinh x}{\cosh x - \sinh x} = \dfrac{e^x}{e^{-x}} = e^{2x}.$

Therefore $\quad e^{2x} = \dfrac{1+\sin\phi}{1-\sin\phi} = \dfrac{1+2t/(1+t^2)}{1-2t/(1+t^2)},$

where $t = \tan\frac{1}{2}\phi$.

$$e^{2x} = \frac{(1+t)^2}{(1-t)^2}$$

and so $\quad e^x = +\dfrac{1+t}{1-t}.$

$$x = \log_e\left[\frac{1+t}{1-t}\right] = \log_e\left[\frac{\tan\frac{1}{4}\pi + \tan\frac{1}{2}\phi}{1-\tan\frac{1}{4}\pi\tan\frac{1}{2}\phi}\right]$$
$$= \log_e\tan(\tfrac{1}{4}\pi + \tfrac{1}{2}\phi).$$

C Trigonometry: Problems

Answers and hints will be found on pp. 33–6.

1

(i) Verify that the equation $\sin 3\theta = 2\cos 2\theta$ is satisfied by $\theta = 30°$. Find all other angles between 0° and 360° which satisfy this equation.

(ii) Three angles are in arithmetic progression, and their sines (not all equal), taken in the same order, are in geometric progression. Find the common ratio of the geometric progression. [L, J 1968, P I, Q 4]

2 A flat watch is fixed face upwards on a plane which is inclined at an angle $\alpha°$ to the horizontal. At noon the hands point up a line of greatest slope. Find

(a) the inclination of the minute hand to the horizontal at t minutes past each hour,

(b) the angle between the vertical plane through the minute hand at that time and the vertical plane through a line of greatest slope. [L, J 1968, P II, Q 5]

3 A vertical mast stands on the north bank of a river with straight parallel banks running from east to west. The angle of elevation of the top of the mast is α when measured from a point A on the south bank distant $3a$ to the east of the mast, and β when measured from another point B on the south bank distant $5a$ to the west of the mast. Prove that the height of the mast is

$$\frac{4a}{(\cot^2\beta-\cot^2\alpha)^{\frac{1}{2}}}$$

and that the angle of elevation θ measured from a point midway between A and B is given by the equation

$$2\cot^2\theta = 3\cot^2\alpha-\cot^2\beta.$$ [L, S 1967, P I, Q 4]

4

(i) Find the general solution of

(a) $\cos 2x-\cos 4x = \sin 2x$,

(b) $2\cos^2 x-4\sin x\cos x-\sin^2 x = 1$,

giving your answer to (a) in radians and to (b) in degrees and minutes.

(ii) Solve the equation $\tan^{-1}\left[\frac{1-x}{1+x}\right] = \frac{1}{2}\tan^{-1}x$.

[L, S 1967, P II, Q 4]

5 With the usual notation for a triangle ABC prove that

$$\tan\frac{B-C}{2} = \frac{b-c}{b+c}\cot\frac{A}{2}.$$

To ascertain the distance between two inaccessible points X and Y on level ground a base line AB, of length 100 metres, is laid out with X and Y on different sides of it. Angles are measured as follows: $\angle$ ABX = 60°, $\angle$ ABY = 46°, $\angle$ BAX = 30°, $\angle$ BAY = 67°. Find the acute angle between AB and XY and the distance XY. [L, J 1967, P I, Q 6]

6 A round tower, standing on a level plain, has a steeple built symmetrically on top of it. From a point on the ground a man observes that P, the nearest point of the top of the tower, and Q, the top of the steeple, are in line, and that PQ makes an angle $\tan^{-1}3$ with the horizontal. He then moves back from the tower a distance of 60 m in the vertical plane through P and Q, and observes that the angles of elevation of P and Q are $\tan^{-1}\frac{3}{4}$ and 45° respectively. Find the height of the tower, the height of the steeple (above the ground) and the radius of the top of the tower. [O, S 1968, M & PM I, Q 4]

7 If $\cos a+\sin b = 0$, prove that $a = 2n\pi\pm(\frac{1}{2}\pi+b)$, where n is any positive or negative integer or zero, and hence find the general solution of the equation $\cos 3x+\sin 2x = 0$.

Express

$$\frac{\cos 3x+\sin 2x}{\cos x}$$

in terms of $\sin x$ and, using the results of the first part of this question, prove that

$$4\sin^2 x-2\sin x-1 \equiv 4(\sin x-\sin\tfrac{3}{10}\pi)(\sin x+\sin\tfrac{1}{10}\pi).$$

[O, S 1966, P II, Q 1]

8 Prove that $\cot\theta-\cot 2\theta = \operatorname{cosec} 2\theta$.

Replace θ by 2θ in this identity and deduce that if $0° < \theta < 45°$ then $\cot\theta-\cot 4\theta > 2$.

The acute-angled triangle ABC is such that $B = 4C$. By using the above result, or otherwise, prove that

$$b^2-c^2 < 4\triangle,$$

where $\triangle$ is the area of the triangle. [C, S 1967, P I, Q 8]

9 The points P_1 and P_2 on the surface of the earth (assumed to be a sphere of radius r and centre O) are at the same (Northern) latitude λ, and their respective longitudes are L_1 and L_2. Show that the length C of the route from P_1 to P_2 via a circle centre O (called a *great circle*) satisfies equation

$$\cos\left[\frac{C}{r}\right] = \sin^2\lambda+\cos^2\lambda\cos(L_2-L_1).$$

If the separation in longitude is 90° show that either

$$C = 2r\sin^{-1}\left[\frac{\cos\lambda}{\sqrt{2}}\right],\quad\text{or}\quad C = 2\pi r-2r\sin^{-1}\left[\frac{\cos\lambda}{\sqrt{2}}\right].$$

[C, S 1967, Sp O, Q 3]

10 Express $\tan 3\theta$ and $\tan 4\theta$ as rational functions of $\tan\theta$.

By considering the roots of the equation $\tan 3\theta+\tan 4\theta = 0$, or otherwise, obtain a cubic equation whose roots are

$$\tan^2\left[\frac{\pi}{7}\right],\quad \tan^2\left[\frac{2\pi}{7}\right]\quad\text{and}\quad \tan^2\left[\frac{3\pi}{7}\right].$$

Deduce (or prove otherwise) that

(i) $\tan\left[\frac{\pi}{7}\right]\tan\left[\frac{2\pi}{7}\right]\tan\left[\frac{3\pi}{7}\right] = \sqrt{7}$.

(ii) $\tan^4\left[\frac{\pi}{7}\right]+\tan^4\left[\frac{2\pi}{7}\right]+\tan^4\left[\frac{3\pi}{7}\right] = 371$.

[C, S 1966, P S, Q 5]

11 Prove that

$$\sin\alpha+\sin(\alpha+\beta)+\sin(\alpha+2\beta)+\ldots+\sin\{\alpha+(n-1)\beta\} = \sin\{\alpha+\tfrac{1}{2}(n-1)\beta\}\sin\tfrac{1}{2}n\beta\operatorname{cosec}\tfrac{1}{2}\beta.$$

A regular polygon $A_1A_2\ldots A_n$ is inscribed in a circle of radius R, and P is a point on the circle, between A_1 and A_n, such that A_1P subtends an angle 2α at the centre. Prove that the sum of the lengths of the chords $PA_1, PA_2, \ldots, PA_n$ is

$$2R\cos\left[\alpha-\frac{\pi}{2n}\right]\operatorname{cosec}\frac{\pi}{2n}.$$

Prove that this sum lies between $2R\cot(\pi/2n)$ and $2R\operatorname{cosec}(\pi/2n)$. [O & C, S 1967, M & HM I, Q 4]

12 Prove that
$\cos\theta+\cos 2\theta+\ldots+\cos n\theta = \cos\frac{1}{2}(n+1)\theta \sin\frac{1}{2}n\theta \operatorname{cosec}\frac{1}{2}\theta.$

Deduce that

$$\cos\frac{\pi}{2n}+\cos\frac{2\pi}{2n}+\ldots+\frac{\cos(n-1)\pi}{2n} = \frac{1}{2}\left(\cot\frac{\pi}{4n}-1\right).$$

Show that $\cot\frac{1}{8}\pi = 1+\sqrt{2}$, and find a similar expression for $\cot\frac{1}{12}\pi$. [O & C, S 1966, M & HM I, Q 4]

13 H is the top of a vertical tower of height h metres and K is the base. A is a point due South of K and at the same level. B is a point due East of the tower but at a level d metres higher than A and K. The angles of elevation of H from A and B are α and β respectively, and the length of AB is l metres. Show that h is given by a root of the equation.

$$h^2(\cot^2\alpha+\cot^2\beta)-2hd\cot^2\beta+d^2\operatorname{cosec}^2\beta-l^2 = 0.$$

If squares and higher powers of d may be neglected, show that one root of the equation is negative and h is given by the formula

$$h = \frac{l}{(\cot^2\alpha+\cot^2\beta)^{\frac{1}{2}}}+\frac{d\cot^2\beta}{\cot^2\alpha+\cot^2\beta}.$$

[O & C, S 1967, M for Sc I, Q 11]

14 In the triangle ABC, BC $= a$, AC $= b$, AB $= c$. Prove that
$a = b\cos C+c\cos B.$

By using this and two similar results show that

(i) $\dfrac{\cos A}{a}+\dfrac{\cos B}{b}+\dfrac{\cos C}{c} = \dfrac{a^2+b^2+c^2}{2abc}$,

(ii) $\cos A = \dfrac{b^2+c^2-a^2}{2bc}$. [AEB, S 1968, P & A I, Q 4]

15 If $\tan\theta = \frac{3}{4}$ calculate, without the use of tables,

(i) $\sin 2\theta$,
(ii) $\sin 4\theta$.

In Figure 8, O is the centre of a circle of radius $5r$. The diameter TOP meets the circumference at P and an arc of a circle of radius $8r$, centre P, is drawn to cut the given circle at R and S. Prove that the area of the crescent RTS is $2r^2(12-7\tan^{-1}\frac{3}{4})$. [AEB, S 1968, P & A II, Q 2]

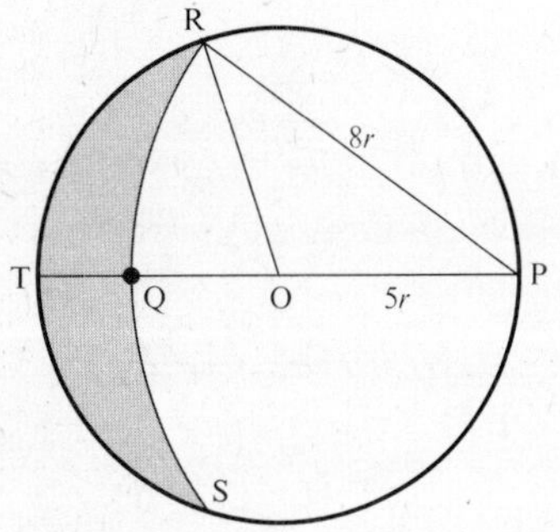

Figure 8

16

(a) Obtain an expression for $\sin(A+B)$ in terms of sines and cosines of A and B, showing that it is true for angles of any magnitude. Deduce an expression for $\sin 3A$ in terms of $\sin A$.

(b) If $\lambda > 0$ and $0 < \phi < \frac{1}{2}\pi$, prove that

$$\sin^{-1}\frac{\lambda\sin\phi}{\sqrt{(1+2\lambda\cos\phi+\lambda^2)}}+\sin^{-1}\frac{\sin\phi}{\sqrt{(1+2\lambda\cos\phi+\lambda^2)}} = \phi.$$

[You may assume that the sum of the angles on the left-hand side is less than $\frac{1}{2}\pi$.] [W, S 1969, P II, Q 1]

17 ABCD and $A_1B_1C_1D_1$ are two opposite faces of a cube of side 6 cm, such that AA_1 etc. are the remaining edges. Points P, Q and R are chosen on AD, BC and C_1D_1 respectively such that AP = 1 cm, BQ = 2 cm and C_1R = 5 cm. Find the area of the triangle PQR as accurately as your tables allow. [W, S 1968, P II, Q 1]

D Complex numbers: Theory summary

1 *Rectangular and polar form of a complex number*

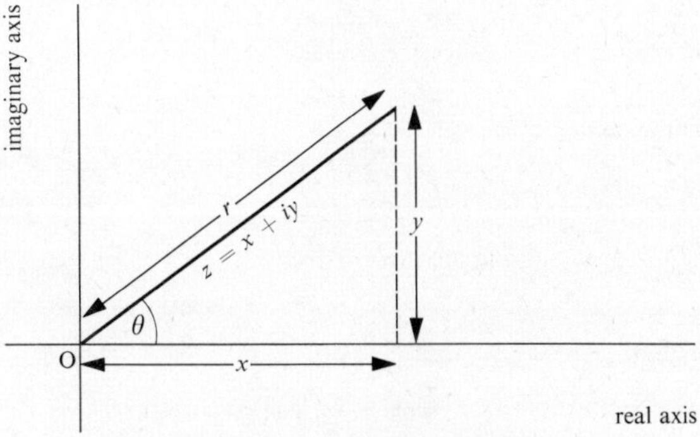

Figure 9 Argand diagram (complex plane) showing the geometrical interpretation of the complex number $z = x+iy = re^{i\theta}$

$z = x+iy$
$\quad = r(\cos\theta+i\sin\theta) = re^{i\theta}.$

$x = r\cos\theta$ = real part of z, Re z;

$y = r\sin\theta$ = imaginary part of z, Im z.

The numbers $\pm i$ are roots of the equation $i^2 = -1$.

$r = \sqrt{(x^2+y^2)} = |z|$ is known as the *modulus* of z

$\theta = \tan^{-1}\dfrac{y}{x}$ is known as the *argument* of z

2 *Powers of i*

$i^2 = -1$; $\quad i^3 = -i$;

$i^4 = 1$; $\quad \dfrac{1}{i} = -i.$

3 *Addition and subtraction*

If $z_1 = x_1+iy_1$, $z_2 = x_2+iy_2$, $z_3 = x_3+iy_3$, then

(a) $z_1+z_2 = (x_1+x_2)+i(y_1+y_2)$
$z_1+z_2+z_3 = (x_1+x_2+x_3)+i(y_1+y_2+y_3)$

(b) $z_1-z_2 = (x_1-x_2)+i(y_1-y_2)$

4 *Multiplication*

(a) $z_1 z_2 = (x_1 + iy_1)(x_2 + iy_2) = x_1 x_2 - y_1 y_2 + i(x_1 y_2 + x_2 y_1)$,
i.e. multiply algebraically and replace i^2 by -1 wherever it appears.

(b) In polar form, $z_1 = r_1 e^{i\theta_1}$ and $z_2 = r_2 e^{i\theta_2}$, and
$z_1 z_2 = r_1 e^{i\theta_1} r_2 e^{i\theta_2} = r_1 r_2 e^{i(\theta_1+\theta_2)}$,
i.e. multiply moduli and add arguments.

5 *Complex conjugate*

(a) The *complex conjugate* of $z = x+iy$ is $z^* = x - iy = re^{-i\theta}$.

(b) $zz^* = |z|^2$; $\quad x = \frac{1}{2}(z+z^*)$, $\quad y = \frac{1}{2i}(z-z^*)$.

N.B. z^* is sometimes written as $\bar{z}$.

6 *Division*

(a)
$$\frac{z_1}{z_2} = \frac{x_1+iy_1}{x_2+iy_2} = \frac{x_1+iy_1}{x_2+iy_2} \times \frac{x_2-iy_2}{x_2-iy_2} = \frac{x_1 x_2 + y_1 y_2 + i(x_2 y_1 - x_1 y_2)}{x_2^2 + y_2^2}.$$

The above procedure of making denominator a real quantity is known as *rationalization*.

(b)
$$\frac{z_1}{z_2} = \frac{r_1 e^{i\theta_1}}{r_2 e^{i\theta_2}} = \frac{r_1}{r_2} e^{i(\theta_1-\theta_2)},$$
i.e. divide moduli and subtract arguments.

7 *De Moivre's theorem*

$(\cos\theta + i\sin\theta)^n = \cos n\theta + i\sin n\theta$,

for all positive or negative integral n. If n is fractional, r.h.s. is only one of the possible values of l.h.s.

8 *n-th root of a number* $N = r(\cos\theta + i\sin\theta)$

$$\sqrt[n]{N} = r^{1/n}\left\{\cos\frac{1}{n}(\theta+2m\pi) + i\sin\frac{1}{n}(\theta+2m\pi)\right\},$$
$(m = 0, 1, 2, 3, \ldots, n-1)$.

e.g. the cube roots of minus one, $-1 = \cos\pi + i\sin\pi$, are
$\sqrt[3]{-1} = \frac{1}{2}(1+i\sqrt{3}),\ -1,\ \frac{1}{2}(1-i\sqrt{3})$.

9 *Euler's relations*

$\cos\theta + i\sin\theta = e^{i\theta}$;
$\cos\theta - i\sin\theta = e^{-i\theta}$.

10 *Exponential definitions of* $\sin\theta$ *and* $\cos\theta$

$$\sin\theta = \frac{1}{2i}(e^{i\theta} - e^{-i\theta});$$
$$\cos\theta = \frac{1}{2}(e^{i\theta} + e^{-i\theta}).$$

E Hyperbolic functions: Theory summary

1 *Exponential definitions*

$\sinh x = \frac{1}{2}(e^x - e^{-x})$;
$\cosh x = \frac{1}{2}(e^x + e^{-x})$;
$$\tanh x = \frac{e^x - e^{-x}}{e^x + e^{-x}} = \frac{\sinh x}{\cosh x}.$$

N.B. $\sinh i\beta = i\sin\beta$, $\quad \cosh i\beta = \cos\beta$,
$\tanh i\beta = i\tan\beta$.

2 *Graphs of* sinh *x*, cosh *x*, *and* tanh *x*

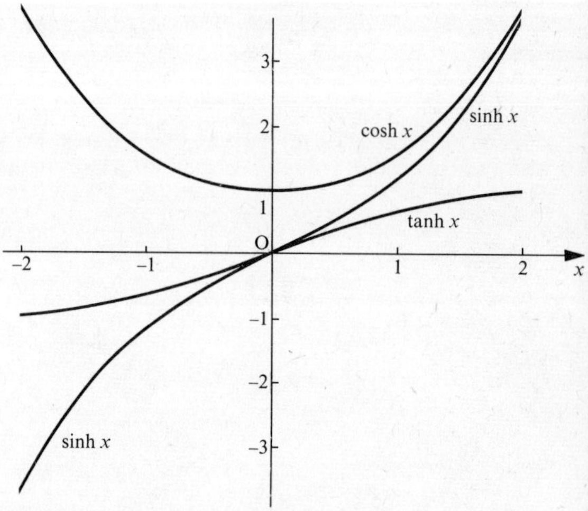

Figure 10 Curves of the hyperbolic functions

3 $\cosh x + \sinh x = e^x$;
$\cosh x - \sinh x = e^{-x}$.

4 $\cosh^2 x - \sinh^2 x = 1$;
$1 - \tanh^2 x = \text{sech}^2 x$;
$\coth^2 x - 1 = \text{cosech}^2 x$.

5

(a) $\sinh(A \pm B) = \sinh A \cosh B \pm \cosh A \cosh B$;
$\sinh(\alpha \pm i\beta) = \sinh\alpha\cos\beta \pm i\cosh\alpha\sin\beta$.

(b) $\cosh(A \pm B) = \cosh A \cosh B \pm \sinh A \sinh B$;
$\cosh(\alpha \pm i\beta) = \cosh\alpha\cos\beta \pm i\sinh\alpha\sin\beta$.

6

(a) $\sinh A + \sinh B = 2\sinh\frac{1}{2}(A+B)\cosh\frac{1}{2}(A-B)$.
(b) $\sinh A - \sinh B = 2\cosh\frac{1}{2}(A+B)\sinh\frac{1}{2}(A-B)$.
(c) $\cosh A + \cosh B = 2\cosh\frac{1}{2}(A+B)\cosh\frac{1}{2}(A-B)$.
(d) $\cosh A - \cosh B = 2\sinh\frac{1}{2}(A+B)\sinh\frac{1}{2}(A-B)$.

N.B. Any identity between sin *x* and cos *x* may be transformed into an identity between sinh *x* and cosh *x* by replacing sin *x* by *i* sinh *x* and cos *x* by cosh *x*.

7 *Power series expansions*

$$\sinh x = x + \frac{x^3}{3!} + \frac{x^5}{5!} + \frac{x^7}{7!} + \ldots;$$
$$\cosh x = 1 + \frac{x^2}{2!} + \frac{x^4}{4!} + \frac{x^6}{6!} + \ldots.$$

8 *Inverse hyperbolic functions*

$x = \sinh^{-1} y$ denotes $y = \sinh x$;

$x = \log_e\{y+\sqrt{(y^2+1)}\}$.

$x = \cosh^{-1} y$ denotes $y = \cosh x$;

$x = \log_e\{y \pm \sqrt{(y^2-1)}\}\quad (y > 1)$.

$x = \tanh^{-1} y$ denotes $y = \tanh x$;

$x = \dfrac{1}{2}\log_e\left[\dfrac{1+y}{1-y}\right]\quad (|y| < 1)$.

F Complex numbers and hyperbolic functions: Illustrative worked problems

1

(i) Given that $2+i$ is a root of the equation

$3x^3-10x^2+7x+10 = 0$,

solve the equation completely.

(ii) If $w = \dfrac{z+1}{z-i}$,

find the locus of the point representing z in an Argand diagram when the locus of the point representing w is

(a) the real axis,
(b) the imaginary axis,
(c) the circle $|w| = 1$. [L, J 1968, FM V, Q 2]

Solution.

(i) If $2+i$ is a root, its complex conjugate $2-i$ is also a root.

Thus $\{x-(2+i)\}\{x-(2-i)\} = x^2-4x+5$ is a factor.

Hence $3x^3-10x^2+7x+10 = (x^2-4x+5)(3x+2) = 0$

and $x = 2+i, 2-i$ and $-\frac{2}{3}$.

(ii) Writing $z = x+iy$,

$$w = \frac{z+1}{z-i} = \frac{(x+1)+iy}{x+i(y-1)},$$

$$\text{i.e.}\quad w = \frac{(x+1)+iy}{x+i(y-1)} \times \frac{x-i(y-1)}{x-i(y-1)}$$

$$= \frac{x^2+x+y^2-y+i(x-y+1)}{x^2+(y-1)^2}.$$

(a) If w is purely real, Im $w = 0$,

thus $x-y+1 = 0$,

i.e. $y = x+1$.

Hence locus of z is a straight line, slope 1, intercept -1 on real axis.

(b) If w is purely imaginary, Re $w = 0$,

thus $x^2+y^2+x-y = 0$.

Hence locus of z is a circle centre $(-\frac{1}{2}, \frac{1}{2})$,

radius $= \dfrac{1}{\sqrt{2}}$.

(c) If $|w| = 1$, then $|z+1|^2 = |z-i|^2$,

i.e. $(z+1)(z^*+1) = (z-i)(z^*+i)$,

$zz^*+(z+z^*)+1 = zz^*+i(z-z^*)+1$,

$2x = i(2iy) = -2y$.

Hence the locus of z in this case is the straight line $y = -x$.

2

(a) Define $\sinh\theta$, and using your definition express $\sinh^3\theta$ in terms of $\sinh 3\theta$ and $\sinh\theta$. If $\sinh\theta = x$, use your result to express $e^{3\theta}$ as a function of x, and hence find the real value of $(38+\sqrt{1445})^{\frac{1}{3}}$ in simple surd form.

(b) Show that the equation $\sqrt{3}\sinh\phi+\cosh\phi = 2$ has a unique real solution

$\phi = \log\dfrac{2+\sqrt{6}}{1+\sqrt{3}}$. [L, S 1967, FM VI, Q 6]

Solution.

(a) $\sinh\theta = \frac{1}{2}(e^\theta-e^{-\theta})$.

$\sinh^3\theta = \frac{1}{8}(e^\theta-e^{-\theta})^3 = \frac{1}{8}(e^{3\theta}-e^{-3\theta})+\frac{1}{8}(-3e^\theta+3e^{-\theta})$

$= \frac{1}{4}\sinh 3\theta-\frac{3}{4}\sinh\theta$.

If $\sinh\theta = x$, $\sinh^3\theta = x^3$

and $\sinh 3\theta = 4x^3+3x = \frac{1}{2}(e^{3\theta}-e^{-3\theta})$

$\equiv \dfrac{1}{2}\left[y-\dfrac{1}{y}\right]$.

Hence $y^2-(8x^3+6x)y-1 = 0$

and $y \equiv e^{3\theta} = 4x^3+3x\pm\sqrt{\{(4x^3+3x)^2+1\}}$.

Let $e^\theta = (38+\sqrt{1445})^{\frac{1}{3}}$, $x = \sinh\theta$,

then $e^{3\theta} = 38+\sqrt{1445} = 4x^3+3x\pm\sqrt{\{(4x^3+3x)^2+1\}}$ **2.5**

and by inspection it is seen that equation **2.5** is satisfied when $x = 2$.

Thus $\sinh\theta = \frac{1}{2}(e^\theta-e^{-\theta}) = 2$ or $e^{2\theta}-4e^\theta-1 = 0$.

Hence $e^\theta = 2\pm\sqrt{5}$ and, since $e^\theta > 0$,

$e^\theta = (38+\sqrt{1445})^{\frac{1}{3}} = 2+\sqrt{5}$.

(b) If $\sqrt{3}\sinh\phi+\cosh\phi = 2$,

then $\sqrt{3}(e^\phi-e^{-\phi})+e^\phi+e^{-\phi} = 4$,

or $(\sqrt{3}+1)e^{2\phi}-4e^\phi-(\sqrt{3}-1) = 0$.

Thus $e^\phi = \dfrac{4\pm\sqrt{(16+4\times 2)}}{2(\sqrt{3}+1)} = \dfrac{2\pm\sqrt{6}}{1+\sqrt{3}}$.

For ϕ real, $e^\phi > 0$,

$e^\phi = \dfrac{2+\sqrt{6}}{1+\sqrt{3}}$ and $\phi = \log_e\dfrac{2+\sqrt{6}}{1+\sqrt{3}}$.

3 If $\sinh y = x$, prove that

$y = \log_e\{x+\sqrt{(x^2+1)}\}$,

explaining carefully why the positive square root must be taken.

If $\sinh y = \cot\theta$, where $0 < \theta < \pi$, prove that $y = \log_e\cot\frac{1}{2}\theta$, and find a similar expression for y when $\pi < \theta < 2\pi$. [O & C, S 1966, M & HM V Sp, Q 10]

Solution.

If $\sinh y = x$, then $\frac{1}{2}(e^y-e^{-y}) = x$,

thus $e^y-2x-e^{-y} = 0$,

$e^{2y}-2xe^y-1 = 0$.

This is a quadratic equation in e^y. On solving,

$e^y = \frac{1}{2}\{2x\pm\sqrt{(4x^2+4)}\} = x\pm\sqrt{(x^2+1)}$.

Now $e^y > 0$ for all y, hence the positive sign only can be taken in this solution:

$$e^y = x+\sqrt{(x^2+1)},$$
$$y = \log_e\{x+\sqrt{(x^2+1)}\}.$$

If $\sinh y = \cot\theta, 0 < \theta < \pi$, then using the above, with $x = \cot\theta$, we obtain

$$y = \log_e\{\cot\theta+\sqrt{(\cot^2\theta+1)}\} = \log_e(\cot\theta+\operatorname{cosec}\theta),$$

since $\operatorname{cosec}\theta > 0$ for the given θ, so that

$$y = \log_e\left[\frac{\cos\theta+1}{\sin\theta}\right]$$
$$= \log_e\left[\frac{2\cos^2\frac{1}{2}\theta}{2\sin\frac{1}{2}\theta\cos\frac{1}{2}\theta}\right]$$
$$= \log_e\cot\tfrac{1}{2}\theta \quad (0 < \theta < \pi).$$

With $\pi < \theta < 2\pi$, $\operatorname{cosec}\theta < 0$, so that

$$y = \log_e(\cot\theta-\operatorname{cosec}\theta)$$
$$= \log_e\left[\frac{\cos\theta-1}{\sin\theta}\right]$$
$$= \log_e\left[\frac{-2\sin^2\frac{1}{2}\theta}{2\sin\frac{1}{2}\theta\cos\frac{1}{2}\theta}\right]$$
$$= \log_e(-\tan\tfrac{1}{2}\theta) \quad (\pi < \theta < 2\pi).$$

G Complex numbers and hyperbolic functions: Problems

Answers and hints will be found on pp. 36–8.

1 If $(w-1)/(w+1) = \tan z$, where $w = u+iv$ and $z = x+iy$, and u, v, x, y are all real, prove that the modulus of w is

$$\sqrt{\left(\frac{\cosh 2y+\sin 2x}{\cosh 2y-\sin 2x}\right)}$$

and find the argument of w. [L, S 1967, FM V, Q 2]

2 If $\sin\theta = n\sin(\theta+\alpha)$, where n and α are real constants, prove that

$$e^{2i\theta} = \frac{1-ne^{-i\alpha}}{1-ne^{i\alpha}}.$$

Hence, or otherwise, prove that

$$\theta = \sum_{k=1}^{\infty}\frac{n^k}{k}\sin k\alpha.$$

State the range of values of n for which the series is valid. [L, S 1967, FM V, Q 3]

3

(i) The complex numbers z and w are related by the formula

$$w = z+\frac{2a^2}{z},$$

where a is a positive constant. If, on an Argand diagram, points P and Q correspond to z and w respectively, and if P describes each of the following curves in turn

(a) $|z| = a$, (b) $\arg z = \frac{1}{4}\pi$,

show that the corresponding curves described by Q are conics and find their eccentricities.

Draw two separate diagrams showing the corresponding z- and w-curves in case (a) and in case (b).

(ii) The points on the Argand diagram which represent the complex numbers z_1, z_2, z_3 lie on a circle passing through the origin. Prove that the points which represent $1/z_1, 1/z_2, 1/z_3$ are collinear. [L, J 1967, FM V, Q 3]

4 By using the exponential form of $\cos\theta$, or otherwise, show that the sum of the series

$$1+n\cos\theta+\ldots+\frac{n!}{r!(n-r)!}\cos r\theta+\ldots+\cos n\theta$$

is $(2\cos\frac{1}{2}\theta)^n\cos\frac{1}{2}n\theta$.

Prove that, as n tends to infinity, the sum of the series tends to zero if $\frac{2}{3}\pi < \theta < \frac{4}{3}\pi$. [O, S 1967, PM II, Q 5]

5 If z is a complex number and

$$f(z) \equiv \frac{z+i}{z-i},$$

denoting $f(z)$ by z_1, $f(z_1)$ by z_2 and $f(z_2)$ by z_3, prove that $z_3 = z$.

If the point in the Argand diagram which represents z moves on the axis of x, i.e. if z is real, show that the points which represent z_1 and z_2 move respectively on the circle of unit radius with the origin as centre and on the axis of y. [C, S 1966, P S, Q 7]

6 Show that the square of the distance between the two points which correspond in the Argand diagram to the complex numbers z_1 and z_2 is $(z_1-z_2)(\bar{z}_1-\bar{z}_2)$, where $\bar{z}$ is conjugate to z, i.e. if $z = x+iy$, then $\bar{z} = x-iy$.

Show that the equation of any circle in the Argand diagram may be written in the form

$$z\bar{z}-\bar{a}z-a\bar{z}+b = 0,$$

where b is real.

If two circles have the equations

$$z\bar{z}-\bar{a}_1z-a_1\bar{z}+b_1 = 0,$$
$$z\bar{z}-\bar{a}_2z-a_2\bar{z}+b_2 = 0,$$

find the condition that they should cut orthogonally. [O & C, S 1967, M & HM V (PS), Q 3]

7 Prove that the equation

$$a\cosh x+b\sinh x = c,$$

where a, b, c are constants, and $c > 0$, has

(i) two real roots if $a > 0$ and $b^2+c^2 > a^2 > b^2$,
(ii) no real root if $a < 0$ and $b^2+c^2 > a^2 > b^2$ or if $a^2 > b^2+c^2$,
(iii) one real root if $a^2 < b^2$.

Solve the equation

$$5\cosh x+2\sinh x = 11.$$

[O & C, S 1967, M & HM V (P S), Q 11]

8

(i) The point P on the Argand diagram, which represents the complex number z, lies on the circle of radius unity and centre the origin. Show that z can be written in the form $\cos\theta+i\sin\theta$, and indicate θ on a diagram. Prove that the complex number z^2+2 can be written as

$$2+\cos 2\theta+i\sin 2\theta$$

and find expressions of the form $a+ib$ for the numbers $-2z$ and $1/(z+1)$.

Indicate on a diagram the loci of the points representing the numbers $-2z$, z^2+2 and $1/(z+1)$ as P moves anticlockwise round the circle, starting at the point $\theta = 0$, and ending at the point $\theta = \pi$; indicate directions on the loci.

(ii) In the equation $z^2+2\lambda z+1 = 0$, λ is a parameter which can take any *real* value. Show that, if $-1 < \lambda < 1$, the roots of this equation lie on a certain circle in the Argand diagram, but that if $\lambda > 1$, one root lies inside the circle and one outside.

Prove that for very large values of λ the roots are approximately -2λ and $-1/2\lambda$.

[O & C, S 1966, M for Sc I, Q 3]

9

(a) Solve for x the equation

$$(x-1)^2+(x-2)^2+\ldots+(x-n)^2 = \tfrac{1}{12}n(n^2-1).$$

$\left[\text{You may use the formula } \sum_1^n r^2 = \tfrac{1}{6}n(n+1)(2n+1)\right].$

(b) If x is the root of the equation $x \cosh x = \frac{1}{2}$, show that

$$0 < xe^x < 1 \quad \text{and} \quad 0 < xe^{-x} < 1,$$

and that the sum of infinity of the series

$$1+x\cosh x+x^2\cosh 2x+x^3\cosh 3x+\ldots = \frac{1}{2x^2}.$$

[JMB, S 1967, P II, Q 1]

10 Express the roots of the equation $z^7-1 = 0$ in the form $\cos\theta+i\sin\theta$. Hence or otherwise show that the roots of the equation

$$u^3+u^2-2u-1 = 0$$

are $2\cos\frac{2\pi}{7}$, $2\cos\frac{4\pi}{7}$ and $2\cos\frac{6\pi}{7}$, and find the roots of

$$8w^3+4w^2-4w-1 = 0.$$

[W, S 1969, P S, Q 1]

11

(a) Sketch the region in the complex plane in which z can lie:

(i) when $0 < \arg(z+i) < \frac{1}{4}\pi$;

(ii) when $|z-1| < |z|$.

(b) The complex numbers z_1, z_2 are represented in the Argand diagram by points P_1, P_2.

If $z_2 = z_1+\dfrac{4}{z_1}$,

show that when P_1 describes a circle of radius 4 and centre the origin, the locus of P_2 is an ellipse.

(c) Find all real values of α such that the number

$$\frac{3+e^{2i\alpha}}{2-e^{i\alpha}}$$

is real.

[JMB, S 1967, FM II, Q 2]

12

(a) Express $(6+7i)/(4-2i)$ in the form $A+iB$, where A and B are real.

(b) Express $\cos^4 x$ as a sum of cosines of multiples of x.

(c) Express $\cos 6x$ as a polynomial in $\cos x$ and $\sin x$.

[W, S 1968, P I, Q 8]

H Trigonometry: Answers and hints

1

(i) $\sin 3\theta-2\cos 2\theta = 3\sin\theta-4\sin^3\theta-2(1-2\sin^2\theta) = 0$

i.e. $4\sin^3\theta-4\sin^2\theta-3\sin\theta+2 = 0$

and, on factorizing,

$(2\sin\theta-1)(2\sin^2\theta-\sin\theta-2) = 0.$

Hence $\sin\theta = \frac{1}{2}$,

$\theta = 30°, 150°$ are solutions;

or $\sin\theta = \frac{1}{4}(1-\sqrt{17})$,

$\theta = 308°\,40', 231°\,30'$ are also solutions.

(ii) Let the three angles be $\alpha-\delta, \alpha, \alpha+\delta$, then, since their sines are in geometrical progression,

$$\sin^2\alpha = \sin(\alpha-\delta)\sin(\alpha+\delta)$$

or $\frac{1}{2}(1-\cos 2\alpha) = \frac{1}{2}(\cos 2\delta-\cos 2\alpha)$.

Hence $\cos 2\delta = 1$ and $\delta = \pi$.

$$\text{Common ratio of G.P.} = \frac{\sin\alpha}{\sin(\alpha-\pi)} = \frac{1}{\cos\pi} = -1.$$

2

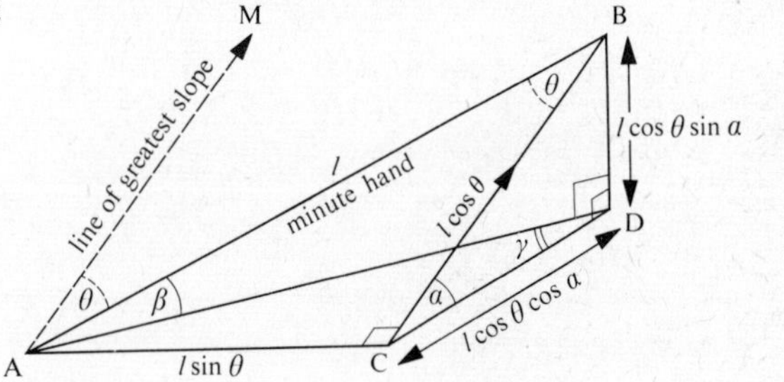

Figure 11

(a) In Figure 11 AM//CB are parallel to the line of greatest slope. At t min past each hour, angle θ, which the minute hand AB makes with AM, is

$$\theta = \frac{t}{60}\times 360° = 6t°.$$

∠ BAD, which AB makes with horizontal is given by

$$\sin\beta = \frac{\text{BD}}{\text{AB}} = \cos\theta\sin\alpha,$$

$$\beta = \sin^{-1}(\cos 6t°\sin\alpha).$$

(b) Angle between vertical planes ABD and CBD is

$$\gamma = \tan^{-1}\frac{\text{AC}}{\text{CD}} = \tan^{-1}\frac{\sin\theta}{\cos\theta\cos\alpha} = \tan^{-1}(\tan 6t°\sec\alpha)$$

3 Let the width of river be c, and the height of mast be h, then, applying Pythagoras' theorem to △BPQ and △QPA in Figure 12, we have

$$c^2 = h^2\cot^2\beta-25a^2 = h^2\cot^2\alpha-9a^2$$

hence $h = \dfrac{4a}{\cot^2\beta-\cot^2\alpha}.$

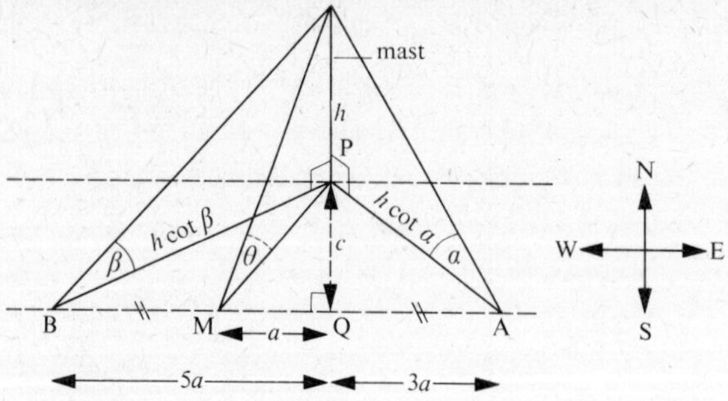

Figure 12

If M is midpoint of AB,

$$\mathrm{PM}^2 = c^2 + a^2 = h^2 \cot^2\beta - 24a^2,$$

$$\cot^2\theta = \frac{\mathrm{PM}^2}{h^2} = \frac{1}{h^2}\left[h^2\cot^2\beta - \frac{24h^2}{16}(\cot^2\beta - \cot^2\alpha)\right]$$

and hence $2\cot^2\theta = 3\cot^2\alpha - \cot^2\beta$.

4

(i)

(a) $$\cos 2x - \cos 4x = \sin 2x,$$

i.e. $2\sin 3x \sin x - 2\sin x\cos x = 0,$

$$2\sin x(\sin 3x - \cos x) = 0.$$

Thus $\sin x = 0$ and $x = n\pi \quad (n = 0, 1, 2, 3, \ldots).$

Also $\sin 3x - \cos x = \sin 3x - \sin(\frac{1}{2}\pi - x) = 0,$

therefore $3x = n\pi + (-1)^n(\frac{1}{2}\pi - x);$

if $n = 2m$ (n even) $x = \dfrac{\pi}{8}(4m+1),$

if $n = 2m+1$ (n odd) $x = \dfrac{\pi}{4}(4m+1).$

(b) Dividing the equation by $\cos^2 x$ we obtain

$2\tan^2 x + 4\tan x - 1 = 0.$

Thus $\tan x = -1 \pm \sqrt{1{\cdot}5}, \qquad x = 12°\,41'$ or $-65°\,48'.$

Hence general solution is $x = 180n° + 12°\,41', 180n° - 65°\,48'$

(ii) $\tan^{-1}\left[\dfrac{1-x}{1+x}\right] = \dfrac{1}{2}\tan^{-1}x = y,$ say,

then $\tan y = \dfrac{1-x}{1+x}$ and $\tan 2y = x = \dfrac{2\tan y}{1-\tan^2 y} = \dfrac{2(1-x^2)}{4x}.$

Hence $4x^2 = 2(1-x^2)$

and $x = \pm\dfrac{1}{\sqrt{3}}.$

5 From the sine rule, $\dfrac{\sin B}{b} = \dfrac{\sin C}{c},$

thus $$\frac{\sin B + \sin C}{b+c} = \frac{\sin B - \sin C}{b-c}.$$

Hence $$\frac{b-c}{b+c} = \frac{\sin B - \sin C}{\sin B + \sin C} = \frac{2\sin\frac{1}{2}(B-C)\cos\frac{1}{2}(B+C)}{2\cos\frac{1}{2}(B-C)\sin\frac{1}{2}(B+C)}$$

$$= \tan\tfrac{1}{2}(B-C)\cot\tfrac{1}{2}(B+C),$$

but $\frac{1}{2}(B+C) = 90° - \frac{1}{2}A,$

$\cot\frac{1}{2}(B+C) = \tan\frac{1}{2}A,$ hence required result.

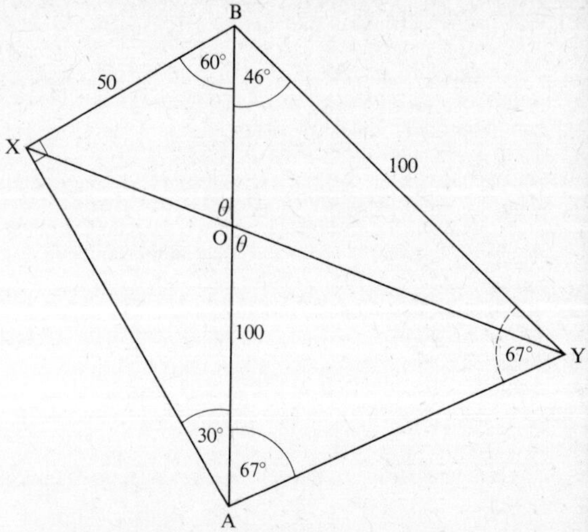

Figure 13

From Figure 13 it is seen that $\mathrm{BX} = 100\cos 30° = 50,$
$\mathrm{BY} = \mathrm{AB} = 100.$

Applying cosine rule to $\triangle\mathrm{BYX}$,

$$\mathrm{XY}^2 = 100^2 + 50^2 - 100^2\cos 106° = 100^2(1{\cdot}25 + \cos 74°),$$

$$\mathrm{XY} = 123{\cdot}6 \text{ metres}.$$

Also $\mathrm{XO}\sin\theta = 50\sin 60° = 43{\cdot}30,$

$\mathrm{OY}\sin\theta = 100\sin 46° = 71{\cdot}93,$

hence $(\mathrm{XO}+\mathrm{OY})\sin\theta = \mathrm{XY}\sin\theta = 115{\cdot}23$

and therefore $\sin\theta = \dfrac{115{\cdot}23}{\mathrm{XY}} = 0{\cdot}9323,$

$$\theta = 68°\,48'.$$

6

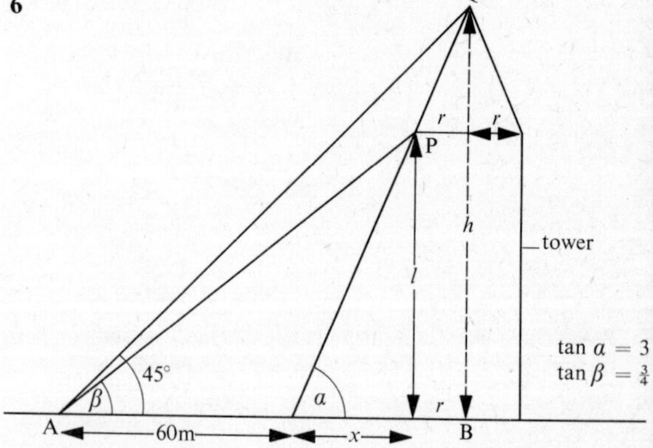

Figure 14

Height of tower $l = x\tan\alpha = 3x$
$= (60+x)\tan\beta = \frac{3}{4}(60+x),$

hence $12x = 180 + 3x,$
$x = 20\,\mathrm{m}$ and $l = 60\,\mathrm{m}.$

Height of steeple $h = (x+r)\tan\alpha = 3(20+r)$
$= 60 + x + r$ (as AB = QB),

therefore, radius of tower $r = 10\,\mathrm{m}$ and $h = 90\,\mathrm{m}.$

7 If $\cos a+\sin b = 0$,

then $\cos a = -\sin b = \cos\{\pm(b+\frac{1}{2}\pi)\}$
$= \cos\{2n\pi\pm(b+\frac{1}{2}\pi)\}$

and $a = 2n\pi\pm(\frac{1}{2}\pi+b)$.

Using the above result the general solution of $\cos 3x = -\sin 2x$ is

$3x = 2n\pi\pm(\frac{1}{2}\pi+2x)$,

hence $x = \frac{1}{2}\pi(4n+1)$ and $x = \frac{1}{10}\pi(4n-1)$.

$$\frac{\cos(x+2x)+\sin 2x}{\cos x}$$

$$= \frac{\cos x\cos 2x-(\sin x)2\sin x\cos x+2\sin x\cos x}{\cos x}$$

$$= (1-2\sin^2 x)-2\sin^2 x+2\sin x = -(4\sin^2 x-2\sin x-1).$$

Now as $4\sin^2 x-2\sin x-1 = -\dfrac{\cos 3x+\sin 2x}{\cos \alpha}$,

the first expression has the same roots as $\cos 3x+\sin 2x = 0$. Thus, on selecting $x = \frac{1}{10}\pi(4n-1)$ with $n = 0, 1$, i.e. $x = -\frac{1}{10}\pi, x = \frac{3}{10}\pi$,

$$4\sin^2 x-2\sin x-1 \equiv 4\{\sin x-\sin(-\tfrac{1}{10}\pi)\}(\sin x-\sin\tfrac{3}{10}\pi)$$
$$\equiv 4(\sin x-\sin\tfrac{3}{10}\pi)(\sin x+\sin\tfrac{1}{10}\pi).$$

8 Work in terms of sines and cosines, and use the appropriate addition formula. Add the two identities and obtain

$\cot\theta-\cot 4\theta = \operatorname{cosec} 2\theta+\operatorname{cosec} 4\theta$.

Each cosecant term exceeds 1, for θ as given, and the inequality follows.

In the acute-angled triangle, draw the height h from A and show that

$$b^2-c^2 = h^2(\operatorname{cosec}^2 C-\operatorname{cosec}^2 B) = h^2(\cot^2 C-\cot^2 B)$$
$$= ha(\cot C-\cot B).$$

Deduce that this is less than $2ha$, using the above, and so less than $4\triangle$.

9

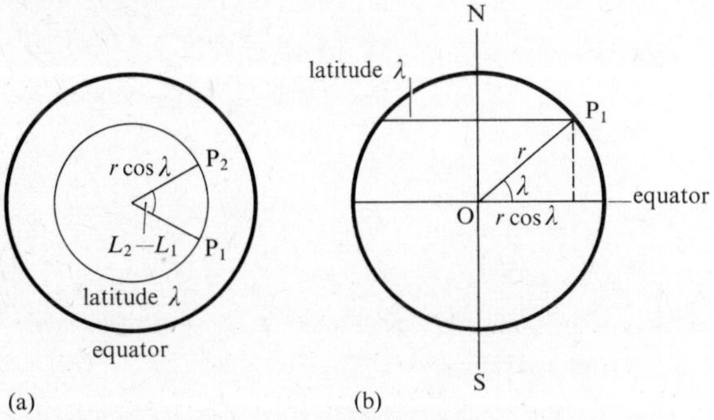

Figure 15 (a) View along earth's axis. (b) Section across earth's axis, through P

Show that $\cos\dfrac{C}{r} = \dfrac{r^2+r^2-(P_1P_2)^2}{2r^2}$,

where $(P_1P_2)^2 = (r\cos\lambda)^2+(r\cos\lambda)^2-$
$-2(r^2\cos^2\lambda)\cos(L_1-L_2)$.

The equation follows.

If $L_1-L_2 = \pm 90°$, the equation reduces to

$$\cos\frac{C}{r} = \sin^2\lambda.$$

Convert to the half-angle $C/2r$, and hence show that

$$\sin\frac{C}{2r} = \frac{\cos\lambda}{\sqrt{2}}.$$

The results for $C/2r$ are now immediate.

10 If $t \equiv \tan\theta$, then

$$\tan 3\theta = \frac{3t-t^3}{1-3t^2},\qquad \tan 4\theta = \frac{4t-4t^3}{1-6t^2+t^4}$$

and using the latter

$\tan 3\theta+\tan 4\theta = 0$ **2.6**

becomes $t^6-21t^4+35t^2-7 = 0$,

i.e. $x^3-21x^2+35x-7 = 0 \quad (x \equiv t^2)$. **2.7**

By inspection, the three roots of equation **2.6** are

$$\frac{\pi}{7},\ \frac{2\pi}{7},\ \frac{3\pi}{7},$$

so the roots of equation **2.7** are

$$\tan^2\frac{\pi}{7},\ \tan^2\frac{2\pi}{7},\ \tan^2\frac{3\pi}{7}$$

and hence equation **2.7** is the required cubic.

(i) From equation **2.7**, product of roots is $+7$, hence result.

(ii) Write l.h.s. as

$$\left[\sum_{r=1}^{3}\tan^2\frac{r\pi}{7}\right]^2$$
$$-2\left[\tan^2\frac{\pi}{7}\tan^2\frac{2\pi}{7}+\tan^2\frac{2\pi}{7}\tan^2\frac{3\pi}{7}+\tan^2\frac{3\pi}{7}\tan^2\frac{\pi}{7}\right]$$
$$= (-21)^2-2(35) = 371.$$

11 Begin by proving that

$$2\sin\tfrac{1}{2}\beta\sin(\alpha+r\beta) = \cos\{\alpha+(r-\tfrac{1}{2})\beta\}-\cos\{\alpha+(r+\tfrac{1}{2})\beta\}.$$

Sum such results for $r = 0, 1, \ldots, n-1$. After a series of cancellations, followed by use of the factorization formula for the difference of two cosines, the formula is proved.

In the polygon, each

$$PA_r = 2R\sin\left[\alpha+\frac{(r-1)\pi}{n}\right].$$

Sum for $r = 1, \ldots, n$, using the result above with $\beta = \pi/n$. This gives the sum of the lengths of the chords. Use the fact that for all α,

$$\cos\frac{\pi}{2n} \leqslant \cos\left(\alpha-\frac{\pi}{2n}\right) \leqslant 1,$$

to deduce the last result.

12 Prove first that

$$2\sin\tfrac{1}{2}\theta\cos r\theta = \sin(r+\tfrac{1}{2})\theta-\sin(r-\tfrac{1}{2})\theta.$$

Sum such results for $r = 1, \ldots, n$, and obtain

$$2 \sin \tfrac{1}{2}\theta \sum_{r=1}^{n} \cos r\theta = \sin(n+\tfrac{1}{2})\theta - \sin \tfrac{1}{2}\theta.$$

Deduce the first formula from this. Put $\theta = \pi/2n$ and prove the second formula.

$n = 2$ gives $\cot \frac{1}{8}\pi$; $n = 3$ is used to prove $\cot \frac{1}{12}\pi = 2+\sqrt{3}$.

13 In the Pythagorean result $AK^2+KL^2 = 1^2-d^2$, substitute $AK = h \cot \alpha$, and $KL = (h-d)\cot \beta$. This gives the quadratic equation in h. Neglect the term in d^2 and solve for the positive root. This gives h, after again neglecting a term in d^2 in the discriminant. The other root of the quadratic is negative because the discriminant is greater than $d \cot^2 \beta$.

14 Drop perpendicular from A to BC and prove by simple trigonometry that

$$a = b \cos C + c \cos B.$$

(i) From above, $\dfrac{a}{bc} = \dfrac{\cos C}{c} + \dfrac{\cos B}{b}$,

and two similar results provide

$$\frac{b}{ac} = \frac{\cos A}{a} + \frac{\cos C}{c}, \qquad \frac{c}{ab} = \frac{\cos B}{b} + \frac{\cos A}{a}.$$

Hence on adding all three equations,

$$\frac{a}{bc}+\frac{b}{ac}+\frac{c}{ab} = 2\left[\frac{\cos A}{a}+\frac{\cos B}{b}+\frac{\cos C}{c}\right],$$

i.e. $$\frac{a^2+b^2+c^2}{2abc} = \frac{\cos A}{a}+\frac{\cos B}{b}+\frac{\cos C}{c}.$$

(ii) Using above results,

$$\frac{\cos A}{a} = \frac{a^2+b^2+c^2}{2abc} - \left(\frac{\cos B}{b}+\frac{\cos C}{c}\right)$$
$$= \frac{a^2+b^2+c^2}{2abc} - \frac{a}{bc}$$
$$= \frac{b^2+c^2-a^2}{2abc};$$

Thus $$\cos A = \frac{b^2+c^2-a^2}{2bc}.$$

15 If $\tan \theta = \frac{3}{4}$, $\sin \theta = \frac{3}{5}$ and $\cos \theta = \frac{4}{5}$.

(i) $\sin 2\theta = 2 \sin \theta \cos \theta = 2.\frac{3}{5}.\frac{4}{5} = \frac{24}{25}$.

(ii) $\sin 4\theta = 2 \sin 2\theta \cos 2\theta = 2 \sin 2\theta(2\cos^2\theta - 1) = \frac{336}{625}$.
Referring to Figure 8,

$\frac{1}{2}$ crescent RTS = Area sector OTR –
– (Area sector PQR – Area △ORP)
$= 25r^2\theta - [32r^2\theta - \frac{1}{2}.5r.5r \sin 2\theta]$,

where $\theta = \angle RPO = \cos^{-1}\frac{4}{5} = \tan^{-1}\frac{3}{4}$. Hence result follows.

16

(a) $\sin 3A = 3 \sin A - 4 \sin^3 A$.

(b) The two inverse sines are acute angles; their sum can be written, more simply, as

$$\tan^{-1}\left(\frac{\lambda \sin \phi}{1+\lambda \cos \phi}\right) + \tan^{-1}\left(\frac{\sin \phi}{\lambda + \cos \phi}\right).$$

Now use the formula for $\tan^{-1} m_1 + \tan^{-1} m_2$, to simplify the sum to $\tan^{-1}(\tan \phi) = \phi$.

17 By Pythagoras' theorem in three dimensions, $PQ = \sqrt{37} \approx 8{\cdot}083$ cm, $PR = \sqrt{62} \approx 7{\cdot}874$ cm, $QR = \sqrt{77} \approx 8{\cdot}875$ cm. Then use Hero's formula and show that area $\triangle PQR \approx 543{\cdot}4\ \text{cm}^2$.

I Complex numbers and hyperbolic functions: Answers and hints

1 If $\dfrac{w-1}{w+1} = \tan z$,

then $$w = \frac{1+\tan z}{1-\tan z} = \frac{\cos z+\sin z}{\cos z-\sin z}.$$

i.e. $$w = \frac{\cos(x+iy)+\sin(x+iy)}{\cos(x+iy)-\sin(x+iy)}$$
$$= \frac{\cos x \cos iy - \sin x \sin iy + \sin x \cos iy + \cos x \sin iy}{\cos x \cos iy - \sin x \sin iy - \sin x \cos iy - \cos x \sin iy}$$
$$= \frac{\cosh y(\cos x+\sin x) + i \sinh y(\cos x - \sin x)}{\cosh y(\cos x-\sin x) - i \sinh y(\cos x + \sin x)}$$
$$\equiv \frac{r_1 e^{i\theta_1}}{r_2 e^{i\theta_2}},$$

where

$$r_1 = \sqrt{\{\cosh^2 y(\cos x+\sin x)^2 + \sinh^2 y(\cos x - \sin x)^2\}}$$
$$= \sqrt{\{(\cosh^2 y+\sinh^2 y)(\cos^2 x+\sin^2 x) + 2 \sin x \cos x(\cosh^2 y - \sinh^2 y)\}}$$
$$= \sqrt{(\cosh 2y + \sin 2x)},$$

$$\theta_1 = \tan^{-1}\frac{\sinh y(\cos x - \sin x)}{\cosh y(\cos x+\sin x)}$$
$$= \tan^{-1}\left[\tanh y \frac{\cos x - \sin x}{\cos x + \sin x}\right],$$

$$r^2 = \sqrt{(\cosh 2y - \sin 2x)},$$

$$\theta_2 = -\tan^{-1}\left[\tanh y \frac{\cos x + \sin x}{\cos x - \sin x}\right].$$

Hence $$|w| = \frac{r_1}{r_2} = \sqrt{\left[\frac{\cosh 2y + \sin 2x}{\cosh 2y - \sin 2x}\right]}$$

and $\arg w = \theta_1 - \theta_2$,
$\tan(\theta_1 - \theta_2) = 2 \sinh y \sec 2x$.

2 Using the exponential definition for $\sin \theta$, $\sin \theta = n \sin(\theta+\alpha)$ becomes

$$\frac{1}{2i}(e^{\theta} - e^{-\theta}) = \frac{n}{2i}[e^{i(\theta+\alpha)} - e^{-i(\theta+\alpha)}]$$

and multiplying by $e^{\theta}/2i$,

$$e^{2i\theta} = n(e^{2i\theta+i\alpha} - e^{-i\alpha}),$$

hence $$e^{2i\theta} = \frac{1-ne^{-i\alpha}}{1-ne^{i\alpha}}. \qquad \textbf{2.8}$$

By taking logarithms of equation **2.8**,

$$2i\theta = \log_e(1-ne^{-i\alpha}) - \log_e(1-ne^{i\alpha}).$$

i.e. $2i\theta = -\left[ne^{-i\alpha}+\frac{n^2}{2}e^{-2i\alpha}+\frac{n^3}{3}e^{-3i\alpha}+\ldots\right]+$

$$+\left[ne^{i\alpha}+\frac{n^2}{2}e^{2i\alpha}+\frac{n^3}{3}e^{3i\alpha}+\ldots\right].$$

$$\theta = \frac{1}{2i}\left[n(e^{i\alpha}-e^{-i\alpha})+\frac{n^2}{2}(e^{2i\alpha}-e^{-2i\alpha})+\ldots\right]$$

$$= \sum_{k=1}^{\infty}\frac{n^k}{k}\sin k\alpha.$$

The logarithmic expansions are valid provided $|ne^{\pm i\alpha}| < 1$, hence the series is valid for $|n| < 1$.

3

(i) $w = u+iv = z+\dfrac{2a^2}{z}$

$$= x+iy+\frac{2a^2}{x+iy}$$

$$= x+iy+\frac{2a^2(x-iy)}{x^2+y^2},$$

hence $u = x\left[1+\dfrac{2a^2}{x^2+y^2}\right]$, $\quad v = y\left[1-\dfrac{2a^2}{x^2+y^2}\right]$.

(a) If $|z| = a$, $x^2+y^2 = a^2$ and thus $u = 3x$, $v = -y$.

Thus $\dfrac{u^2}{9}+v^2 = x^2+y^2 = a^2$,

i.e. Q describes an ellipse of semi-axes $3a$, a and eccentricity e found from

$$a^2 = 9a^2(1-e^2),$$

i.e. $e = \dfrac{2\sqrt{2}}{3}$.

(b) If $\arg z = \frac{1}{4}\pi$,

then $\tan\frac{1}{4}\pi = 1 = \dfrac{x}{y}$

and hence $x = y = t$, say.

Thus $u = t+\dfrac{a^2}{t}$, $\quad v = t-\dfrac{a^2}{t}$,

giving $u+v = 2t$, $\quad u-v = \dfrac{2a^2}{t}$ and $u^2-v^2 = 4a^2$.

I.e. Q describes a rectangular hyperbola of eccentricity e given by $4a^2 = 4a^2(e^2-1)$, i.e. $e = \sqrt{2}$.

(ii) Here we require the locus of the transform,

$$w = u+iv = \frac{1}{z} = \frac{1}{x+iy} = \frac{x}{x^2+y^2}-\frac{iy}{x^2+y^2},$$

but as the points $z_i (i = 1, 2, 3)$ lie on a circle through the origin, $x^2+y^2+2fx+2gy = 0$,

$$u = \frac{+x}{x^2+y^2} = \frac{-x}{2(fx+gy)} \quad\text{and}\quad v = \frac{y}{2(fx+gy)}.$$

Thus $-2fu+2gv = 1$ and the points $1/z_i$ are collinear.

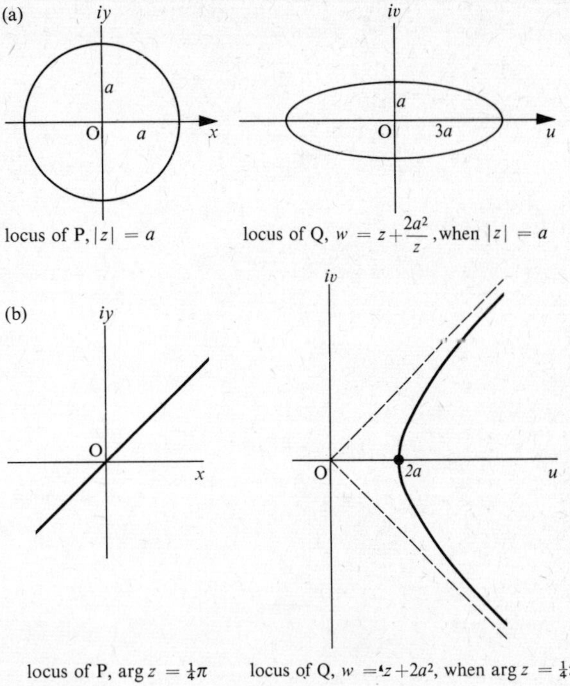

Figure 16

4 Using the form $\cos r\theta = \frac{1}{2}(e^{ir\theta}+e^{-ir\theta})$, series becomes

$$\sum_{r=1}^{n} {}^nC_r \cos r\theta = \tfrac{1}{2}(1+e^{i\theta})^n+\tfrac{1}{2}(1+e^{-i\theta})^n$$
$$= \tfrac{1}{2}e^{i\frac{1}{2}n\theta}(e^{-i\frac{1}{2}\theta}+e^{i\frac{1}{2}\theta})^n+\tfrac{1}{2}e^{-i\frac{1}{2}n\theta}(e^{i\frac{1}{2}\theta}+e^{-i\frac{1}{2}\theta})^n$$
$$= \tfrac{1}{2}(e^{i\frac{1}{2}n\theta}+e^{-i\frac{1}{2}n\theta})(e^{i\frac{1}{2}\theta}+e^{-i\frac{1}{2}\theta})^n$$
$$= \cos\tfrac{1}{2}n\theta(2\cos\tfrac{1}{2}\theta)^n.$$

Now $(2\cos\frac{1}{2}\theta)^n \cos n\theta \to 0$ if $(2\cos\frac{1}{2}\theta)^n \to 0$, i.e. if $\cos\frac{1}{2}\theta < \pm\frac{1}{2}$; but $\cos\frac{1}{3}\pi = \frac{1}{2}$ and $\cos\frac{2}{3}\pi = -\frac{1}{2}$, hence $\frac{1}{3}\pi < \frac{1}{2}\theta < \frac{2}{3}\pi$ or $\frac{2}{3}\pi < \theta < \frac{4}{3}\pi$.

5 To prove that $z_3 = z$, prove first that

$$z_3 = \frac{i(z_1+1)}{z_1-1}.$$

If the locus of z is the x-axis, then $z_1 = e^{i2\theta}$, where $\theta = \arg(z+i)$ $(-\pi < \theta < \pi)$, whence locus of z_1 is the unit circle. On an Argand diagram, show that

$$\arg z_2 \equiv \arg(z_1+i)-\arg(z_1-i) = 90°,$$

whence z_2 lies on the axis of y.

6 The first part follows from the fact that $|z|^2 = z\bar{z}$ for any z. The equation of a standard circle can be written

$$|z|^2+g(z+\bar{z})+\frac{f}{i}(z-\bar{z})+c = 0,$$

i.e. $z\bar{z}-\bar{a}z-a\bar{z}+b = 0$,

where $-a = g+if$ and $b = c$. The condition for orthogonality of the two circles, which is $2(g_1g_2+f_1f_2) = c_1+c_2$, becomes in complex numbers, $b_1+b_2 = a_1\bar{a}_2+\bar{a}_1a_2$.

7 Write the equation as $(a+b)y^2-2cy+(a-b)=0$, where $y=e^x$. The roots y are

$$\frac{c \pm \sqrt{\{c^2-(a^2-b^2)\}}}{a+b} \quad \text{(real or complex).}$$

Conclusions (i—iii) follow from consideration of this formula. The last equation has $a=5, b=2, c=11$. Then $y=\frac{1}{7}, 3$, so $x=-\log_e 7$, and $x=\log_e 3$.

8

(i) The whole of (i) is straightforward. $1/(z+1)$ simplifies to $\frac{1}{2}(1-i\tan\frac{1}{2}\theta)$. The locus of the three points as P moves anticlockwise round the semicircle is given in Figure 17.

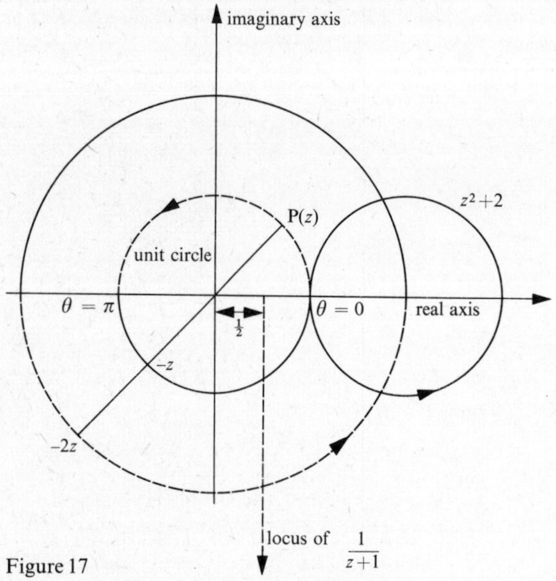

Figure 17

(ii) If $|\lambda|<1$, the roots of the quadratic equation are $z=-\lambda \pm i\sqrt{(1-\lambda^2)}$; both values are such that $|z|=1$. If $\lambda>1$, $z=-\lambda\pm\sqrt{(\lambda^2-1)}$; the lesser root lies outside the unit circle, the larger root lies inside since $z_+z_-=1$. For large λ, use the binomial theorem to find z_+.

9

(a) $x=\frac{1}{2}(1+n)$.

(b) $\text{Sum}=\dfrac{1-x\cosh x}{1-2x\cosh x+x^2}=\dfrac{1}{2x^2}$ (on using $x\cosh x=\frac{1}{2}$).

10 The roots of the equations are

$$\cos\frac{2\pi k}{7}+i\sin\frac{2\pi k}{7} \quad (k=0, \pm 1, \pm 2, \pm 3).$$

Remove the root $z=1$ and show that

$$z^6+z^5+\ldots+z+1=(z^2-\alpha z+1)(z^2-\beta z+1)(z^2-\gamma z+1),$$

where $\alpha=2\cos\frac{2\pi}{7}$, $\beta=2\cos\frac{4\pi}{7}$, $\gamma=2\cos\frac{6\pi}{7}$.

Expand and show that

$$\alpha+\beta+\gamma=-1, \quad \alpha\beta+\beta\gamma+\gamma\alpha=-2, \quad \alpha\beta\gamma=1,$$

so that the equation $u^3+u^2-2u-1=0$ has the roots α, β, γ. In the final part, put $u=2w$, making the roots of the cubic in w $\cos\frac{2\pi}{7}, \cos\frac{4\pi}{7}, \cos\frac{6\pi}{7}$.

11

(a)

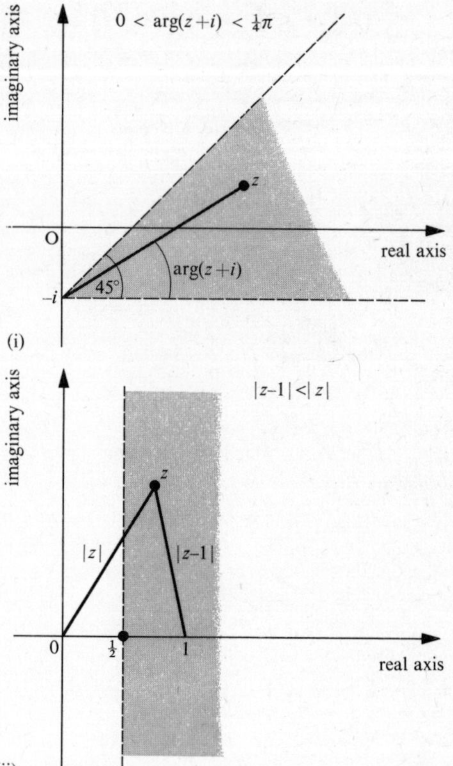

(ii)

Figure 18

(b) Locus of P_2 is the ellipse

$$\left(\frac{x}{5}\right)^2+\left(\frac{y}{3}\right)^2=1.$$

(c) $\alpha=n\pi, 2n\pi\pm\frac{2}{3}\pi$ (n integral).

12

(a) Make the denominator real. Show that $A=\frac{1}{2}, B=2$.

(b) Use $\cos^2 x=\frac{1}{2}(1+\cos 2x)$; square this and repeat the use of the formula to find $\cos^4 x=\frac{1}{8}(3+4\cos 2x+\cos 4x)$.

(c) $\cos 6x=\text{Re}\{(\cos x+i\sin x)^6\}$.
$=\cos^6 x-15\cos^4 x\sin^2 x+15\cos^2 x\sin^4 x-\sin^6 x$.

Chapter Three
Further Topics in Algebra

A Determinants: Theory summary

1 *Definition*

An nth order *determinant* consists of $n \times n$ numbers arranged in a square array thus:

$$\Delta = \begin{vmatrix} a_{11} & a_{12} & \dots & a_{1n} \\ a_{21} & a_{22} & \dots & a_{2n} \\ \vdots & \vdots & & \vdots \\ a_{n1} & a_{n2} & \dots & a_{nn} \end{vmatrix} \equiv |a_{ij}|.$$

Δ has a value which is obtained by following the rules given in 3.A.4 below. The numbers a_{ij} are called *elements*, and the subscripts define the position of the element in the array, e.g. a_{ik} is located in the ith row and kth column.

2 *Minor*

A *minor* of a_{ik} in a given determinant is the determinant D_{ik} formed by deleting all the elements in the ith row and kth column, e.g. the minor D_{12} of the element a_{12} in the third-order determinant

$$\Delta = \begin{vmatrix} a_{11} & a_{12} & a_{13} \\ a_{21} & a_{22} & a_{23} \\ a_{31} & a_{32} & a_{33} \end{vmatrix} \qquad \textbf{3.1}$$

is $D_{12} = \begin{vmatrix} a_{21} & a_{23} \\ a_{31} & a_{33} \end{vmatrix}$.

3 *Cofactor*

The signed minor $(-1)^{i+k} D_{ik}$ is called the *cofactor* of a_{ik} and is usually denoted by Δ_{ik},

i.e. $\Delta_{ik} = (-1)^{i+k} D_{ik}$.

4 *The numerical value of a determinant*

(a) The numerical value of a second-order determinant

$$\begin{vmatrix} a_{11} & a_{12} \\ a_{21} & a_{22} \end{vmatrix} = a_{11}a_{22} - a_{12}a_{21}.$$

(b) The value of an nth order determinant Δ is equal to the sum of the n products obtained by multiplying each element of any row or column in Δ by its corresponding cofactor; e.g. on expanding down the second column of the determinant **3.1**, we have

$$\Delta = a_{12}\Delta_{12} + a_{22}\Delta_{22} + a_{32}\Delta_{32}$$
$$= -a_{12}\begin{vmatrix} a_{21} & a_{23} \\ a_{31} & a_{33} \end{vmatrix} + a_{22}\begin{vmatrix} a_{11} & a_{13} \\ a_{31} & a_{33} \end{vmatrix} - a_{32}\begin{vmatrix} a_{11} & a_{13} \\ a_{21} & a_{23} \end{vmatrix}.$$

5 *General properties of determinants*

(a) The value of determinant is unchanged if all its rows are systematically interchanged with all its columns, i.e. row 1 becomes column 1, row 2 becomes column 2, etc.

(b) The interchange of any two rows or any two columns will produce a determinant of the same magnitude as the original but of opposite sign.

(c) If any two rows or columns are identical the value of the determinant is zero.

(d) The value of a determinant is unchanged if to a row (or column) is added a multiple of another row (or column),

e.g. $\begin{vmatrix} a_{11} & a_{12} & a_{13} \\ a_{21} & a_{22} & a_{23} \\ a_{31} & a_{32} & a_{33} \end{vmatrix} = \begin{vmatrix} a_{11}+\lambda a_{13} & a_{12} & a_{13} \\ a_{21}+\lambda a_{23} & a_{22} & a_{23} \\ a_{31}+\lambda a_{33} & a_{33} & a_{33} \end{vmatrix}$.

B Matrices: Theory summary

1

Definitions

(a) A *matrix* is a rectangular array of numbers set out in rows and columns and enclosed in a pair of brackets. A matrix consisting of m rows and n columns has the form

$$\begin{bmatrix} a_{11} & a_{12} & \dots & a_{1n} \\ a_{21} & a_{22} & \dots & a_{2n} \\ \vdots & \vdots & & \vdots \\ a_{m1} & a_{m2} & \dots & a_{mn} \end{bmatrix} \equiv [a_{ij}]$$

and is said to be of order $m \times n$. If $m = n$ the matrix is called a *square matrix*; if $m = 1$, a *row matrix*; if $n = 1$, a *column matrix*.

(b) The *unit matrix* I is a square matrix in which each element of the principal diagonal is unity and every other element is zero.

The *zero* or *null matrix* O is a matrix in which every element is zero.

e.g. the 3×3 unit matrix and the 3×1 null matrix are given respectively by:

$$\mathrm{I} = \begin{bmatrix} 1 & 0 & 0 \\ 0 & 1 & 0 \\ 0 & 0 & 1 \end{bmatrix}, \qquad \mathrm{O} = \begin{bmatrix} 0 \\ 0 \\ 0 \end{bmatrix}.$$

(c) If the rows and columns of a matrix A are interchanged the resulting matrix $\tilde{A}$ is known as the *transpose* of A. If $A = \tilde{A}$, A must be square and symmetrical about its principal diagonal, i.e. $a_{ij} = a_{ji}$.

2 *Addition and subtraction*

(a) $A \pm B = [a_{ij}] \pm [b_{ij}] = [a_{ij} \pm b_{ij}]$,

where $A = [a_{ij}]$, $B = [b_{ij}]$, and A and B must be of the same order.

(b) The associative law holds for the addition of matrices,

i.e. $A+B+C = (A+B)+C = A+(B+C)$.

3 *Multiplication*

(a) *Scalar multiplication.* To multiply a matrix A by a scalar λ we multiply all elements a_{ij} by λ,

i.e. $\lambda A = \lambda[a_{ij}] = [\lambda a_{ij}]$.

(b) Multiplication of two matrices is possible only if the number of columns in the 'pre-multiplier' matrix is equal to the number of rows in the 'post-multiplier' matrix.

The product of matrices $[a_{ij}]$ and $[b_{ij}]$ of respective orders $m_1 \times r$ and $r \times n_2$ is $[a_{ij}][b_{ij}] = [c_{ij}]$,

where $c_{ij} = \sum_{k=1}^{r} a_{ik} b_{kj}$;

e.g. $$\begin{bmatrix} a_{11} & a_{12} & a_{13} \\ a_{21} & a_{22} & a_{23} \\ a_{31} & a_{32} & a_{33} \end{bmatrix} \begin{bmatrix} b_{11} \\ b_{21} \\ b_{31} \end{bmatrix} = \begin{bmatrix} a_{11}b_{11}+a_{12}b_{21}+a_{13}b_{31} \\ a_{21}b_{11}+a_{22}b_{21}+a_{23}b_{31} \\ a_{31}b_{11}+a_{32}b_{21}+a_{33}b_{31} \end{bmatrix},$$

$$\begin{bmatrix} a_{11} & a_{12} \\ a_{21} & a_{22} \end{bmatrix} \begin{bmatrix} b_{11} & b_{12} \\ b_{21} & b_{22} \end{bmatrix} = \begin{bmatrix} a_{11}b_{11}+a_{12}b_{21} & a_{11}b_{12}+a_{12}b_{22} \\ a_{21}b_{11}+a_{22}b_{21} & a_{21}b_{12}+a_{22}b_{22} \end{bmatrix}.$$

(c) Matrix multiplication is in general non-commutative,

i.e. $AB \neq BA$.

(d) The associative law holds for the multiplication of matrices,

i.e. $ABC = (AB)C = A(BC)$.

4 *Inversion*

(a) The inverse matrix A^{-1} of a square matrix A is such that $AA^{-1} = A^{-1}A = I$.

(b) To find the inverse of A:

(i) Interchange rows and columns of A, i.e. find the transposed matrix $\tilde{A}$.

(ii) Replace each element in $\tilde{A}$ by its cofactor. The resulting matrix is known as the *adjugate* or *adjoint matrix*, adj A.

(iii) The inverse matrix A^{-1} is then formed by dividing each element in adj A by the determinant $|A|$ of the original matrix.

Thus $A^{-1} = \frac{1}{|A|} \text{adj } A$.

C Systems of linear equations: Theory summary

1 *Solution of linear equations*

E.g. $$a_{11}x+a_{12}y+a_{13}z = d_1,$$
$$a_{21}x+a_{22}y+a_{23}z = d_2,$$
$$a_{31}x+a_{32}y+a_{33}z = d_3,$$

or $$\begin{bmatrix} a_{11} & a_{12} & a_{13} \\ a_{21} & a_{22} & a_{23} \\ a_{31} & a_{32} & a_{33} \end{bmatrix} \begin{bmatrix} x \\ y \\ z \end{bmatrix} = \begin{bmatrix} d_1 \\ d_2 \\ d_3 \end{bmatrix}.$$

(a) *By determinants (Cramer's rule).* The solution is

$$x = \frac{\Delta_x}{\Delta}, \quad y = \frac{\Delta_y}{\Delta}, \quad z = \frac{\Delta_z}{\Delta};$$

provided $\Delta \neq 0$, and where

$$|A| \equiv \Delta = \begin{vmatrix} a_{11} & a_{12} & a_{13} \\ a_{21} & a_{22} & a_{23} \\ a_{31} & a_{32} & a_{33} \end{vmatrix}, \quad \Delta_x = \begin{vmatrix} d_1 & a_{12} & a_{13} \\ d_2 & a_{22} & a_{23} \\ d_3 & a_{32} & a_{33} \end{vmatrix},$$

$$\Delta_y = \begin{vmatrix} a_{11} & d_1 & a_{13} \\ a_{21} & d_2 & a_{23} \\ a_{31} & d_3 & a_{33} \end{vmatrix}, \quad \Delta_z = \begin{vmatrix} a_{11} & a_{12} & d_1 \\ a_{21} & a_{22} & d_2 \\ a_{31} & a_{32} & d_3 \end{vmatrix}$$

$\Delta_x, \Delta_y, \Delta_z$ being derived from Δ by replacing the elements in the first, second and third columns respectively, by d_1, d_2, d_3.

(b) *By inversion.*

$$\begin{bmatrix} x \\ y \\ z \end{bmatrix} = A^{-1} \begin{bmatrix} d_1 \\ d_2 \\ d_3 \end{bmatrix}, \quad A^{-1} = \frac{1}{\Delta} \begin{bmatrix} \Delta_{11} & \Delta_{21} & \Delta_{31} \\ \Delta_{12} & \Delta_{22} & \Delta_{32} \\ \Delta_{13} & \Delta_{23} & \Delta_{33} \end{bmatrix},$$

$\Delta_{11}, \Delta_{21}, \ldots, \Delta_{33}$ being the cofactors of $a_{11}, a_{21}, \ldots, a_{33}$ in $\Delta = |A|$.

2 *Linear dependence and independence*

The columns $C_1, C_2, \ldots, C_n$ of a matrix A of order $m \times n$ are linearly dependent if there are numbers $\lambda_1, \lambda_2, \ldots, \lambda_n$ which are not all zero such that

$\lambda_1 C_1 + \lambda_2 C_2 + \ldots + \lambda_n C_n = O$ of order $m \times 1$.

However, if the above matrix equation can only be satisfied when $\lambda_1 = \lambda_2 = \ldots = \lambda_m = 0$, then the columns are linearly independent. The same argument applies to the rows of A.

3 *The rank r of a matrix*

The rank of a matrix, rectangular or square, is equal to the order *r* of the highest-order non-zero minor contained in the matrix array. Or equivalently *r* is the number of linearly independent rows.

4 *Consistency or compatibility of sets of equations*

Consider a set of *m* equations in *n* unknowns:

$$a_{11}x_1+a_{12}x_2+\ldots+a_{1n}x_n = d_1,$$
$$a_{21}x_1+a_{22}x_2+\ldots+a_{2n}x_n = d_2,$$
$$\vdots$$
$$a_{m1}x_1+a_{m2}x_2+\ldots+a_{mn}x_n = d_m,$$

i.e. $$\begin{bmatrix} a_{11} & a_{12} & \ldots & a_{1n} \\ a_{21} & a_{22} & \ldots & a_{2n} \\ \vdots & \vdots & & \vdots \\ a_{m1} & a_{m2} & \ldots & a_{mn} \end{bmatrix} \begin{bmatrix} x_1 \\ x_2 \\ \vdots \\ x_n \end{bmatrix} = \begin{bmatrix} d_1 \\ d_2 \\ \vdots \\ d_m \end{bmatrix},$$

or in abbreviated form AX = D, where A in the $m \times n$ matrix formed by the coefficients $a_{11}, a_{12}, \ldots a_{mn}$; X is the column matrix of the unknowns $x_1, x_2, \ldots, x_n$; D is the column matrix composed of $d_1, d_2, \ldots, d_m$.

(a) *Homogeneous equations.* AX = O,

i.e. $d_1 = d_2 = \ldots = d_m = 0$.

(i) If $m = n$, a set of n homogeneous equations in n variables has a non-trivial solution (viz. not all $x_1, x_2, \ldots, x_n = 0$) if and only if the determinant $|A| = 0$.

(ii) $m \neq n$, the necessary and sufficient condition for the set to be consistent with a non-zero solution is that the rank of A is less than n.

(b) *Non-homogeneous equations.* AX = D. The condition for consistency of the set is that the rank of A and the rank of the augmented matrix [A, D] are equal, where the augmented matrix is defined as

$$[A, D] = \begin{bmatrix} a_{11} & a_{12} & \ldots & a_{1n} & d_1 \\ a_{21} & a_{22} & \ldots & a_{2n} & d_2 \\ \vdots & \vdots & & \vdots & \vdots \\ a_{m1} & a_{m2} & \ldots & a_{mn} & d_m \end{bmatrix}.$$

D Determinants and matrices: Illustrative worked problems

1 *Solve the equations*

$$ax+y+z = b,$$
$$x+ay+z = 2b,$$
$$x+y+az = 1$$

in terms of a and b.

Show that when a has either of two particular values, the solution obtained is inapplicable. Show further that when a has one of these values, the equations are inconsistent for all values of b; and that when a has the other value, the equations are inconsistent except when $b = -\frac{1}{3}$. Find y and z in terms of x in the latter case. [L, J 1968, FM V, Q 1]

Solution.

The determinant of the coefficients of x, y, z in the set of equations is

$$\Delta = \begin{vmatrix} a & 1 & 1 \\ 1 & a & 1 \\ 1 & 1 & a \end{vmatrix} = \begin{vmatrix} a-1 & 1 & 1 \\ 1-a & a & 1 \\ 0 & 1 & a \end{vmatrix} = (a-1)^2(a+2).$$

The solutions of the equations are (using Cramer's rule)

$$x = \frac{1}{\Delta}\begin{vmatrix} b & 1 & 1 \\ 2b & a & 1 \\ 1 & 1 & a \end{vmatrix} = \frac{(a-1)(ab-b-1)}{(a-1)^2(a+2)} = \frac{ab-b-1}{(a-1)(a+2)},$$

$$y = \frac{1}{\Delta}\begin{vmatrix} a & b & 1 \\ 1 & 2b & 1 \\ 1 & 1 & a \end{vmatrix} = \frac{(a-1)(2ab+b-1)}{(a-1)^2(a+2)} = \frac{2ab+b-1}{(a-1)(a+2)},$$

$$z = \frac{1}{\Delta}\begin{vmatrix} a & 1 & b \\ 1 & a & 2b \\ 1 & 1 & 1 \end{vmatrix} = \frac{(a-1)(a-3b+1)}{(a-1)^2(a+2)} = \frac{a-3b+1}{(a-1)(a+2)},$$

provided $\Delta \neq 0$, i.e. $a \neq 1$ and $a \neq -2$.

If $a = 1$ or $a = -2$, $\Delta = 0$ and the above solutions are clearly inapplicable. If $a = 1$, the set of equations becomes

$$x+y+z = b, \qquad \textbf{3.2}$$
$$x+y+z = 2b, \qquad \textbf{3.3}$$
$$x+y+z = 1. \qquad \textbf{3.4}$$

Clearly equations **3.2** and **3.4** are only consistent if $b = 1$, in which case equation **3.3** is inconsistent. Thus for $a = 1$, the equations are inconsistent for all values of b.

If $a = -2$, then on adding together the three equations, we obtain

$$(-2+1+1)x+(1-2+1)y+(1+1-2)z = 3b+1,$$
i.e. $0 = 3b+1$. **3.5**

Equation **3.5** can only be satisfied if $b = -\frac{1}{3}$. Thus on substituting $a = -2, b = -\frac{1}{3}$ the set becomes

$$-2x+y+z = -\tfrac{1}{3}, \qquad \textbf{3.2a}$$
$$x-2y+z = -\tfrac{2}{3}, \qquad \textbf{3.3a}$$
$$x+y-2z = 1, \qquad \textbf{3.4a}$$

and since **3.2a** + **3.3a** = −**3.4a**, equation **3.4a** adds no further information and may be ignored. Solving equations **3.2a** and **3.3a** yields

$$y = \tfrac{1}{9}+x, \qquad z = -\tfrac{4}{9}+x, \qquad x \text{ arbitrary.}$$

2

(i) Prove that

$$\begin{vmatrix} \sin^2 x & \sin 2x & \cos 2x \\ \sin^2 y & \sin 2y & \cos 2y \\ \sin^2 z & \sin 2z & \cos 2z \end{vmatrix} = -2\sin(y-z)\sin(z-x)\sin(x-y).$$

(ii) If A, B, C are matrices such that AB = C and

$$A = \begin{bmatrix} 2 & 3 \\ 5 & 4 \end{bmatrix}, \qquad C = \begin{bmatrix} -1 & 7 \\ -6 & 7 \end{bmatrix},$$

find B.

(iii) Use matrices to solve

$$\begin{bmatrix} 3 & 2 & -2 \\ 2 & -1 & -4 \\ -1 & 1 & 5 \end{bmatrix}\begin{bmatrix} x \\ y \\ z \end{bmatrix} = \begin{bmatrix} -5 \\ -4 \\ 7 \end{bmatrix}.$$

[L, J 1968, FM VI, Q 1]

Solution

(i)
$$\begin{vmatrix} \sin^2 x & \sin 2x & \cos 2x \\ \sin^2 y & \sin 2y & \cos 2y \\ \sin^2 z & \sin 2z & \cos 2z \end{vmatrix}$$

$$= \tfrac{1}{2}\begin{vmatrix} 1-\cos 2x & \sin 2x & \cos 2x \\ 1-\cos 2y & \sin 2y & \cos 2y \\ 1-\cos 2z & \sin 2z & \cos 2z \end{vmatrix}$$

$$= \tfrac{1}{2}\begin{vmatrix} 1 & \sin 2x & \cos 2x \\ 1 & \sin 2y & \cos 2y \\ 1 & \sin 2z & \cos 2z \end{vmatrix}$$

$$= \tfrac{1}{2}\begin{vmatrix} 1 & \sin 2x & \cos 2x \\ 0 & \sin 2y-\sin 2z & \cos 2y-\cos 2z \\ 0 & \sin 2z-\sin 2x & \cos 2z-\cos 2x \end{vmatrix}$$

$$= \tfrac{1}{2}\begin{vmatrix} 1 & \sin 2x & \cos 2x \\ 0 & 2\sin(y-z)\cos(y+z) & -2\sin(y+z)\sin(y-z) \\ 0 & 2\sin(z-x)\cos(z+x) & -2\sin(z+x)\sin(z-x) \end{vmatrix}$$

$$= \tfrac{1}{2}(-4)\sin(y-z)\sin(z-x)\{\cos(y+z)\sin(z+x) - \sin(y+z)\cos(z+x)\}$$

$$= -2\sin(y-z)\sin(z-x)\sin(x-y).$$

(ii) If AB = C, then B = $A^{-1}C$.

$$\text{where}\quad A^{-1} = \begin{bmatrix} 2 & 3 \\ 5 & 4 \end{bmatrix}^{-1}$$

$$= -\tfrac{1}{7}\begin{bmatrix} 4 & -3 \\ -5 & 2 \end{bmatrix}.$$

$$\text{Thus}\quad B = -\tfrac{1}{7}\begin{bmatrix} 4 & -3 \\ -5 & 2 \end{bmatrix}\begin{bmatrix} -1 & 7 \\ -6 & 7 \end{bmatrix}$$

$$= -\tfrac{1}{7}\begin{bmatrix} -4+18 & 28-21 \\ 5-12 & -35+14 \end{bmatrix}$$

$$= \begin{bmatrix} -2 & -1 \\ 1 & 3 \end{bmatrix}.$$

(iii) Write given equation as AX = C, then $X = A^{-1}C$,

$$\text{where}\quad A^{-1} = \begin{matrix} 3 & 2 & -2 \\ 2 & -1 & -4 \\ -1 & 1 & 5 \end{matrix}^{\,-1}$$

$$= -\tfrac{1}{17}\begin{bmatrix} -1 & -12 & -10 \\ -6 & 13 & 8 \\ 1 & -5 & -7 \end{bmatrix}.$$

$$\text{Hence}\quad X = \begin{bmatrix} x \\ y \\ z \end{bmatrix} = -\tfrac{1}{17}\begin{bmatrix} -1 & -12 & -10 \\ -6 & 13 & 8 \\ 1 & -5 & -7 \end{bmatrix}\begin{bmatrix} -5 \\ -4 \\ 7 \end{bmatrix}$$

$$= \begin{bmatrix} 1 \\ -2 \\ 2 \end{bmatrix}.$$

I.e. $x = 1, \quad y = -2, \quad z = 2.$

3 By reduction to echelon form find the general solution of the following system of equations:

$$\begin{aligned} w+2x+y+z &= 7, \\ w \qquad +y-z &= 1, \\ 2x+y+z &= 5, \\ -2w+\ x+y \qquad &= -2. \end{aligned}$$

Solution. On subtracting the third equation from the first, adding the third to the second, and then interchanging the third and fourth, we obtain

$$\begin{aligned} w \qquad\qquad &= 2, && \textbf{3.6} \\ w+2x+2y \quad &= 6, && \textbf{3.7} \\ -2w+\ x+\ y \quad &= -2, && \textbf{3.8} \\ 2x+\ y+z &= 5. && \textbf{3.9} \end{aligned}$$

Subtracting twice equation **3.8** from equation **3.7** gives $5w = 10$. Since this is a multiple of equation **3.6** this gives no new information and hence the system reduces to

$$\begin{aligned} w \qquad\qquad &= 2, && [\textbf{3.6}] \\ -2w+\ x+y \quad &= -2, && [\textbf{3.8}] \\ 2x+y+z &= 5. && [\textbf{3.9}] \end{aligned}$$

On eliminating w from equation **3.8** and y from equation **3.9**, we obtain the echelon system:

$$\begin{aligned} w \qquad &= 2, \\ x+y \quad &= 2, \\ x \quad +z &= 3, \end{aligned}$$

and the general solution:

$$\begin{aligned} w &= 2, \\ y &= 2-x = 2-t, \\ z &= 3-x = 3-t, \end{aligned}$$

where $x = t$ can take any value.

E Determinants and matrices: Problems

Answers and hints will be found on pp. 50–51.

1

(i) Use determinants to solve the following linear equations for x, y, z,

$$\begin{aligned} kx+\ y+\ z &= 1, \\ x+ky+\ z &= 1, \\ x+\ y+kz &= -2, \end{aligned}$$

where k is a constant.

Find the values of k for which the equations fail to have a unique solution and explain the nature of the solution, if any, for each of these values of k.

(ii) If A is the matrix $\begin{bmatrix} 1, & 1 \\ -1, & 2 \end{bmatrix}$, prove that

$$A^2 - 3A + 3I = O,$$

where I is the identity (unit) two-by-two matrix and O is the zero matrix. [L, S 1967, F M V, Q 1]

2 Express

$$\begin{vmatrix} a & b & c \\ b & c & a \\ c & a & b \end{vmatrix},$$

where a, b, c are real, finite and not all zero, as a product of two real factors.

Discuss the consistency of the simultaneous equations in x, y, z,

$$\begin{aligned} ax+by+cz &= a, \\ bx+cy+az &= b, \\ cx+ay+bz &= c, \end{aligned}$$

solving the equations as far as possible for the different cases which may arise. [L, J 1967, F M V, Q 1]

3

(i) If α, β, γ are the roots of the equation

$$t^3 - lt^2 + mt - n = 0$$

and the equations

$$\begin{aligned} (k+\alpha)^2x + k(k+2\alpha)(y+z) &= 0, \\ (k+\beta)^2y + k(k+2\beta)(z+x) &= 0, \\ (k+\gamma)^2z + k(k+2\gamma)(x+y) &= 0 \end{aligned}$$

are simultaneously true when x, y, z are not all zero, show that k is a root of the equation

$$(m^2 - 2nl)t^2 + 2mnt + n^2 = 0.$$

(ii) If ω is a complex cube root of unity,

(a) express the roots of the equation $\sum_{r=0}^{8} t^r = 0$ in terms of ω,

(b) simplify

$$\begin{vmatrix} x & \omega^n y & \omega^{2n} z \\ \omega^{2n} y & z & \omega^n x \\ \omega^n z & \omega^{2n} x & y \end{vmatrix},$$

where n is an integer. [L, S 1966, F M V, Q 1]

4

(i) When are two matrices conformable for multiplication? Express

$$[1 \quad 2 \quad -3]\begin{bmatrix}2 & 1 & 4\\1 & 0 & 3\\4 & 3 & 5\end{bmatrix}\begin{bmatrix}1\\2\\-3\end{bmatrix}$$

as a single matrix.

(ii) If $A = \begin{bmatrix}a_1 & b_1 & c_1\\a_2 & b_2 & c_2\\a_3 & b_3 & c_3\end{bmatrix}$ and X and B are matrices such that AX = B, prove that if this equation is solved as $X = A^{-1}B$, then

$$A^{-1} = \frac{1}{\Delta}\begin{bmatrix}A_1 & A_2 & A_3\\B_1 & B_2 & B_3\\C_1 & C_2 & C_3\end{bmatrix},$$

where Δ is the determinant of A and A_1 is the cofactor of a_1, etc., in Δ.

Solve for x, y and z, by this method,

$$\begin{bmatrix}1 & \omega & \omega^2\\\omega^2 & \omega^2 & 1\\1 & \omega & \omega\end{bmatrix}\begin{bmatrix}x\\y\\z\end{bmatrix} = \begin{bmatrix}1\\0\\\omega\end{bmatrix},$$

where ω is a complex cube root of unity.

[L, S 1966, FM VI, Q 1]

5

(i) If A is the matrix

$$\begin{bmatrix}\cos\theta & \sin\theta\\-\sin\theta & \cos\theta\end{bmatrix},$$

proved by induction or otherwise that

$$A^n = \begin{bmatrix}\cos n\theta & \sin n\theta\\-\sin n\theta & \cos n\theta\end{bmatrix},$$

where n is a positive integer.

(ii) If B, X and C are the matrices

$$B = \begin{bmatrix}2 & -1\\1 & 2\end{bmatrix}, \quad X = \begin{bmatrix}x_1 & y_1\\x_2 & y_2\end{bmatrix}, \quad C = \begin{bmatrix}10 & 5\\-5 & -10\end{bmatrix},$$

solve the matrix equation BX = C.

(iii) If 1, ω and ω^2 are the three cube roots of unity, simplify the determinant

$$\begin{vmatrix}\lambda+1 & \omega & \omega^2\\\omega & \lambda+\omega^2 & 1\\\omega^2 & 1 & \lambda+\omega\end{vmatrix}.$$

[L, J 1966, FM V, Q 1]

6

(i) Show that, if $abc \neq 0$, there are two distinct values of k for which the equations

$$a^2x+aby+acz = kx,$$
$$abx+b^2y+bcz = ky,$$
$$acx+bcy+c^2z = kz$$

are consistent and show that for one of these values the ratios $x:y:z$ are unique.

(ii) Solve the following simultaneous equations as completely as possible:

$$x+2y+2z = 2,$$
$$2x+5y+3z = 4,$$
$$px+8y+4z = q,$$

in the two cases (a) $p = 1, q = 1$, (b) $p = 3, q = 6$.

[L, J 1966, FM V, Q 2]

7 If A is the matrix $\begin{bmatrix}a & b\\c & d\end{bmatrix}$ and X is the non-zero matrix $\begin{bmatrix}x\\y\end{bmatrix}$, show that, if $AX = \lambda X$, where λ is a number, real or complex, then $\lambda^2-(a+d)\lambda+ad-bc = 0$.

Show that A itself satisfies this quadratic equation, in the sense that

$$A^2-(a+d)A+(ad-bc)I = 0,$$

where I is the matrix $\begin{bmatrix}1 & 0\\0 & 1\end{bmatrix}$ and O is the matrix $\begin{bmatrix}0 & 0\\0 & 0\end{bmatrix}$.

[O, S 1967, P II, Q 4]

8 In the matrices

$$A = \begin{bmatrix}0 & \omega\\\omega^2 & 0\end{bmatrix} \quad \text{and} \quad B = \begin{bmatrix}0 & \omega^2\\\omega & 0\end{bmatrix},$$

ω is a complex cube root of unity. Show that A, B, A^2, AB, ABA and ABAB are six distinct matrices. Show also that any product of any number of successive powers of A and B (e.g. AB^3A^2B) reduces to one of these six matrices.

[O, S 1967, PS, Q 4]

9

(i) Given that no two of the constants a, b, c are equal, solve for x, y, z the simultaneous equations

$$x+y+z = 0,$$
$$ax+by+cz = 0,$$
$$a^2x+b^2y+c^2z = 1,$$

giving your solutions in their simplest factorized form.

If $b = c$, prove (considering all possibilities) that the equations cannot be simultaneously true.

(ii) If the three straight lines

$$x+\alpha y+\alpha^3 = 0,$$
$$x+\beta y+\beta^3 = 0,$$
$$x+\gamma y+\gamma^3 = 0$$

are distinct and concurrent, find a relation between α, β, γ; and show that the point of concurrence is

$(-\alpha\beta\gamma, \beta\gamma+\gamma\alpha+\alpha\beta)$.

[C, S 1966, PS, Q 1]

10 Define the value of the third-order determinant

$$\begin{vmatrix}a_1 & a_2 & a_3\\b_1 & b_2 & b_3\\c_1 & c_2 & c_3\end{vmatrix}$$

and prove from your definition that the value is zero if any two rows or any two columns are equal.

Express the determinant

$$\begin{vmatrix}b^2c^2+a^2 & bc+a & 1\\c^2a^2+b^2 & ca+b & 1\\a^2b^2+c^2 & ab+c & 1\end{vmatrix}$$

as the product of six linear factors.

[C, S 1966, PS, Q 3]

11 If the equations

$$a_1x+b_1y+c_1z = 0,$$
$$a_2x+b_2y+c_2z = 0,$$
$$a_3x+b_3y+c_3z = 0$$

have a solution other than $x = y = z = 0$, show that

$$\begin{vmatrix} a_1 & b_1 & c_1 \\ a_2 & b_2 & c_2 \\ a_3 & b_3 & c_3 \end{vmatrix} = 0.$$

Find the values of λ for which the equations

$$4x-6y-z = \lambda x,$$
$$x-4y-z = \lambda y,$$
$$2x+3y+z = \lambda z$$

have a solution other than $x = y = z = 0$, and find the ratios $x:y:z$ for each of these values of λ.

[O & C, S 1966, M & HM V (PS), Q 9]

12

(a) Express as a product of linear factors

$$\begin{vmatrix} x^2 & ax+y+z & x \\ y^2 & x+ay+z & y \\ z^2 & x+y+az & z \end{vmatrix}.$$

(b) Prove that the equations

$$x-5y+2z = 1,$$
$$x+(k-1)y+4z = 2,$$
$$kx-4y+3kz = 3$$

have a unique solution provided $k \neq 2$ and $k \neq 4$.

If $k = 2$, prove that the three planes which the equations represent have a line in common. Find the coordinates of the points where this line intersects the planes $x = 0$, $y = 0$ and $z = 0$. [JMB, S 1967, FM I, Q 3]

13

(a) Given that the roots of the equation

$$x^3+qx+r = 0$$

are α, β, γ, express the value of the determinant

$$\begin{vmatrix} 1+\alpha & 1 & 1 \\ 1 & 1+\beta & 1 \\ 1 & 1 & 1+\gamma \end{vmatrix}$$

in terms of q and r.

(b) Show that $x = -6$ is the only real root of the equation

$$\begin{vmatrix} x & 4 & 2 \\ 1 & x & 5 \\ 3 & 3 & x \end{vmatrix} = 0.$$

[JMB, S 1967, P I, Q 10]

14 Express the determinant

$$\begin{vmatrix} a^2-a & a^3-a & a^4-a \\ b^2-b & b^3-b & b^4-b \\ c^2-c & c^3-c & c^4-c \end{vmatrix}$$

as the product of factors which are linear in a, b and c. Four points have Cartesian coordinates (a^r, b^r, c^r), $r = 1, 2, 3, 4$. Prove that they all lie in a plane if and only if at least one of the quantities a, b, c is equal to its square, or at least two of them are equal. Deduce that, if the four points are coplanar but not collinear, their plane must be one of nine certain planes.

[JMB, S 1966, FMS, Q 4]

15

(a) Prove that

$$\begin{vmatrix} a_1 & b_1 & c_1 \\ a_2 & b_2 & c_2 \\ a_3 & b_3 & c_3 \end{vmatrix} = \frac{1}{a_1}\begin{vmatrix} a_1b_2-a_2b_1 & a_1c_2-a_2c_1 \\ a_1b_3-a_3b_1 & a_1c_3-a_3c_1 \end{vmatrix},$$

and use this result to solve the system of equations

$$x+2y+z = 3,$$
$$2x-y+z = 2,$$
$$x+y-z = 4.$$

(b) Factorize

$$\begin{vmatrix} a & c & b \\ d & e & d \\ b & c & a \end{vmatrix}.$$

[W, S 1969, P I, Q 6]

16 (a) Prove that the equations

$$a\lambda^2+b\lambda+c = 0,$$
$$a'\lambda^2+b'\lambda+c' = 0$$

have a common solution if and only if

$$\begin{vmatrix} b & c \\ b' & c' \end{vmatrix}\begin{vmatrix} a & b \\ a' & b' \end{vmatrix} = \begin{vmatrix} c & a \\ c' & a' \end{vmatrix}^2.$$

(b) Find the most general solution of the equations

$$3x+4y+z = 1,$$
$$2x-y+3z = 0,$$
$$3x-7y+8z = -1.$$

[W, S 1968, PS, Q 5]

F Permutations, combinations and probability: Theory summary

1 *Definitions*

(a) A permutation is an arrangement of objects, regard being paid to the order of the objects in the arrangement.

(b) A combination is a selection of objects, no regard being paid to the order of selection of the objects.

(c) If an event can happen in a ways and fail to happen in b ways, where each of the $a+b$ ways is equally likely, then the probability of the event happening is $a/(a+b) = p$; that of its failing is $b/(a+b) = 1-p$. The odds on its happening is $a:b$, the odds against its happening is $1-a/b:1$.

2 *The r, s principle*

If one operation is possible in r ways, and for each of these ways a second operation is possible in s ways, then both operations can be performed in $r \times s$ ways.

3

(a) The number of permutations of n dissimilar objects taken r at a time is

$$n(n-1)(n-2)\ldots(n-r+1) = \frac{n!}{(n-r)!} = {}_nP_r \quad (\text{or } {}^nP_r).$$

(b) The number of permutations of n dissimilar objects taken all together (the number of ways of arranging n objects in line amongst themselves) $= n(n-1)(n-2)\ldots2.1 = n! = {}_nP_n$.

(c) The number of ways of arranging n dissimilar objects in a circle is $(n-1)!$

4

(a) The number of combinations of n dissimilar objects taken r at a time (i.e. the number of selections of r objects from the n) is

$$\frac{{}_nP_r}{r!} = \frac{n!}{r!(n-r)!} = {}_nC_r, \quad \text{also written as } {}^nC_r \text{ or } \binom{n}{r}.$$

(b) The number of ways of arranging n objects in line, when p are alike of one kind, q are alike of another kind, . . . , is

$$\frac{n!}{p!q!\dots} \quad (\text{where } p+q+\dots = n).$$

(c) The number of ways of dividing n objects into a number of unequal groups, the first to contain p, the second q, . . . is

$$\frac{n!}{p!q!\dots} \quad (p+q+\dots = n),$$

but if the groups are of equal size p and they number s, then the number of ways of dividing the n objects is

$$\frac{n!}{(p!)^s s!}.$$

5

The number of permutations of n dissimilar objects taken r at a time when repetitions are allowed is n^r.

6 The number of ways of selecting some or all of a given number n of dissimilar objects is

$$2^n - 1 = ({}_nC_1 + {}_nC_2 + \dots + {}_nC_n).$$

7 The number of ways of selecting from n dissimilar objects, p similar objects of one kind and q similar objects of another kind, if any number may be taken is

$$2^n(p+1)(q+1)-1.$$

8 *Probability definitions*
If A and B are events associated with an experiment, and $p(\mathrm{A})$ and $p(\mathrm{B})$ denote the probabilities of A and B happening, respectively, then the notation:

$p(\mathrm{B}|\mathrm{A})$ denotes the conditional probability of the event B, given that the event A has occurred;
$p(\mathrm{A} \cap \mathrm{B})$ denotes the probability of both A and B happening;
$p(\mathrm{A} \cup \mathrm{B})$ denotes the probability of either A or B happening;
$p(\bar{\mathrm{A}})$ denotes the probability of A not happening ($\bar{\mathrm{A}}$ is the complementary event to A).

9 $p(\mathrm{A}) = 1 - p(\bar{\mathrm{A}})$.

10 For any two events A, B

$$p(\mathrm{A} \cup \mathrm{B}) = p(\mathrm{A}) + p(\mathrm{B}) - p(\mathrm{A} \cap \mathrm{B}).$$

For mutually exclusive events, $p(\mathrm{A} \cap \mathrm{B}) = 0$, and
$p(\mathrm{A} \cup \mathrm{B}) = p(\mathrm{A}) + p(\mathrm{B})$.

11 The probability of an event A consisting of r outcomes equals the sum of the probabilities of the various individual outcomes making up A,

$$p(\mathrm{A}) = p_1 + p_2 + \dots + p_r.$$

12 Choosing an object at random from a set of objects means that each object has the same probability of being chosen as any other.

13 *Multiplication theorem of probability*

$$p(\mathrm{A} \cap \mathrm{B}) = p(\mathrm{B}|\mathrm{A})\,p(\mathrm{A}) \quad \text{or} \quad p(\mathrm{A}|\mathrm{B})\,p(\mathrm{B}).$$

14 *Theorem on total probability*
If in an experiment one and only one of the events B_i *occurs* ($i = 1, 2, \dots, k$), *and* A *is an event with respect to the set of all the events* B_i, *then*

$$p(\mathrm{A}) = p(\mathrm{A}|\mathrm{B}_1)\,p(\mathrm{B}_1) + \dots + p(\mathrm{A}|\mathrm{B}_k)\,p(\mathrm{B}_k).$$

15 The events A and B are independent if and only if

$$p(\mathrm{A} \cap \mathrm{B}) = p(\mathrm{A})\,p(\mathrm{B}).$$

G Permutations, combinations and probability: Illustrative worked problems

1 Write down, in factorial form, the number of different permutations which can be made using all the letters of the word

PARALLELEPIPED.

How many different permutations of the letters of this word can be made using

(a) three letters, of which the first and third are consonants and the second is a vowel,
(b) five letters, of which the first, third and fifth are consonants and the second and fourth are vowels? [L, S 1966, P I, Q 2]

Solution. Total number of letters in PARALLELEPIPED is 14, number of Ps, Ls and Es is 3, number of As is 2, and number of Rs, Is and Ds is 1 each. Therefore number of different permutations possible using all letters is

$$\frac{14!}{(3!)^3 2!(1!)^3}.$$

(a) Words will take the form (CVC) where (C) is a consonant, (V) a vowel.
Vowels may be chosen in three ways, i.e. A, E or I. Then words of the form:

P(V) may terminate in P, R, L or D,
i.e. $3 \times 4 = 12$ permutations;
L(V) may terminate in P, R, L or D,
i.e. $3 \times 4 = 12$ permutations;
R(V) may terminate in P, L or D,
i.e. $3 \times 3 = 9$ permutations;
D(V) may terminate in P, R or L,
i.e. $3 \times 3 = 9$ permutations.
Hence total number of permutations is 42.

(b) Words will take the form (CVCVC) where the vowels may appear as AA, EE, AE, AI, IA, EI, IE, that is, in eight different ways. Then if first and third letters are PP, LL, PL, LP, in each case fifth letter may be P, R, L or D, giving $4 \times 8 \times 4 = 128$ permutations. If first and third letters are PR, RP, PD, DP, LR, RL, LD, DL, fifth letter may be chosen in three ways giving $8 \times 8 \times 3 = 192$ permutations. If first and third letters are RD, DR, fifth letter can only be P or L, yielding $2 \times 8 \times 2 = 32$ permutations.
Hence total number of permutations is $128 + 192 + 32 = 352$.

2 A square board game consists of $n+1$ 'vertical' squares and $n+1$ 'horizontal' squares. A piece can move horizontally to the right or vertically upwards, one square at a time, according to the rules of the game. Prove that the total number of paths by which a piece starting from the lower left-hand corner can move to the upper right-hand corner is $(2n)!/(n!)^2$. If $n = 6$, what is the chance that the path traversed passes through the centre square of the board?

Solution. Let H denote a horizontal step to the right and V a vertical step upwards.

The path is described by a sequence of n Hs and n Vs in various juxtapositions since $2n$ steps have to be taken,

e.g. HHVHV . . . VHVV.

There are $(2n)!$ ways of arranging these, of which $n!$ correspond to H permutations and $n!$ to V permutations

Number of distinguishable paths is $\dfrac{(2n)!}{(n!)(n!)}$.

When $n = 6$, number of paths is $\dfrac{12!}{6!6!}$.

If the path has to pass through the centre square, then number of distinguishable paths is the product of the number of paths leading from the initial square to the centre square and the number of paths leading from the centre to the final square,

i.e. $\dfrac{(2\times 3)!}{3!\,3!}\times\dfrac{(2\times 3)!}{3!\,3!} = \dfrac{(6!)^2}{(3!)^4}$.

Probability of the path passing through the centre is

$$\frac{(6!)^2/(3!)^4}{12!/(6!)^2} = \frac{(6!)^4}{(3!)^4 12!}$$

$$= \frac{6\times 5\times 4\times 6\times 5\times 4\times 6\times 5\times 4}{12\times 11\times 10\times 9\times 8\times 7\times 3\times 2} = \frac{100}{231}.$$

3 Four brothers compare the seasons (winter, spring, summer and autumn) of their birthdays. Show that the probability that each of them is born in a different season equals (a) one sixth the probability that only one pair of them shares a birthday in the same season, and equals (b) six times the probability that all their birthdays fall in the same season.

Determine the probability of two brothers sharing a birthday in one season, and the other two sharing a birthday in another season.

Solution. Count the number of possibilities of attaching a season to each brother's birthday.

If no seasons are common, number of ways is $4! = 24$. If only two brothers share one season, number of ways is ${}^4C_2\times 4! = 144$. If two brothers share one season, and the other two share another season, number of ways is ${}^4C_3\times 4\times 3 = 48$. If all four brothers share one season, number of ways is 4. The total number of ways is $256 = 4^4$, as expected.

The ratios $\dfrac{24}{144} = \dfrac{1}{6}$ and $\dfrac{24}{4} = 6$ give the answers to (a) and (b) respectively.

For the final part the probability is $\dfrac{18}{256} = \dfrac{9}{128}$.

H Permutations, combinations and probability: Problems

1 *Answers and hints will be found on p. 52.*

(i) A touring party of 15 cricket players consists of 7 batsmen, 6 bowlers and 2 wicket-keepers. Each team of 11 players must include at least 5 batsmen, 4 bowlers and 1 wicket-keeper. How many different teams can be selected when

(a) 1 batsman and 1 wicket-keeper are injured,
(b) all the players are available for selection?

(ii) If A, B, C are the angles of a triangle, find in its simplest form the ratio of the sum $\sin A+\sin B+\sin C$ to the product

$\cos\frac{1}{2}A\cos\frac{1}{2}B\cos\frac{1}{2}C$. [L, J 1968, P I, Q 3].

2

(i) Find the number of ways in which a committee of 4 can be chosen from 6 boys and 6 girls

(a) if it must contain 2 boys and 2 girls,
(b) if it must contain at least 1 boy and 1 girl,
(c) if either the oldest boy or the oldest girl must be included, but not both.

(ii) If n is an integer, use the *method of induction* to prove that

$1^3+2^3+\ldots+n^3 = \frac{1}{4}n^2(n+1)^2$. [L, S 1970, P I, Q 2]

3 Assuming that marbles of the same colour are not distinguishable from one another, find

(a) the number of distinguishable ways of arranging 7 red marbles and 4 white marbles in a row, if no two white marbles are to be together and if a red marble is to be at each end of the row;
(b) the number of distinguishable ways in which 7 marbles, of which 3 are white, 3 are red and 1 is blue, can be arranged in a row so that no white marble is next to the blue one.

[O, S 1966, P I, Q 3]

4

(i) Expand $\{x-(1/x)\}^7$ by means of the binomial theorem. Use your result to evaluate $(9{\cdot}9)^7$ correct to the nearest thousand.

(ii) A pair of integers is selected from the set of positive integers $1, 2, 3, \ldots, n$. In how many ways may this be done? (In each pair the order of the integers is immaterial, e.g. (2,3) and (3, 2) count as one pair only.)

If the integers in each pair are multiplied together show that, in the case when n is odd, the number of products which will be odd integers is $\frac{1}{8}(n^2-1)$. If n is large show that this number is approximately one-quarter of the total number of products.

[C, S 1966, P I, Q 5]

5

(i) Prove that

$$\sum_{r=1}^{n} r^2(r+1) = \tfrac{1}{12}n(n+1)(n+2)(3n+1).$$

(ii) A party of 4 men, 4 ladies and 6 children seats itself at random at a round table. Determine the chance that a given man will find himself between (a) 2 children, (b) a lady and a child.

[C, S 1966, P III, Q 3]

6

(a) Express $1/\{r(r+2)\}$ in partial fractions. Hence, or otherwise, prove that

$$\sum_{r=n}^{2n}\frac{1}{r(r+2)} = \frac{1}{2n}+\frac{1}{4(n+1)}-\frac{1}{2(2n+1)}.$$

(b) A box contains $n+2$ cards, consisting of 2 identical red cards, and n other cards each of a different colour, none of them red. If n of these cards are to be placed in a line, find the number of different arrangements of the colours (i) when only one red card is used, (ii) when both red cards are used.

Show that, when the choice of the n cards is unrestricted, the *total* number of arrangements of the colours is

$(n+1)!+\frac{1}{4}n(n-1)(n!)$. [JMB, S 1967, FM I, Q 1]

7 A set of n cups and saucers consists of one cup and one saucer in each of n different colours. Show that the number of ways in which r pairs of cups and saucers (not necessarily matching) can be chosen from the set is

$$\frac{1}{r!}\left[\frac{n!}{(n-r)!}\right]^2.$$

Denoting by $N(n, r)$ the number of ways of choosing r non-matching pairs of cups and saucers from the set of n, prove that

$$\frac{1}{r!}\left[\frac{n!}{(n-r)!}\right]^2 = N(n, r)+{}_nC_1N(n-1, r-1)+\ldots+$$
$$+{}_nC_{r-1}N(n-r+1, 1)+{}_nC_r.$$

Determine the numbers of ways of choosing three non-matching pairs of cups and saucers from a set of six matching pairs in six different colours.

[JMB, S 1966, FM Sp, Q 1]

8 Five identical red and five identical white balls are to be placed in three boxes subject to the following conditions:

(a) Of the balls in box 1 at most one is red; of those in box 2 at most two are red and of those in box 3 at most three are red;
(b) each box must contain at least one ball;
(c) every ball must be used.

In how many ways can all these conditions be satisfied simultaneously? [W, S 1969, P I, Q 3]

9

(a) Derive a formula for the number of permutations of $n_1+n_2+\ldots+n_r$ objects, of which n_1 are alike of one kind, n_2 alike of another kind, and so on.

How many different arrangements are there of the letters in 'Abertawe'?

(b) A committee consists of four men and their wives. In how many ways can a sub-committee be formed, each consisting of three people, subject to the condition that a man and his wife may not serve on the same sub-committee?

[W, S 1968, P I, Q 9]

I Convergence of series: Theory summary

1 *Definitions*

With a sequence $\{a_n\}$, associate another sequence $\{A_n\}$ where

$$A_n = \sum_{r=1}^{n} a_r.$$

The infinite series $\sum_1^\infty a_n$ has a_n as the nth term and A_n as the nth partial sum.

If $\{A_n\}$ converges, $\lim_{n\to\infty} A_n$ is the sum of the series $\sum a_n$. A series for which $\sum |a_n|$ is convergent is absolutely convergent.

2 *Theorems and tests for convergence*

(a) If $\sum a_n$ converges, $\sum \lambda a_n = \lambda \sum a_n$ (λ constant).

(b) If $\sum a_n$, $\sum b_n$ each converge,
$\sum (a_n+b_n) = \sum a_n+\sum b_n$.

(c) If $\sum a_n$ converges, then $a_n \to 0$.
(N.B. The converse is not necessarily true.)

(d) If $0 \leqslant a_n \leqslant kb_n$ for all large n, where k is a positive constant, and if $\sum b_n$ converges, so does $\sum a_n$. Important corollaries:

(i) If $a_n \geqslant 0$, $b_n > 0$, and $a_n/b_n \to L \neq 0$, then if $\sum b_n$ is convergent so is $\sum a_n$ and vice versa.

(ii) If $a_n > 0$, $b_n > 0$, $\sum b_n$ converges and $a_{n+1}/a_n \leqslant b_{n+1}/b_n$ for all large n, $\sum a_n$ wlll converge.

(e) *D'Alembert's ratio test*. If $a_n > 0$ and $\lim a_{n+1}/a_n = L$, then if $L < 1$, $\sum a_n$ converges.

(f) If $\sum |a_n|$ converges, so does $\sum a_n$.

(g) *Leibnitz's alternating-series test*. If u_n decreases to zero, $\sum (-1)^{n+1}u_n$ converges to a value between u_1 and u_1-u_2.

(h) *Maclaurin–Cauchy integral test*. If $f(x)$ is positive, continuous and decreasing on $x \geqslant 0$, then $\sum_0^\infty f(n)$ converges if $\int_0^\infty f(x)\,dx$ converges and vice versa. If the series and the integral are both convergent, then

$$0 < \sum_0^\infty f(n) - \int_0^\infty f(x)\,dx < f(0).$$

Also $\sum_0^N f(n) - \int_0^N f(x)\,dx$ decreases to a limit which lies strictly between 0 and $f(0)$.

J Convergence of series: Illustrative worked problems

1

(i) If $a_{n+1} > a_n$ for all positive integers n, and if

$$b_n = \frac{1}{n}\sum_{r=1}^{n} a_r,$$

prove that $b_{n+1} > b_n$.

(ii) For real values of k find

$$\lim_{x\to\infty} \frac{(\log_e x)^k}{x^{k+1}}.$$

(iii) Find for what real values of x and of b the series

$$\sum_{r=1}^{\infty} r^b(x+1)^r$$

is convergent. [L, S 1967, FM VI, Q 1]

Solution.

(i) $b_{n+1} = \dfrac{1}{n+1}\sum_{r=1}^{n+1} a_r$ and $b_n = \dfrac{1}{n}\sum_{r=1}^{n} a_r$.

$$b_{n+1}-b_n = \frac{1}{n(n+1)}\{n(a_1+a_2+\ldots+a_{n+1})-$$
$$-(n+1)(a_1+a_2+\ldots+a_n)\}$$
$$= \frac{na_{n+1}-(a_1+\ldots+a_n)}{n(n+1)}$$
$$\equiv \frac{(a_{n+1}-a_1)+(a_{n+1}-a_2)+\ldots+(a_{n+1}-a_n)}{n(n+1)},$$

which is positive, since each term in the brackets is positive because, since $a_{n+1} > a_n$ for all n, $a_m > a_n$ for all $m > n$.

Hence $b_{n+1} > b_n$.

(ii) $$\lim_{x\to\infty}\frac{(\log_e x)^k}{x^{k+1}}$$
$$= \lim_{y\to\infty}\frac{y^k}{e^{(k+1)y}}, \quad \text{where } y = \log_e x \text{(or } x = e^y\text{)},$$
$$= \lim_{y\to\infty}\frac{y^k}{1+(k+1)y+\ldots+\{(k+1)^k/k!\}y^k}$$
$$= \lim_{y\to\infty}\frac{1}{1/y^k+\ldots+(k+1)^k/k!+\{(k+1)^{k+1}/(k+1)!\}y+\ldots}$$
$$= 0.$$

(iii) Let $a_r = r^b(x+1)^r$ and so $a_{r+1} = (r+1)^b(x+1)^{r+1}$.

Then $$\frac{a_{r+1}}{a_r} = \left[\frac{r+1}{r}\right]^b(x+1).$$

The given series is convergent if

$$\lim_{r\to\infty}\left|\left[\frac{r+1}{r}\right]^b(x+1)\right| < 1,$$

i.e. if $|x+1| < 1$,

i.e. if $-1 < x+1 < 1$,

i.e. if $-2 < x < 0$

If $x = 0$, the series is

$$\sum_{r=1}^{\infty} r^b = 1^b+2^b+3^b+\ldots = 1+\frac{1}{2^{-b}}+\frac{1}{3^{-b}}+\ldots$$

and this series will converge if $-b > 1$, that is, $b < -1$ (from the integral test).

If $x = -2$, the series is $\sum_{r=1}^{\infty}(-1)^r r^b$. This alternating series will converge if the terms decrease in magnitude,

i.e. if $r^b > (r+1)^b$, i.e. if $b < 0$.

2

If $f(x)$ is a positive decreasing continuous function of x (i.e. $0 < f(x+\alpha) \leqslant f(x)$ for all $\alpha > 0$), prove, by considering a suitable area under the curve $y = f(x)$, that

$$f(n) \leqslant \int_{n-1}^{n} f(x)\,dx \leqslant f(n-1),$$

where n is a positive integer.

Deduce that $$S_n - f(1) \leqslant \int_1^n f(x)\,dx \leqslant S_n - f(n),$$

where $S_n = \sum_{r=1}^{n} f(r)$.

Show, further, that if

$\int_1^n f(x)\,dx$ tends to a finite limit, then

$\sum f(n)$ is a convergent series, but that if the limit is infinite, then $\sum f(n)$ is divergent.

Discuss the convergence of the following series:

(i) $\sum \dfrac{1}{n\sqrt{n}}$,

(ii) $\sum \dfrac{1}{n\log_e n}$. [L, J 1966, FM VI, Q 2]

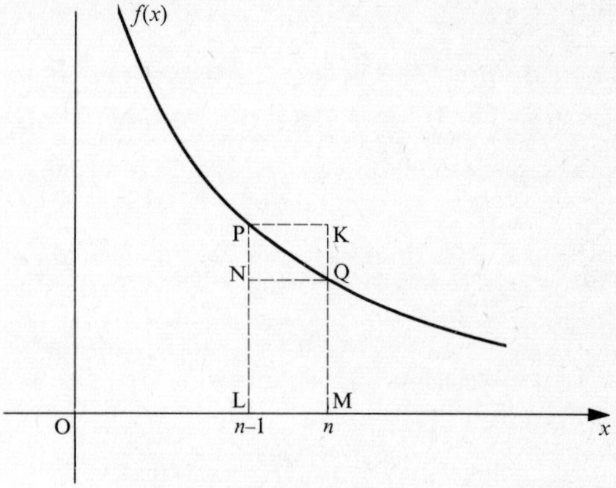

Figure 19

Solution. Consider the area under the curve $y = f(x)$ between the points P and Q at which $x = n-1$ and $x = n$ (see Figure 19). This area lies between the area of rectangle LMQN and the area of rectangle LMKP,

i.e. $$\text{QM}\times\text{LM} \leqslant \int_{n-1}^{n} f(x)\,dx \leqslant \text{PL}\times\text{LM}$$

i.e. $$f(n) \leqslant \int_{n-1}^{n} f(x)\,dx \leqslant f(n-1) \quad \text{since LM} = 1.$$

Replace n by r and sum this result for $r = 2, 3, \ldots, n$, obtaining

$$f(2)+f(3)+\ldots+f(n)$$
$$\leqslant \int_1^2 f(x)\,dx+\int_2^3 f(x)\,dx+\ldots+\int_{n-1}^{n} f(x)\,dx$$
$$\leqslant f(1)+f(2)+\ldots+f(n-1).$$

Now $S_n = \sum_{r=1}^{n} f(v)$,

and the sum of the integrals is

$$\int_1^n f(x)\,dx,$$

so the inequality reads

$$S_n - f(1) \leqslant \int_1^n f(x)\,dx \leqslant S_n - f(n),$$

and so $S_n \leqslant \int_1^n f(x)\,dx + f(1)$ **3.10**

and $S_n \geqslant \int_1^n f(x)\,dx + f(n)$. **3.11**

We also know that $\{S_n\}$ is an increasing sequence, since

$S_{n+1} - S_n = f(n+1) > 0$,

hence, from relatlon **3.10** if

$$\lim_{n\to\infty} \int_1^n f(x)\,dx$$

exists and is finite, $\lim S_n$ exists

$$\left(\text{and is less than or equal to } \int_1^\infty f(x)\,dx + f(1)\right),$$

i.e. $\sum f(n)$ is convergent. But, from relation **3.11**, if

$$\lim_{n\to\infty} \int_1^n f(x)\,dx$$

is infinite, $\{S_n\}$ diverges and therefore $\sum f(n)$ diverges.

In example (i) take $f(x) = 1/(x\sqrt{x})$. This function is positive, continuous and decreasing.

$$\int_1^n \frac{1}{x\sqrt{x}}\,dx = \int_1^n x^{-3/2}\,dx = -2\left[\frac{1}{\sqrt{x}}\right]_1^n = 2\left(1 - \frac{1}{\sqrt{n}}\right) \to 2$$

(finite),

so that $\sum \frac{1}{n\sqrt{n}}$ converges.

In example (ii) take $f(x) = 1/(x \log_e x)$. This function is also positive, continuous and decreasing, and

$$\int_2^n \frac{1}{x \log_e x}\,dx = [\log_e(\log_e x)]_2^n = \log_e(\log_e n) - \log_e(\log_e 2)$$

$$\to \infty,$$

so the series $\sum_{n=2} \frac{1}{n \log_e n}$ diverges also.

(Note that the series cannot start with the term for which $n = 1$.)

K Convergence of series: Problems

1 *Answers and hints will be found on pp. 52–4.*

(i) Find the two conditions that the sequence $a_1, a_2, \ldots a_n, \ldots$,

where $a_n = pn^2 - qn + r$,

p, q, r being real constants, is such that $a_{n+1} \geqslant a_n$ for all n.

(ii) The sequence $b_1, b_2, \ldots b_n, \ldots$ is defined by

$$b_n = \left(1 + \frac{1}{n}\right) \sin \frac{n\pi}{2}.$$

Determine the behaviour of its sub-sequences for the separate cases n odd and n even and find the limits of the sub-sequences.

(iii)

(a) Find $\lim_{x\to 0} (1-2x)^{3/x}$.

(b) Find, when y decreases through positive values to zero,

$$\lim \left[\frac{y^2 \log_e y}{\sin y}\right].$$

[L, J 1967, FM V, Q 2]

2

(i) Discuss the convergence of the infinite geometric series $\sum ar^n$ and deduce that, if $\lim_{n\to\infty} u_n^{1/n} < 1$, a series $\sum u_n$ of positive terms is convergent.

If an infinite series $u_1 + u_2 + \ldots + u_n + \ldots$ is such that

$$u_{2m-1} = a^m, \qquad u_{2m} = k^m a^m,$$

where $1 < k < 1/a$, find

$$\lim_{n\to\infty} \frac{u_{n+1}}{u_n} \quad \text{and} \quad \lim_{n\to\infty} u_n^{1/n}$$

for the cases $n = 2m$, $n = 2m+1$ and test the series for convergence.

(ii) Test for convergence the series for which the nth terms are

(a) $(-1)^{n-1}\left(\sec \frac{\alpha}{n} - 1\right)$,

(b) $\dfrac{x^n}{x^{n-1}+n}$ $(x > 0)$.

[L, J 1967, FM VI, Q 4]

3

(i) In each of the following cases

(a) $u_n = e^{-nx} x^n$ $(x > 0)$,

(b) $u_n = \dfrac{u_{n-1}^2}{u_{n-1}+1}$ and $u_1 > 0$,

find $\lim_{n\to\infty} u_n$.

(ii) Determine the ranges of values of x for which the following series are convergent

(a) $\sum_{n=0}^{\infty} \dfrac{1}{(1+x)^n}$,

(b) $\sum_{n=1}^{\infty} \dfrac{(x-2)^n}{n}$,

(c) $\sum_{n=0}^{\infty} \dfrac{x^n}{1+x^{2n}}$.

[L, S 1966, FM VI, Q 2]

4

(i) $\lim_{x \to \infty} \{\sqrt{(4x^2+3x+1)}-2x\}$,

(ii) $\lim_{x \to 0} \dfrac{\sin^{-1} x - \sinh x}{x^4 \sin x}$,

(iii) $\lim_{x \to 0} \dfrac{(1+x)^{1/x}-e}{x}$. [L, J 1966, FM V, Q 4]

5

(a) Find $\lim_{n \to \infty} \dfrac{(n!)^2}{(2n)!}$.

(b) Find $\lim_{t \to 1} \dfrac{1-t^{2n+1}}{2^{n+1}(1-t)}$

and $\lim_{n \to \infty} \{(1+\cos\theta)(1+\cos^2\theta)\ldots(1+\cos^{2^n}\theta)\}$,

if θ is not an integral multiple of π. [W, S 1969, PS, Q 3]

6 What is meant by the statement '$f(x)$ tends to the limit l as x tends to a'?

Evaluate

(a) $\lim_{x \to \infty} x^2 e^{-x}$,

(b) $\lim_{x \to 0} \dfrac{\sin^3 x + x^4}{x^2}$,

(c) $\lim_{n \to \infty} \left[\dfrac{1^2+2^2+3^3+\ldots+n^2}{n^3+3n+1}\right]$. [W, S 1968, PS, Q 7]

L Determinants and matrices: Answers and hints

1

(i) The determinant of the coefficients, $\Delta = (k-1)^2(k+2)$;

$$x = y = \frac{1}{k-1}, \quad z = \frac{-2}{k-1} \quad \text{provided } k \neq 1.$$

If $k = 1$ equations are inconsistent.

If $k = -2$ equations are consistent but linearly dependent.

Solutions are $x = y = t, z = 1+t$ for all t.

(ii) $$A^2 = \begin{bmatrix} 0 & 3 \\ -3 & 3 \end{bmatrix}, \quad 3A = \begin{bmatrix} 3 & 3 \\ -3 & 6 \end{bmatrix}.$$

$$\text{Thus} \quad A^2 - 3A + 3I = \begin{bmatrix} 0-3+3 & 3-3+0 \\ -3+3+0 & 3-6+3 \end{bmatrix}$$

$$= \begin{bmatrix} 0 & 0 \\ 0 & 0 \end{bmatrix} = O.$$

2 $$\Delta = \begin{vmatrix} a & b & c \\ b & c & a \\ c & a & b \end{vmatrix} = \begin{vmatrix} a+b+c & b & c \\ b+c+a & c & a \\ c+a+b & a & b \end{vmatrix}$$

$$= (a+b+c)\begin{vmatrix} 1 & b & c \\ 1 & c & a \\ 1 & a & b \end{vmatrix}$$

$$= (a+b+c)(ab+bc+ca-a^2-b^2-c^2)$$

$$= \tfrac{1}{2}(a+b+c)[(a-b)^2+(b-c)^2+(c-a)^2].$$

The equations are consistent if the matrix of the set and the augmented matrix have the same rank.

The general solution, when $\Delta \neq 0$, is $x = 1, y = z = 0$.

If $a+b+c = 0$, the equations are linearly dependent and $x = 1+t, y = z = t$ for all t.

If $a = b = c$, then equations all reduce to $x+y+z = 1$ and $x = t, y = u, z = 1-t-u$ for all t and u.

3

(i) Use condition that determinant of the coefficients equals zero

(ii)

(a) $\omega^{\pm\frac{1}{3}}, \omega^{\pm\frac{2}{3}}, \omega^{\pm 1}, \omega^{\pm\frac{4}{3}}$.

(b) $xyz-(x^3+y^3+z^3)$ (by expanding determinant and using $\omega^3 = 1$).

4

(i) Number of columns of pre-multiplier matrix = number of rows of post-multiplier matrix.

$$[1 \quad 2 \quad -3]\begin{bmatrix} 2 & 1 & 4 \\ 1 & 0 & 3 \\ 4 & 3 & 5 \end{bmatrix}\begin{bmatrix} 1 \\ 2 \\ 3 \end{bmatrix} = [1 \quad 2 \quad -3]\begin{bmatrix} 16 \\ 10 \\ 25 \end{bmatrix} = -39.$$

(ii) If $AX = B$, then $A^{-1}AX = A^{-1}B$,

$$\text{but} \quad A^{-1}A = \frac{1}{\Delta}\begin{bmatrix} A_1 & A_2 & A_3 \\ B_1 & B_2 & B_3 \\ C_1 & C_2 & C_3 \end{bmatrix}\begin{bmatrix} a_1 & b_1 & c_1 \\ a_2 & b_2 & c_2 \\ a_3 & b_3 & c_3 \end{bmatrix}$$

$$= \frac{1}{\Delta}\begin{bmatrix} \Delta & 0 & 0 \\ 0 & \Delta & 0 \\ 0 & 0 & \Delta \end{bmatrix} = \begin{bmatrix} 1 & 0 & 0 \\ 0 & 1 & 0 \\ 0 & 0 & 1 \end{bmatrix},$$

i.e. $A^{-1}A = I$.

Thus $A^{-1}AX = IX = X = A^{-1}B$.

$$\text{Let} \quad A = \begin{bmatrix} 1 & \omega & \omega^2 \\ \omega^2 & \omega^2 & 1 \\ 1 & \omega & \omega \end{bmatrix},$$

$$\text{then} \quad A^{-1} = \frac{1}{1-2\omega+\omega^2}\begin{bmatrix} 1-\omega & 1-\omega^2 & 0 \\ 0 & \omega(1-\omega) & -(1-\omega) \\ 1-\omega^2 & 0 & -(1-\omega^2) \end{bmatrix}$$

$$\text{and} \quad \begin{bmatrix} x \\ y \\ z \end{bmatrix} = \frac{1}{(1-\omega)^2}\begin{bmatrix} 1-\omega \\ -\omega(1-\omega) \\ 2-\omega^2-\omega \end{bmatrix}.$$

5

(i) $$\text{Let} \quad A^k = \begin{bmatrix} \cos k\theta & \sin k\theta \\ -\sin k\theta & \cos k\theta \end{bmatrix},$$

$$\text{then} \quad A^{k+1} = \begin{bmatrix} \cos\theta & \sin\theta \\ -\sin\theta & \cos\theta \end{bmatrix}\begin{bmatrix} \cos k\theta & \sin k\theta \\ -\sin k\theta & \cos k\theta \end{bmatrix}$$

$$= \begin{bmatrix} \cos(k+1)\theta & \sin(k+1)\theta \\ -\sin(k+1)\theta & \cos(k+1)\theta \end{bmatrix}.$$

So if result is true when $n = k$, it is true when $n = k+1$. But result is defined when $n = 1$, so result is true for all positive integral n.

(ii) $$X = B^{-1}C = \tfrac{1}{5}\begin{bmatrix} 2 & 1 \\ -1 & 2 \end{bmatrix} 5 \begin{bmatrix} 2 & 1 \\ -1 & -2 \end{bmatrix}$$

$$= \begin{bmatrix} 3 & 0 \\ -4 & -5 \end{bmatrix} \equiv \begin{bmatrix} x_1 & y_1 \\ x_2 & y_2 \end{bmatrix}.$$

(iii) λ^3.

6

(i) Equations are consistent if $\begin{vmatrix} a^2-k & ab & ac \\ ab & b^2-k & bc \\ ac & bc & c^2-k \end{vmatrix} = 0,$

which occurs when $k = 0$ and $k = a^2+b^2+c^2$.

If $k = 0$, $x:y:z$ is not unique; if $k = a^2+b^2+c^2$, $x:y:z = a:b:c$.

(ii)

(a) $x = \frac{5}{2}, \quad y = z = -\frac{1}{8}$.

(b) $x = 2(1-2t), \quad y = z = t, \quad$ for all t.

7 If $AX = \lambda X \equiv \lambda IX$, then

$(A-\lambda I)X = O$, i.e. $\begin{bmatrix} a-\lambda & b \\ c & d-\lambda \end{bmatrix}\begin{bmatrix} x \\ y \end{bmatrix} = \begin{bmatrix} 0 \\ 0 \end{bmatrix}$.

The condition for a non-zero solution for X is

$$\begin{vmatrix} a-\lambda & b \\ c & d-\lambda \end{vmatrix} = 0,$$

hence $\quad (a-\lambda)(d-\lambda)-bc = 0,$

i.e. $\quad \lambda^2-(a+d)\lambda+ad-bc = 0.$

$$A^2 = AA = \begin{bmatrix} a^2+bc & b(a+d) \\ c(a+d) & bc+d^2 \end{bmatrix},$$

$$(a+d)A = \begin{bmatrix} a^2+ad & ab+bd \\ ac+cd & ad+d^2 \end{bmatrix},$$

$$(ad-bc)\begin{bmatrix} 1 & 0 \\ 0 & 1 \end{bmatrix} = \begin{bmatrix} ad-bc & 0 \\ 0 & ad-bc \end{bmatrix};$$

hence $\quad A^2-(a+d)A+(ad-bc)I = O.$

8 $\quad A^2 = \begin{bmatrix} 1 & 0 \\ 0 & 1 \end{bmatrix} = B^2, \quad AB = \omega\begin{bmatrix} \omega & 0 \\ 0 & 1 \end{bmatrix},$

$ABA = \begin{bmatrix} 0 & 1 \\ 1 & 0 \end{bmatrix}, \quad ABAB = \omega\begin{bmatrix} 1 & 0 \\ 0 & \omega \end{bmatrix},$

hence all six matrices are different.

Using $A^2 = B^2 = I$, $A^nB^mA^{n+1}B^{m+1}$ . . . reduces to one of the six matrices.

9

(i) Solve using Cramer's rule, and obtain

$$x = \frac{1}{(a-b)(a-c)}, \quad y = \frac{-1}{(a-b)(b-c)}, \quad z = \frac{1}{(b-c)(a-c)},$$

when no two of a, b and c are equal.

If $b = c$, eliminate $y+z$ from all the equations to give $(a-b)x = 0$ and $(a^2-b^2)x = 1$. Deduce that these cannot hold simultaneously.

(ii) Eliminate x and show that there is a solution for y provided

$$\begin{vmatrix} \alpha-\beta & \alpha^3-\beta^3 \\ \alpha-\gamma & \alpha^3-\gamma^3 \end{vmatrix} \equiv (\alpha+\beta+\gamma)(\gamma-\beta) = 0.$$

Hence $\quad \alpha+\beta+\gamma = 0,$

$y = \alpha\beta-\gamma^2 = \alpha\beta+\beta\gamma+\gamma\alpha$

and $\quad x = \alpha\beta(\alpha+\beta) = -\alpha\beta\gamma.$

10 The first part is straightforward. In the second part, manipulate the given determinant by rows, reduce it to a second-order determinant, and factorize to give $(a-1)(b-1)(c-1)(a-b)(b-c)(c-a)$.

11 Denoting the determinant by Δ, show by elimination of y, z that $\Delta x = 0$, and similarly that $\Delta y = 0$, $\Delta z = 0$. There is a solution other than $x = y = z = 0$ only if $\Delta = 0$.

The given equations can be written in homogeneous form. For a solution other than (0, 0, 0),

$$\begin{vmatrix} 4-\lambda & -6 & -1 \\ 1 & -(4+\lambda) & -1 \\ 2 & 3 & 1-\lambda \end{vmatrix} = 0,$$

giving $\quad (\lambda-3)(\lambda^2+2\lambda+1) = 0.$

Hence $\quad \lambda = 3 \quad$ and $\quad \lambda = -1.$

For $\lambda = 3$, show that $y = 0$ and $x = z$, whence $x:y:z = 1:0:1$. For $\lambda = -1$, show that $x:y:z = 3:4:9$.

12

(a) $(x+y+z)(x-y)(y-z)(z-x)$.

(b) Line intersects:

$x = 0$ at $y = 0$, $z = \frac{1}{2}$;

$y = 0$ at $x = 0$, $z = \frac{1}{2}$;

$z = 0$ at $y = \frac{1}{16}$, $x = \frac{11}{6}$.

13

(a) Determinant $= \alpha\beta+\beta\gamma+\gamma\alpha+\alpha\beta\gamma = q-r$.

(b) The determinantal equation is $(x+6)(x^2-6x+11) = 0$, which has one real root $x = -6$.

14 Determinant $= abc(a-1)(b-1)(c-1)(a-b)(b-c)(c-a)$. The nine planes are $x = 0, y = 0, z = 0, x = 1, y = 1, z = 1, x = y, y = z, z = x$.

15

(a) Factorize a_1 from the first row and alter the second and third rows to find

$$\text{determinant} = a_1\begin{vmatrix} 1 & \dfrac{b_1}{a_1} & \dfrac{c_1}{a_1} \\ 0 & b_2-a_2\dfrac{b_1}{a_1} & c_2-a_2\dfrac{c_1}{a_1} \\ 0 & b_3-a_3\dfrac{b_1}{a_1} & c_3-a_3\dfrac{c_1}{a_1} \end{vmatrix}$$

$$= a_1\begin{vmatrix} \dfrac{a_1b_2-a_2c_1}{a_1} & \dfrac{a_1c_2-a_2c_1}{a_1} \\ \dfrac{a_1b_3-a_3b_1}{a_1} & \dfrac{a_1c_3-a_3c_1}{a_1} \end{vmatrix},$$

which gives the result.

In the problem, use Cramer's rule, expanding the numerators and the denominator by the method indicated. These come to 18, 9, −9 and 9; hence $x = 2, y = 1, z = -1$.

(b) The determinant equals $a^2e+2bcd-b^2e-2acd = (a-b)\{(a+b)e-2cd\}$.

16

(a) Consider the equations to be in two unknowns λ^2, λ. Solve for them and relate them to give

$$(bc'-cb')(ab'-ba') = (ca'-ac')^2.$$

(b) The determinant of the set of coefficients on the l.h.s. is zero. One equation (say the third) is redundant. Put $z = \lambda$ and show that

$$x = \frac{1-13\lambda}{11}, \quad y = \frac{2+7\lambda}{11}.$$

M Permutations, combinations and probability: Answers and hints

1

(i)

(a) With 6 batsmen, 6 bowlers, 1 wicket-keeper available, team can be made up with 5 batsmen, 5 bowlers, 1 wicket-keeper or 6 batsmen, 4 bowlers, 1 wicket-keeper only. In first case, number of ways is ${}_6C_5 \times {}_6C_5 = 6 \times 6 = 36$, in second case, number of ways is $1 \times {}_6C_4 = 15$; Total number $= 36+15 = 51$.

(b) With 7 batsmen, 6 bowlers, 2 wicket-keepers, team can be made up with:
5 batsmen, 4 bowlers, 2 wicket-keepers in
${}_7C_5 \times {}_6C_4 \times 1 = 21 \times 15 = 315$ ways;
5 batsmen, 5 bowlers, 1 wicket-keeper in
${}_7C_5 \times {}_6C_5 \times 2 = 21 \times 6 \times 2 = 252$ ways;
6 batsmen, 4 bowlers, 1 wicket-keeper in
${}_7C_6 \times {}_6C_4 \times 2 = 7 \times 15 \times 2 = 210$ ways.
Total number of ways is 777.

(ii) 4:1.

2

(i)

(a) 2 boys can be chosen in ${}_6C_2 = 6!/2!4! = 15$ ways.
2 girls can also be chosen in ${}_6C_2 = 15$ ways.
Therefore total number of ways for committee to contain 2 boys + 2 girls $= 15 \times 15 = 225$.

(b) Number of ways for committee with

1 boy, 3 girls $= 6 \times {}_6C_3 = 6 \times 20 = 120$,
2 boys, 2 girls $= {}_6C_2 \times {}_6C_2 = 15 \times 15 = 225$,
3 boys, 1 girl $= {}_6C_3 \times 6 = 20 \times 6 = 120$

Therefore total number of ways $= 120+225+120 = 465$.

(c) Number of ways for committee with

oldest boy + 3 boys $= 1 \times {}_5C_3 = 10$,
oldest boy + 2 boys + 1 girl (not including oldest) $1 \times {}_5C_2 \times 5 = 50$,
oldest boy + 1 boy + 2 girls (not including oldest) $1 \times 5 \times {}_5C_2 = 50$,
oldest boy + 3 girls (not including oldest) $= 1 \times {}_5C_3 = 10$.

Therefore total number of ways with oldest boy $= 120$.
Similarly there are 120 with oldest girl. Hence 240 ways in all.

3

(a) Rows must be of the form R[WRWRWRW]R where 2R can be inserted within brackets in any distinguishable way. Total number of ways is 15.

(b) If BR starts or ends row, i.e. BR . . . or . . . RB, there are $5!/(3!2!) = 10$ ways each of arranging 3W and 2R. If . . . RBR . . . there are $(4!/3!)5 = 20$ ways of arranging 3W and 1R. Thus total number of distinguishable ways is $10+10+20 = 40$.

4

(i) $\left(x-\dfrac{1}{x}\right)^7 = x^7-7x^5+21x^3-35x+\dfrac{35}{x}-\dfrac{21}{x^3}+\dfrac{7}{x^5}-\dfrac{1}{x^7}$;

$9{\cdot}9^7 = 9\,321\,000$ to the nearest thousand.

(ii) Number of pairs of integers selected from 1,2,3, . . . , n is ${}_nC_2 = \frac{1}{2}n(n-1)$.

When n is odd, the number of odd products obtained is the number of pairs of integers which can be selected from the set 1, 3, 5, . . . , n. This is ${}_{\frac{1}{2}(n+1)}C_2 = (n^2-1)/8$. If n is large, this number $\approx n^2/8$, but the total number of products ${}_nC_2 \approx \frac{1}{2}n^2$, whence last result.

5

(i) Try induction.

(ii) Number of distinguishable ways of sitting round a table relative to a given man is

$$\frac{13!}{3!\,4!\,6!} = A, \text{ say.}$$

(a) Number of ways with one child on either side is

$$\frac{11!}{4!\,3!\,4!} = B, \text{ say.}$$

Man's chance is $B/A = 5/26$.

(b) Number of ways with one lady and one child on either side is

$$\frac{2 \times 11!}{3!\,3!\,5!} = C, \text{ say.}$$

Man's chance is $C/A = 4/13$.

6

(b) (i) $n.n!$ (ii) ${}_nC_2\dfrac{n!}{2!}$.

7 $N(6, 3) = 1420$.

8 First allocate the red balls in the boxes, labelled A, B, C. There are three possibilities, corresponding to 0, 2, 3; 1, 1, 3; 1, 2, 2. Now distribute the white balls amongst these reds, in the boxes. In the first case there are $1+2+3+4+5 = 15$ ways in the second case there are $1+2+3+4+5+6 = 21$ ways; in the third case also 21 ways. Total number of ways of satisfying requirements is $15+2 \times 21 = 57$.

9

(a) The number of permutations is

$$\frac{(n_1 + \ldots + n_r)!}{n_1!\,n_2! \ldots n_r!};$$

in ABERTAWE the number is $8!/(2!\,2!) = 10080$.

(b) Number of different sub-committees is number with 3W + number with (2M, 1W) + number with (1M, 2W) + number with 3M $= 4+2\times\binom{4}{2}+2\times\binom{4}{2}+4 = 32$.

N Convergence of series: Answers and hints

1

(i) $a_{n+1}-a_n = p(2n+1)-q \geqslant 0$ for all n, provided either $p = 0$ and $q \leqslant 0$ or $p > 0$ and $q \leqslant 3p$.

(ii) $b_{2m+1} = (-1)^m\left[1+\dfrac{1}{2m+1}\right]$. There is no limit as $m \to \infty$.

$b_{2m} = \left[1+\dfrac{1}{2m}\right]\sin m\pi = 0$, so $b_{2m} \to 0$ as $m \to \infty$.

(iii)

(a) Consider $\log_e(1-2x)^{3/x} = \frac{3}{x}\log_e(1-2x) = -6-6x-\ldots,$

on expanding, when $|2x| < 1$; and series tends to -6, as $x \to 0$.

Thus $\lim_{x\to 0}(1-2x)^{3/x} = e^{-6}$.

(b) $$\lim_{y\to 0+}\frac{y^2\log_e y}{\sin y} = \lim_{y\to 0+}\left[\frac{y}{\sin y}\times y\log_e y\right] = 1\times 0 = 0,$$

since $\lim_{y\to 0}\frac{y}{\sin y} = 1$ and $\lim_{y\to 0+}(y\log_e y) = 0$.

2

(i) $\sum ar^n$ converges if $|r| < 1$.

If $\lim u_n^{1/n} = r < 1$, put $r < \rho < 1$, and show for sufficiently large m,

$$\sum u_n \leqslant u_1+u_2+\ldots+u_{m-1}+(\rho^m+\rho^{m+1}+\ldots)$$

and hence deduce the convergence of $\sum u_n$.
In given series,

$$\frac{u_{2m+1}}{u_{2m}} = \frac{a}{k^m} \to 0 \quad \text{as } m \to \infty,$$

and

$$\frac{u_{2m+2}}{u_{2m+1}} = k^{m+1} \to \infty \text{ as } m \to \infty.$$

Also $u_{2m}^{1/2m} = \sqrt{(ka)}$ and is constant as $m \to \infty$, while $u_{2m+1}^{1/(2m+1)} = a^{(m+1)/(2m+1)} \to \sqrt{a}$ as $m \to \infty$ ($\sqrt{a} < 1$).
The series will converge, by latter two results; with a sum

$$\frac{a}{1-a}+\frac{ka}{1-ka}.$$

(ii)

(a) Series is alternating; terms decrease numerically to zero since

$$\sec\frac{\alpha}{n+1}-\sec\frac{\alpha}{n} < 0$$

for all large n and

$$\lim_{n\to\infty}\sec\frac{\alpha}{n} = 1.$$

Therefore, by Leibnitz's test, series converges.

(b) If $x > 1$,

$$\frac{x^n}{x^{n-1}+n} = \frac{x}{1+nx^{1-n}} \to x \quad \text{as } n \to \infty,$$

hence terms do not decrease to zero, therefore series diverges.
If $x = 1$,

$$\frac{x^n}{x^{n-1}+n} = \frac{1}{1+n} > \frac{\frac{1}{2}}{n};$$

the harmonic series diverges, so, therefore, does the given series. If $x < 1$, the ratio of successive terms is

$$x\frac{x^{n-1}+n}{x^n+(n+1)} \to x < 1$$

and the series converges.

3

(i)

(a) $u_n = (xe^{-x})^n$. Max xe^{-x} occurs when $x = 1$, and equals e^{-1}, $0 < u_n < e^{-n}$ and as $n \to \infty$, $u_n \to 0$.

(b) $u_1 > 0$; so $u_2, u_3, \ldots$ are all > 0;

and $u_n - u_{n-1} = -\frac{u_{n-1}}{u_{n-1}+1} < 0$.

Thus $\{u_n\}$ decreases. But sequence is bounded below by zero so that $\lim u_n$ exists, $= l$, say, and l satisfies

$$l = \frac{l^2}{l+1} \quad \text{i.e. } l = 0.$$

(ii)

(a) is a geometric series. Convergent if

$$\frac{1}{|1+x|} < 1 \quad \text{hence if } x > 0 \text{ or if } x < -2.$$

(b) Use d'Alembert's ratio test:

$$\left|\frac{a_{n+1}}{a_n}\right| = \frac{n}{n+1}|x-2| \to |x-2|.$$

Hence series converges if $1 < x < 3$.

If $x = 3$, series is $\sum 1/n$, which diverges.

If $x = 1$, series is $\sum(-1^n/n$ which, by Leibnitz's test, converges.

Summarizing: given series converges only for $1 \leqslant x < 3$.

(c) Use d'Alembert's test again:

$$\left|\frac{a_{n+1}}{a_n}\right| = |x|.\left|\frac{x^{2n}+1}{x^{2n+2}+1}\right|.$$

If $|x| < 1$, the ratio $\to |x| < 1$, so the series converges, and if $|x| > 1$, the ratio $\to |x|^{-1} < 1$ and again the series converges. If $x = 1$, the series is $\sum\frac{1}{2}$ which diverges, and if $x = -1$, the series is $\sum(-1)^n/2$ which diverges. Hence given series converges only if $|x| < 1$ or $|x| > 1$.

4

(i) $$\lim_{x\to\infty}[\sqrt{(4x^2+3x+1)}-2x] = \lim_{x\to\infty}\frac{3x+1}{\sqrt{(4x^2+3x+1)}+2x}$$
$$= \lim_{x\to\infty}\frac{3+1/x}{\sqrt{(4+3/x+1/x^2)}+2}$$
$$= \tfrac{3}{4}.$$

(ii) Obtain first

$$\sin^{-1}x = x+\tfrac{1}{6}x^3+\tfrac{3}{40}x^5+\tfrac{5}{112}x^7+\ldots$$

by integrating the series for $(1-x^2)^{-\frac{1}{2}}$. Then required limit is

$$\lim_{x\to 0}\frac{(x+x^3/6+3x^5/40+\ldots)-(x+x^3/3!+x^5/5!+\ldots)}{x^5-x^7/3!+\ldots}$$
$$= \lim_{x\to 0}\frac{(3/40-1/120)x^5+\ldots}{x^5-\ldots}$$
$$= \tfrac{1}{15}.$$

(iii) Required limit is $\lim \dfrac{e^{(1/x)\log(1+x)} - e}{x}$,

which, by l'Hopital's rule, is

$$\lim_{x\to 0} e^{(1/x)\log(1+x)}\left[\frac{x/(1+x)-\log(1+x)}{x^2}\right]$$

$$= \lim_{x\to 0}(1+x)^{1/x}\lim_{x\to 0}\frac{1}{x}\left[\frac{1}{1+x}-\frac{\log(1+x)}{x}\right]$$

$$= e\lim_{x\to 0}\frac{1}{x}\left[(1-x+x^2-\ldots)-\left(1-\frac{x}{2}+\frac{x^2}{3}-\ldots\right)\right]$$

$$= -\tfrac{1}{2}e.$$

5

(a) $\dfrac{(n!)^2}{(2n)!}$

$$= \frac{n(n-1)\ldots 2.1}{2n(2n-1)\ldots(n+2)(n+1)}$$

$$= \frac{(1-1/n)(1-2/n)\ldots\{1-(n-2)/n\}\{1-(n-1)/n\}}{2^n(1-1/2n)(1-2/2n)\ldots\{1-(n-2)/2n\}\{1-(n-1)/2n\}},$$

which is positive and less than $1/2^n$, which tends to zero. Hence the required limit is zero.

(b) $$\lim_{t\to 1}\frac{1-t^{2^{n+1}}}{2^{n+1}(1-t)} = \lim_{t\to 1}\frac{1+t+t^2+\ldots+t^{2n+1}-1}{2^{n+1}}$$

$$= \frac{2^{n+1}}{2^{n+1}} = 1.$$

Take the logarithm of the expression in the curly brackets to get

$$\sum_{r=0}^{\infty}\log_e(1+\cos^{2^r}\theta)$$

and expand each term in a power series. The sum comes to

$\cos\theta+\frac{1}{2}\cos^2\theta+\frac{1}{3}\cos^3\theta+\ldots,$

which is $-\log_e(1-\cos\theta)$. The required limit is therefore $1/(1-\cos\theta)$.

6

(a) $$x^2e^{-x} = \frac{x^2}{1+x+x^2/(2!)+\ldots}$$

$$= \frac{1}{1/x^2+1/x+1/(2!)+x/(3!)+\ldots} \to 0 \quad \text{as } x\to\infty.$$

(b) $$\frac{\sin^3 x+x^4}{x^2} = \frac{\{x-x^3/(3!)+\ldots\}^3+x^4}{x^2}$$

$$= \frac{x^3+\ldots}{x^2} \to 0 \quad \text{as } x\to 0.$$

(c) The ratio is

$$\frac{(1/6)n(n+1)(2n+1)}{n^3+3n+1} = \frac{(1/3)n^3+\ldots}{n^3+3n+1} \to \frac{1}{3} \quad \text{as } n\to\infty.$$

Chapter Four
Coordinate Geometry

A Coordinate geometry: Theory summary

1 *Rectangular Cartesian coordinates: Distances, gradients, areas*

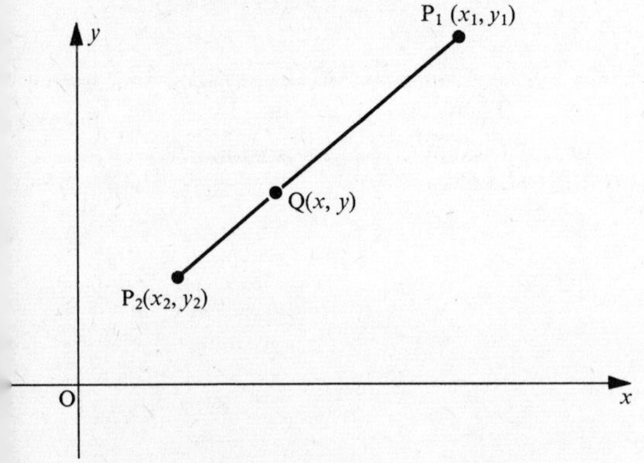

Figure 20

(a) In Figure 20 the points P_1 and P_2 are defined by their Cartesian coordinates $(x_1, y_1), (x_2, y_2)$ respectively.

(i) Distance $P_1P_2 = \sqrt{\{(x_1-x_2)^2+(y_1-y_2)^2\}}$.

(ii) Gradient of line P_1P_2, $\tan\theta = \dfrac{y_1-y_2}{x_1-x_2}$.

(iii) The point Q (x, y) dividing P_1P_2 in the ratio $P_1Q/QP_2 = m/n$ is given by

$$x = \frac{nx_1+mx_2}{n+m}, \qquad y = \frac{ny_1+my_2}{n+m}.$$

Thus any point on the line P_1P_2 may be written in the parametric form:

$$x = ax_1+(1-a)x_2, \qquad y = ay_1+(1-a)y_2.$$

In particular, the midpoint of P_1P_2 corresponds to $n = m = 1$ or $a = \frac{1}{2}$ and is $x = \frac{1}{2}(x_1+x_2), y = \frac{1}{2}(y_1+y_2)$.

(b)

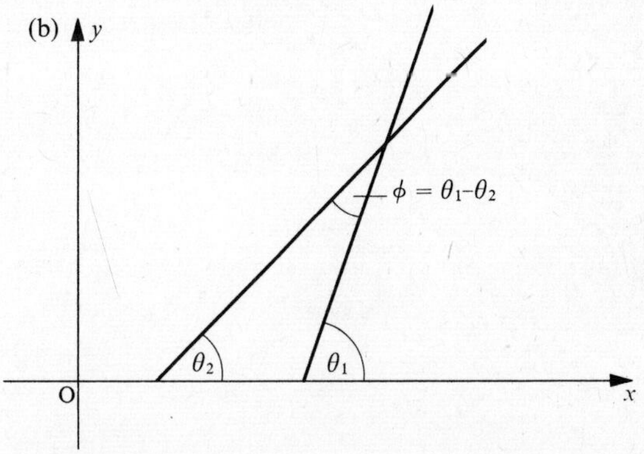

$\tan\theta_1 = m_1, \quad \tan\theta_2 = m_2$

Figure 21

The angle ϕ between two straight lines of gradients m_1 and m_2 is given by

$$\tan\phi = \frac{m_1-m_2}{1+m_1m_2}.$$

In particular, if the lines are parallel, $m_1 = m_2, \phi = 0$, and if the lines are perpendicular, $m_1m_2 = -1, \phi = 90°$.

(c) The area of $\triangle P_1P_2P_3$, where the coordinates of the points P_1, P_2 and P_3 are $(x_1, y_1), (x_2, y_2)$ and (x_3, y_3) respectively, is

$$\triangle = \tfrac{1}{2}(x_1y_2-x_2y_1+x_2y_3-x_3y_2+x_3y_1-x_1y_3)$$
$$= \frac{1}{2}\begin{vmatrix} x_1 & y_1 & 1 \\ x_2 & y_2 & 1 \\ x_3 & y_3 & 1 \end{vmatrix}.$$

2 *Equations and properties of straight lines*

(a) The equation of a straight line can be written in the following forms:

(i) $y = mx + c$,

where m is the gradient of the line and c is the intercept on the y-axis (see Figure 22a).

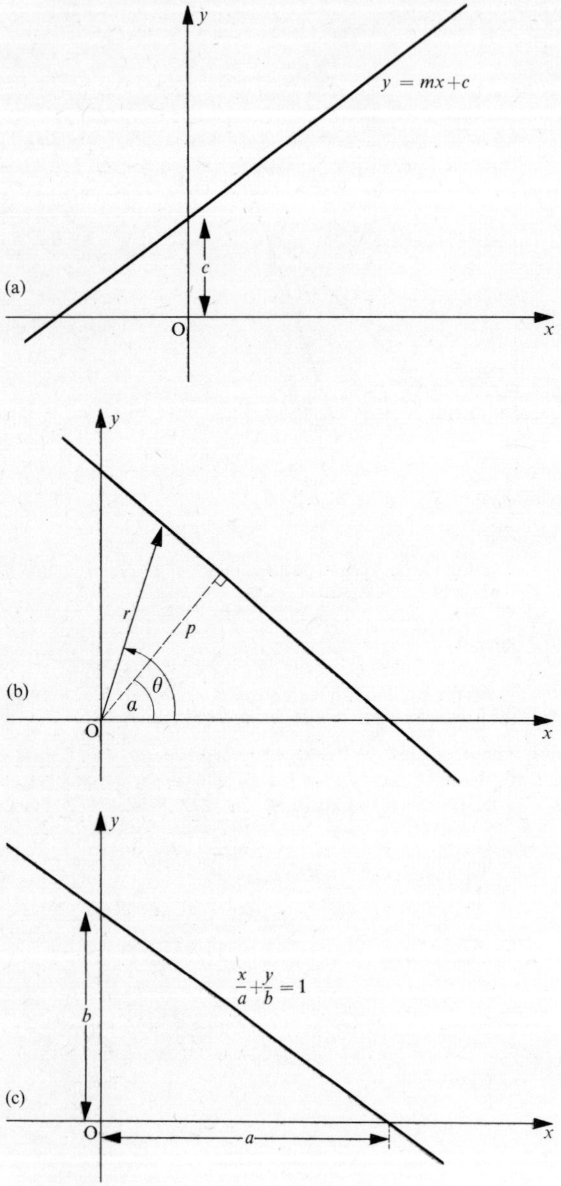

Figure 22

(ii) $y - y_1 = m(x - x_1)$

for a line of gradient m passing through the point $P_1(x_1, y_1)$.

(iii) $y - y_1 = \dfrac{y_1 - y_2}{x_1 - x_2}(x - x_1)$

for a line passing through two points $P_1(x_1, y_1)$ and $P_2(x_2, y_2)$.

(iv) $p = x \cos \alpha + y \sin \alpha$,

where p is the perpendicular distance from origin to line and α is the angle perpendicular makes with x-axis (see Figure 22b) (In polar coordinates (r, θ) the above equation becomes $r \cos(\theta - \alpha) = p$.)

(v) $\dfrac{x}{a} + \dfrac{y}{b} = 1$,

where a is the intercept on the x-axis and b is the intercept on the y-axis (see Figure 22c).

(vi) In the general case any first-degree equation, $lx + my + n = 0$, represents a straight line.

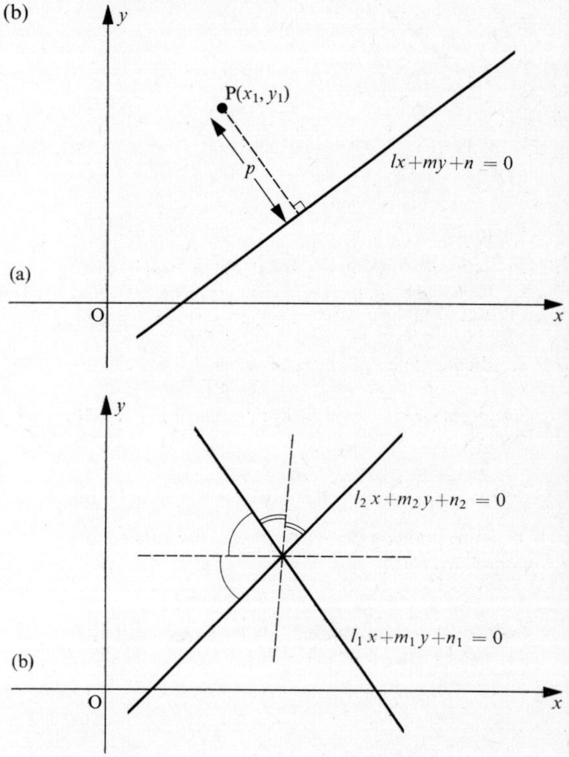

Figure 23

(i) The perpendicular distance from any point $P(x_1, y_1)$ to the line $lx + my + n = 0$ is

$$p = \pm \frac{lx_1 + my_1 + n}{\sqrt{(l^2 + m^2)}}$$

(see Figure 23a).

(ii) The equations of the two angle bisectors of the two straight lines of Figure 23b are

$$\frac{l_1x+m_1y+n_1}{\sqrt{(l_1^2+m_1^2)}} = \pm\frac{l_2x+m_2y+n_2}{\sqrt{(l_2^2+m_2^2)}}.$$

(c)

(i) The point of intersection of lines $L_1 \equiv l_1x+m_1y+n_1 = 0$ and $L_2 \equiv l_2x+m_2y+n_2 = 0$ is found by solving these two equations.

(ii) A general line passing through the point of intersection of $L_1 = 0, L_2 = 0$ is

$$L_1+\lambda L_2 = 0.$$

(iii) The equation

$$ax^2+2hxy+by^2 = 0$$

represents a pair of lines $y = m_1x$ and $y = m_2x$ where the gradients m_1 and m_2 satisfy

$$m_1+m_2 = -\frac{2h}{b} \quad \text{and} \quad m_1m_2 = \frac{a}{b}.$$

The angle between the lines is $\tan^{-1}2\sqrt{(h^2-ab)}/(a+b)$ and if the lines are perpendicular $a+b = 0$.

(iv) The condition for three lines $L_1 = 0, L_2 = 0, L_3 = 0$ to be concurrent is

$$\begin{vmatrix} l_1 & m_1 & n_1 \\ l_2 & m_2 & n_2 \\ l_3 & m_3 & n_3 \end{vmatrix} = 0.$$

3 *Equations and properties of circles*

(a)

(i) The equation of a circle with centre (x_0, y_0) and radius r is

$$(x-x_0)^2+(y-y_0)^2 = r^2.$$

(ii) The general equation of a circle is

$$x^2+y^2+2gx+2fy+c = 0.$$

It has its centre at $(-g, -f)$ and a radius $r = \sqrt{(g^2+f^2-c)}$

(iii) The tangent to the above circle at the point (x_1, y_1) is

$$xx_1+yy_1+g(x+x_1)+f(y+y_1)+c = 0.$$

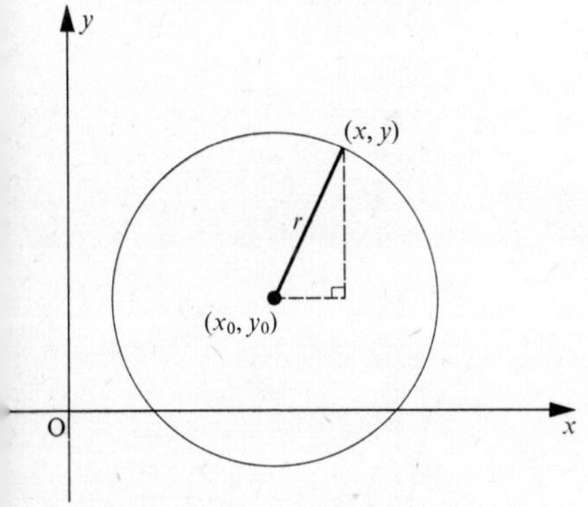

Figure 24

(iv) The tangent length from an external point (x_1, y_1) is

$$t = \sqrt{(x_1^2+y_1^2+2gx_1+2fy_1+c)}.$$

(b) *Systems of circles.*

If $S_1 \equiv x^2+y^2+2g_1x+2f_1y+c_1 = 0$

and $S_2 \equiv x^2+y^2+2g_2x+2f_2y+c_2 = 0$

are the equations of two circles, then:

(i) $S_1+\lambda S_2 = 0 \quad (\lambda \neq -1)$

is a general circle coaxal with S_1 and S_2 (i.e. the centre of this circle, $S_1 = 0$ and $S_2 = 0$ are in the same straight line).

If $S_1 = 0$ intersects $S_2 = 0$, then the above equation gives the general circle through their intersection points.

(ii) The equation of the radical axis of the two circles is

$$S_1-S_2 = 0,$$

i.e. $2(g_1-g_2)x+2(f_1-f_2)y+c_1-c_2 = 0.$

It is a line perpendicular to the line joining the centres of $S_1 = 0$ and $S_2 = 0$.

If the circles intersect then this equation is that of the common chord.

(iii) If the circles cut at right angles (i.e. are orthogonal), then

$$2g_1g_2+2f_1f_2 = c_1+c_2.$$

(iv)

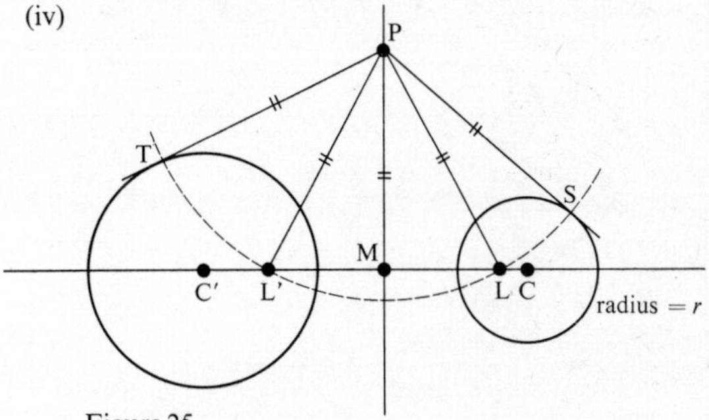

Figure 25

The tangents from any point P on the radical axis to any circle of the coaxal family (shown in Figure 25) are equal,

i.e. PT = PS = ... = PL = PL′,

where L and L′ are point circles (circles of zero radius) in the system and are known as the limiting points. As a consequence of this property,

$$CM^2-r^2 = LM^2, \quad \text{a constant},$$

where r is the radius and C is the centre of any circle in the coaxal set and M is the point where the radical axis cuts the line joining the centres.

(c) *Circle of Apollonius.* If A and B are fixed points and P is a variable point such that PA/PB = k, then the locus of P is a circle (Apollonius' circle) with diameter CD (see Figure 26), where CP is the internal bisector of $A\hat{P}B$ and DP is the external bisector of $A\hat{P}B$, so that $AC/CB = AD/BD = k$.

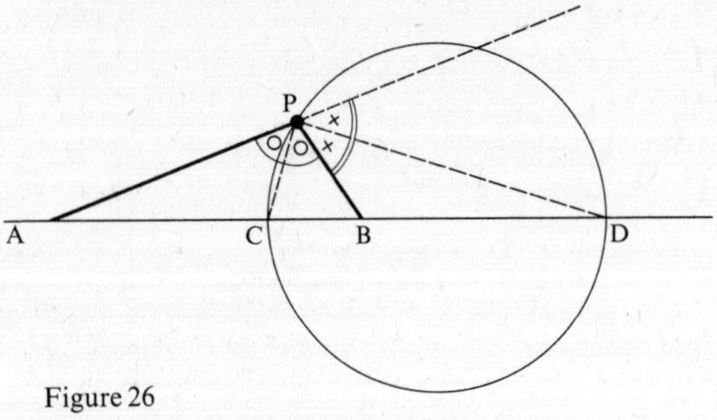

Figure 26

4 *Conics: Equations and general properties*

(a) *Definition.*

(i) A conic is the locus of a point, say P, which moves in a plane so that its distance from a fixed point S (known as the focus) is in a constant ratio to its distance from a fixed line (known as the directrix). The constant ratio $e = SP/PN$ (see Figure 27) is known as the eccentricity of the conic.

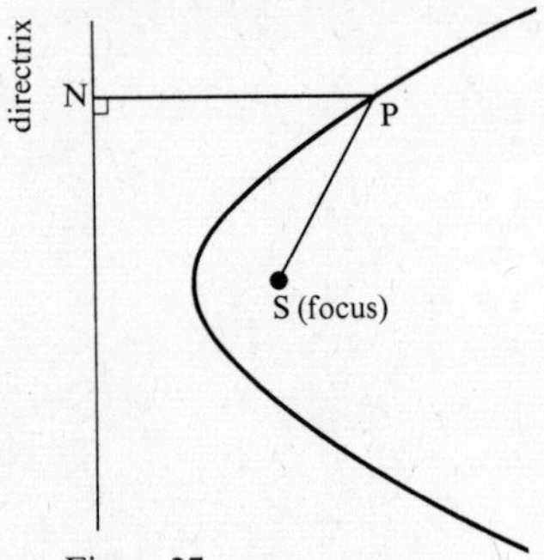

Figure 27

(ii) When $e = 1$ the conic is known as a *parabola*; when $e < 1$ the conic is an *ellipse*; when $e > 1$ the conic is a *hyperbola.*

(iii) Geometrically, we may define a conic as the curve in which a plane cuts a right circular cone. If the plane is perpendicular to the axis of the cone the conic is a circle and if the plane cuts the cone through the vertex the conic is a pair of straight lines. For any other plane the conic is either a parabola, an ellipse or a hyperbola (see Figure 28).

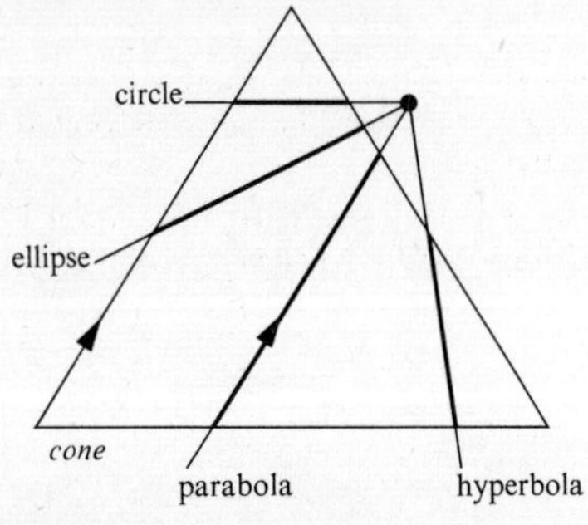

Figure 28

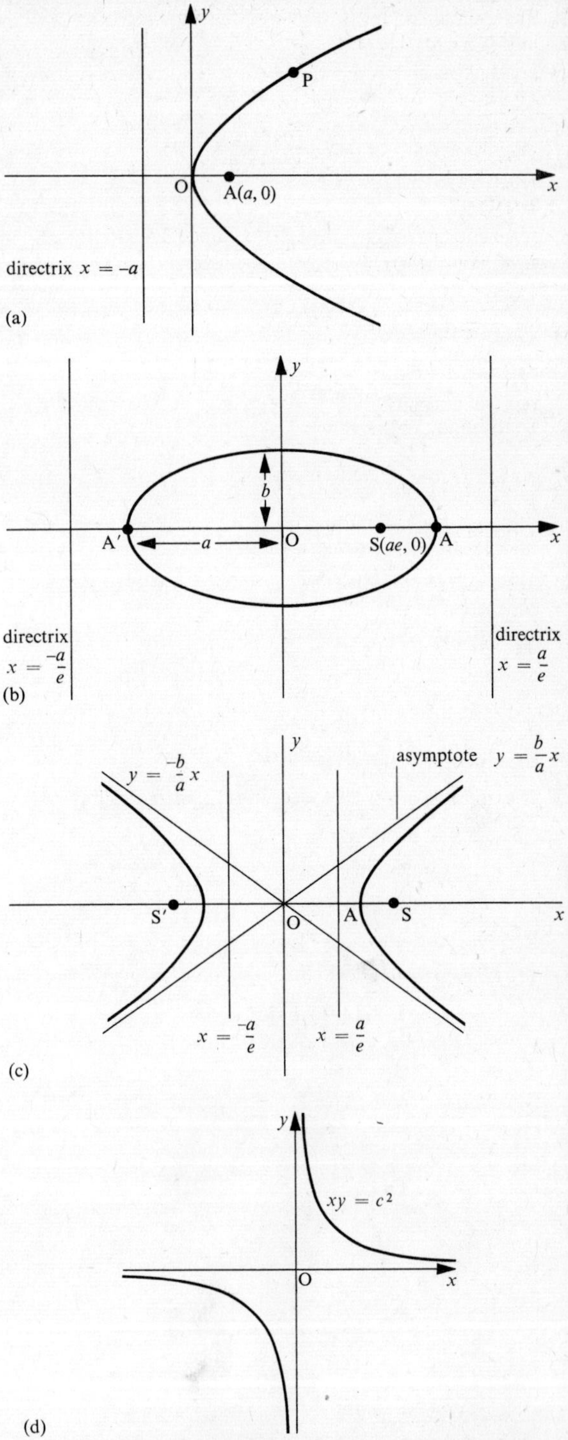

Figure 29 (a) Parabola. (b) Ellipse. (c) Hyperbola. (d) Rectangular hyperbola

(b) *Equations of conics.*

(i) Referred to axes as in Figure 29 the Cartesian equations are:

Parabola $y^2 = 4ax$.

Ellipse $\dfrac{x^2}{a^2}+\dfrac{y^2}{b^2} = 1$, where $b^2 = a^2(1-e^2)$.

Hyperbola $\dfrac{x^2}{a^2}-\dfrac{y^2}{b^2} = 1$, where $b^2 = a^2(e^2-1)$

Rectangular hyperbola $x^2-y^2 = a^2$.

By rotation of axes through 45° it can be written

$xy = c^2(=\tfrac{1}{2}a^2)$ (see Figure 29d).

(ii) Parametric forms:

Parabola $x = at^2$, $y = 2at$.
Ellipse $x = a\cos\theta$, $y = b\sin\theta$.
Hyperbola $x = a\sec u$, $y = b\tan u$.
Rectangular hyperbola (in $xy = c^2$ form)

$x = cu$, $y = \dfrac{a}{u}$.

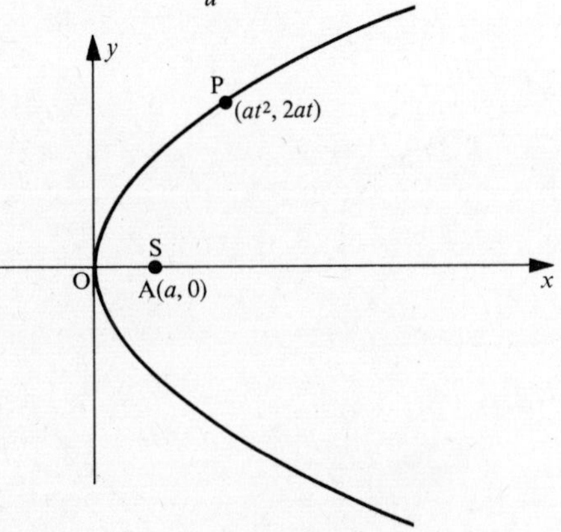

(a)

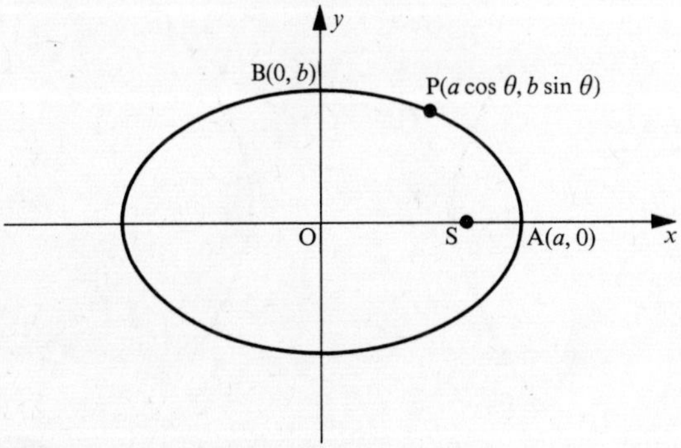

(b)

Figure 30 (a) Parabola. (b) Ellipse

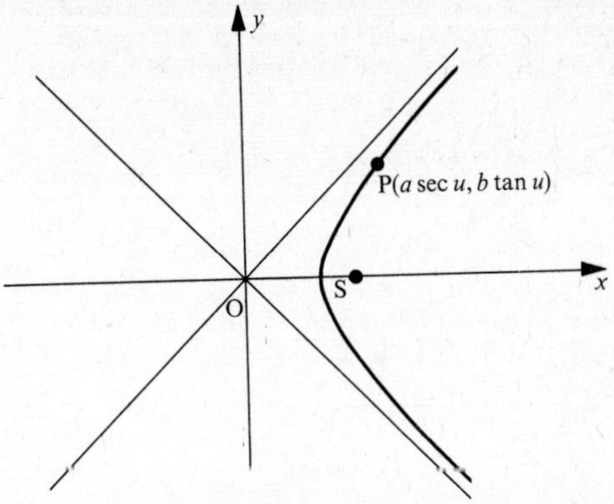

(c)

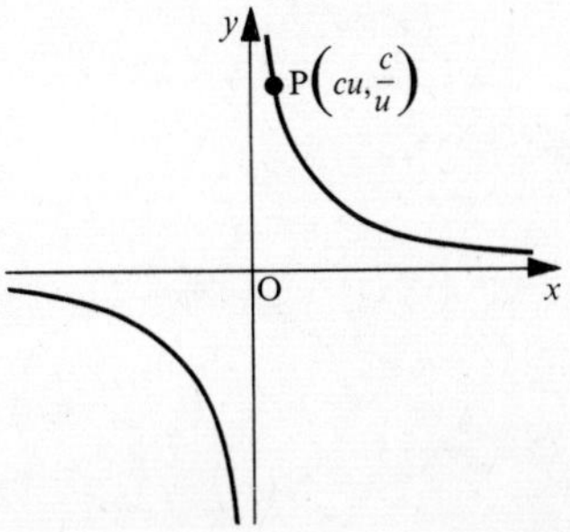

(d)

Figure 30 (c) Hyperbola. (d) Rectangular hyperbola

(c) *Tangents to conics.*

(i) To touch at (x', y'):

Parabola $yy' = 2a(x+x')$.

Ellipse $\dfrac{xx'}{a^2}+\dfrac{yy'}{b^2} = 1$.

Hyperbola $\dfrac{xx'}{a^2}-\dfrac{yy'}{b^2} = 1$.

Rectangular hyperbola (in $xy = c^2$ form)

$xy'+x'y = 2c^2$.

(ii) Of gradient m:

Parabola $y = mx+\dfrac{a}{m}$.

Ellipse $y = mx\pm\sqrt{(a^2m^2+b^2)}$.
Hyperbola $y = mx\pm\sqrt{(a^2m^2-b^2)}$.

(d) *Normals to conics.*

Parabola, at $(at_1^2, 2at_1)$ $y+t_1x = 2at_1+at_1^3$.

Ellipse, at $(a\cos\theta_1, b\sin\theta_1)$ $\dfrac{ax}{\cos\theta_1}-\dfrac{by}{\sin\theta_1} = a^2-b^2$.

Hyperbola, at $(a\sec u_1, b\tan u_1)$
$ax\cos u_1+by\cot u_1 = a^2+b^2$.

(e) *Equations of chords of conics.*
Parabola: chord joining $(at_1^2, 2at_1)$ to $(at_2^2, 2at_2)$ is

$$(t_1+t_2)y-2x = 2at_1t_2.$$

Ellipse: chord joining $(a\cos\theta_1, b\sin\theta_1)$ to $(a\cos\theta_2, b\sin\theta_2)$ is

$$\frac{x}{a}\cos(\theta_1+\theta_2)+\frac{y}{b}\sin(\theta_1+\theta_2) = \cos\tfrac{1}{2}(\theta_1-\theta_2).$$

Rectangular hyperbola (in $xy = c^2$ form): chord joining $(cu_1, c/u_1)$ to $(cu_2, c/u_2)$ is

$$x+u_1u_2y = c(u_1+u_2).$$

(f) *Loci properties.*

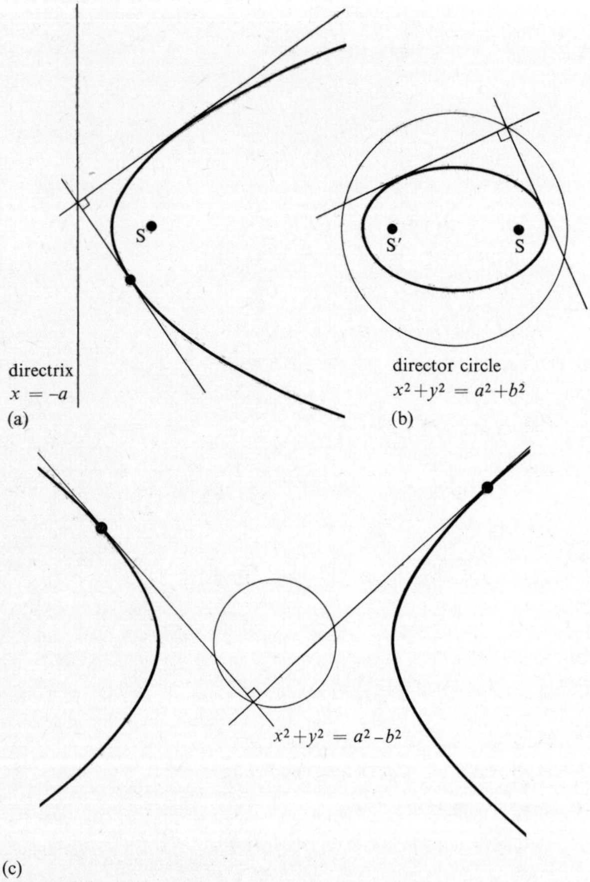

Figure 31 (a) Parabola. (b) Ellipse. (c) Hyperbola

(i) Perpendicular tangents to a parabola meet on the directrix, $x = -a$.

Perpendicular tangents to an ellipse meet on the director circle,

$$x^2+y^2 = a^2+b^2.$$

Perpendicular tangents to a hyperbola meet on the director circle,

$$x^2+y^2 = a^2-b^2.$$

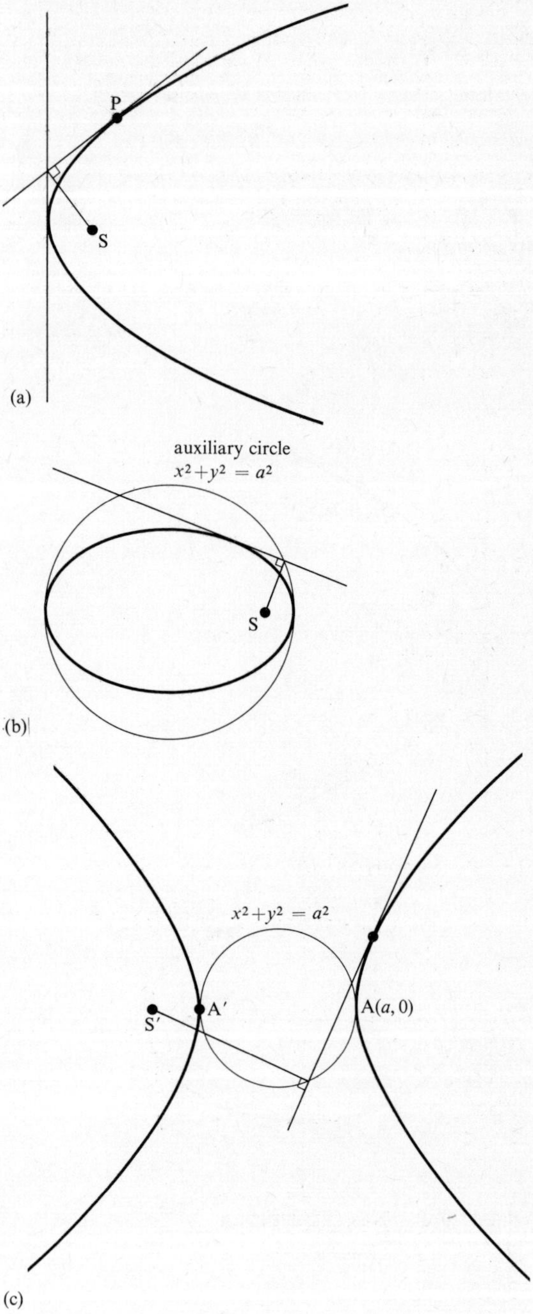

Figure 32 (a) Parabola. (b) Ellipse. (c) Hyperbola

(ii) The locus of the foot of the perpendicular from the focus of a conic to any tangent is:

for the parabola, $x = 0$;
for the ellipse, the auxiliary circle, $x^2+y^2 = a^2$;
for the hyperbola, the auxiliary circle, $x^2+y^2 = a^2$.

B Coordinate geometry: Illustrative worked problems

1 Find the equation of the circle which has its centre at the point of intersection of the common tangents to the circles $x^2+y^2 = 9, x^2+y^2-8x+12 = 0$, and which passes through the points of intersection of the given circles.

[L, J 1968, P I, Q 6]

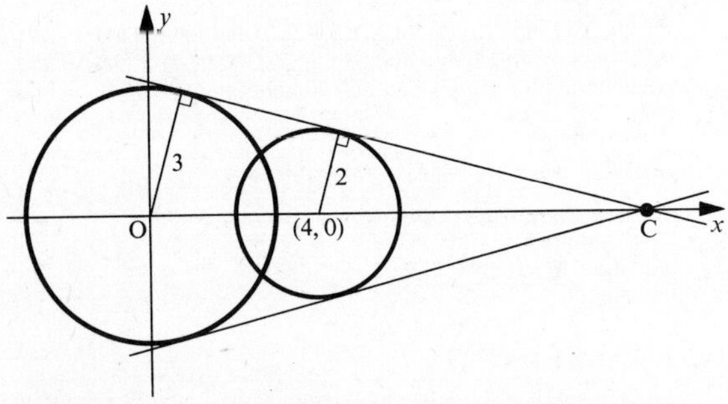

Figure 33

Solution. $x^2+y^2 = 9$ has centre at (0, 0), radius 3.
$x^2+y^2-8x+12 = 0$ has centre at $(-g, -f) = (4, 0)$, radius $\sqrt{(g^2+f^2-c)} = \sqrt{(16-12)} = 2$. These two circles are drawn in Figure 33, together with their common tangents. It is seen that the latter intersect at C, where

$$\frac{3}{2} = \frac{OC}{OC-4}.$$

Hence OC = 12 and point C is (12, 0).

Points of intersection $(x_0, \pm y_0)$ of two circles are found from

$$x^2+y^2-9-(x^2+y^2-8x+12) = 0,$$

i.e. $x = x_0 = \frac{21}{8}, \quad y_0 = \sqrt{(9-x_0^2)} = \frac{\sqrt{135}}{8}.$

Thus equation of required circle centre C and passing through points of intersection is

$$(x-12)^2+(y-0)^2 = AC^2 = (12-x_0)^2+y_0^2 = 90,$$

i.e. $x^2+y^2-24x+54 = 0$.

2 Show that the coordinates of any point P on the ellipse $x^2/a^2+y^2/b^2 = 1 (a > b)$ can be expressed as $(a \cos \phi, b \sin \phi)$.

P and Q are two points on the ellipse such that $\phi = \alpha$ for P and $\phi = \alpha+\frac{1}{2}\pi$ for Q; prove that the equation of the chord PQ is

$$b(\sin \alpha-\cos \alpha)x-a(\sin \alpha+\cos \alpha)y+ab = 0.$$

If PQ passes through the focus $(ae, 0)$ of the ellipse prove that $e\sqrt{2} \geqslant 1$.

[O & C, S 1966, M for Sc I, Q 5]

Solution. From $\frac{x^2}{a^2}+\frac{y^2}{b^2} = 1$

and since $y^2/b^2 \geqslant 0$, we have $x^2/a^2 \leqslant 1$ and hence $|x| \leqslant a$. Thus if we put $x = \cos \phi$, then

$$\frac{y^2}{b^2} = 1-\frac{x^2}{a^2} = 1-\cos^2\phi = \sin^2\phi$$

and so $y = b \sin \phi$ with the sign of ϕ chosen so that $\sin \phi$ has the same sign as y. Hence any point P on the ellipse can be expressed by the coordinates $(x, y) \equiv (a \cos \phi, b \sin \phi)$.

P has coordinates $(a \cos \alpha, b \sin \alpha)$ and Q has coordinates $(a \cos(\alpha+\frac{1}{2}\pi), b \sin(\alpha+\frac{1}{2}\pi))$, that is $(-a \sin \alpha, b \cos \alpha)$. The gradient of PQ is

$$\frac{b \sin \alpha-b \cos \alpha}{a \cos \alpha-(-a \sin \alpha)} = \frac{b(\sin \alpha-\cos \alpha)}{a(\cos \alpha+\sin \alpha)}$$

and so the equation of chord PQ is

$$y-b \sin \alpha = \frac{b(\sin \alpha-\cos \alpha)}{a(\cos \alpha+\sin \alpha)}(x-a \cos \alpha)$$

or $b(\sin \alpha-\cos \alpha)x-a(\sin \alpha+\cos \alpha)y$
$= ab\{\cos \alpha(\sin \alpha-\cos \alpha)-\sin \alpha(\sin \alpha+\cos \alpha)\} \equiv -ab,$

which is the desired result.

If PQ passes through $(ae, 0)$ then substituting $x = ae$ and $y = 0$ into the equation of the chord we have

$$aeb(\sin \alpha-\cos \alpha)+ab = 0.$$

Thus $e(\sin \alpha-\cos \alpha) = -1$
or $e\sqrt{2} \cos(\alpha+\frac{1}{4}\pi) = +1$
and as $|\cos(\alpha+\frac{1}{4}\pi)| \leqslant 1$ for all α, then $e\sqrt{2} \geqslant 1$ for all α.

3 The chord AB joining points $A(ct_1, c/t_1)$ and $B(ct_2, c/t_2)$ on the rectangular hyperbola $xy = c^2$ is of constant length l. Show that, as the position of the chord varies, the centroid G of the triangle AOB, where O is the origin, moves on the curve

$$(9xy-4c^2)(x^2+y^2) = l^2xy.$$

Find the area of the triangle AOB when the coordinates of G are $(c, 2c)$.

[L, S 1967, P II, Q 7]

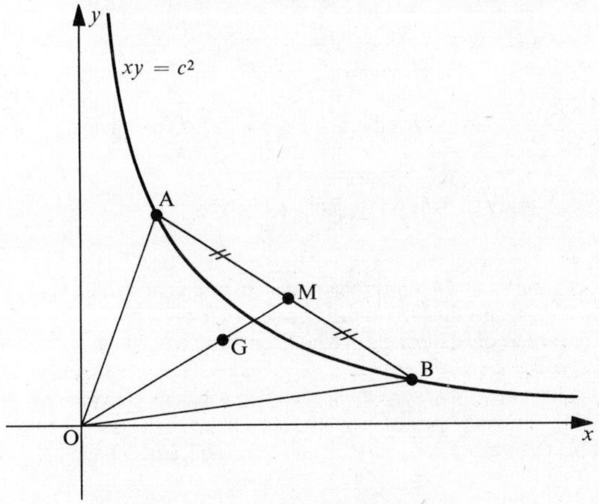

Figure 34

Solution. $AB^2 = l^2 = c^2\left[(t_1-t_2)^2+\left(\frac{1}{t_1}-\frac{1}{t_2}\right)^2\right]$. **4.1**

The midpoint M of AB has coordinates

$$\left(\tfrac{1}{2}c(t_1+t_2), \tfrac{1}{2}c\left[\frac{1}{t_1}+\frac{1}{t_2}\right]\right).$$

Since $OG = \frac{2}{3}OM$, G has coordinates

$$x = \tfrac{2}{3}.\tfrac{1}{2}c(t_1+t_2) = \frac{c}{3}(t_1+t_2),$$

$$y = \tfrac{2}{3}.\tfrac{1}{2}c\left[\frac{1}{t_1}+\frac{1}{t_2}\right] = \frac{c}{3}\left[\frac{1}{t_1}+\frac{1}{t_2}\right].$$

Hence $9x^2 = c^2(t_1+t_2)^2$,

$$9y^2 = c^2\left[\frac{1}{t_1}+\frac{1}{t_2}\right]^2$$

and $\dfrac{x}{y} = t_1t_2$

and thus $c^2(t_1-t_2)^2 = c^2(t_1+t_2)^2-4c^2t_1t_2$

$$= 9x^2-4c^2\frac{x}{y},$$

$$c^2\left[\frac{1}{t_1}-\frac{1}{t_2}\right]^2 = c^2\left[\frac{1}{t_1}+\frac{1}{t_2}\right]^2-4c^2\frac{1}{t_1t_2}$$

$$= 9y^2-4c^2\frac{y}{x}.$$

Thus substituting these results in equation **4.1** we have

$$l^2 = 9(x^2+y^2)-4c^2\frac{1}{xy}(x^2+y^2)$$

or $(9xy-4c^2)(x^2+y^2) = l^2xy$,

which is the locus of G.

Area $\triangle AOB = \frac{1}{2}\begin{vmatrix} 0 & 0 & 1 \\ ct_1 & c/t_1 & \\ ct_2 & c/t_2 & 1 \end{vmatrix} = \frac{1}{2}c^2\left[\frac{t_1}{t_2}-\frac{t_2}{t_1}\right]$

$$= \tfrac{1}{2}c^2\frac{t_1^2-t_2^2}{t_1t_2}$$

$$= \tfrac{1}{2}c^2\frac{(t_1+t_2)(t_1-t_2)}{t_1t_2},$$

and on using relations for the case of $G = (c, 2c)$ we have

$t_1+t_2 = 3, \quad t_1t_2 = 2, \quad t_1-t_2 = \sqrt{7}$

and thus Area $\triangle AOB = 3\sqrt{7}c^2$.

4 Show that the equation $x^2+y^2+2\lambda x+c = 0$, where c is a constant and λ varies, represents a system of coaxal circles and find the condition that real limiting points exist.

At each pair of real points in which the fixed line $y = k$ intersects a circle of the system $x^2+y^2+2\lambda x+4 = 0$, tangents to that circle are drawn and meet at P. Show that as λ varies P lies on a certain parabola, and find the equation of this parabola in terms of k. [L, S 1967, FM VI, Q 5]

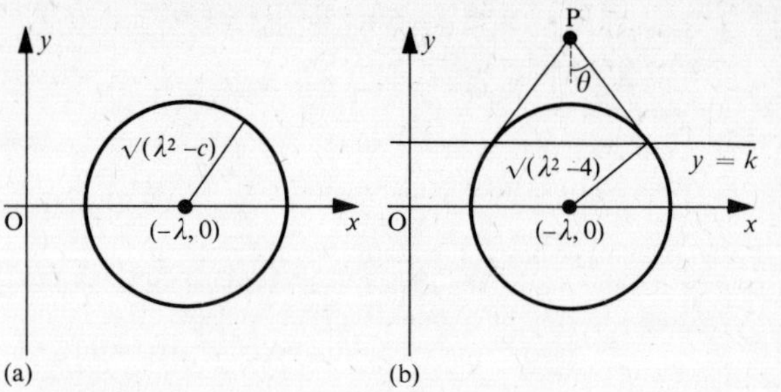

Figure 35

Solution. Each circle of the system has its centre at $(-\lambda, 0)$, and the corresponding radius, $\sqrt{(\lambda^2-c)}$ (see Figure 35a). The centres lie along the x-axis. Consider any two circles

$$S \equiv x^2+y^2+2\lambda x+c = 0$$

and $S' \equiv x^2+y^2+2\lambda' x+c = 0$

of the system. The radical axis will be the line

$$S-S' \equiv 2(\lambda-\lambda')x = 0$$

i.e. $x = 0$,

for all the circles.

The circles of zero radius are at the limiting points $(\pm\sqrt{c}, 0)$, provided

$c \geqslant 0$,

and the circles of the given system are real if $|\lambda| \geqslant \sqrt{c}$.
In the problem, put $c = 4$ (see Figure 35b). The point of intersection P of the tangents has coordinates $(-\lambda, \mu)$, say, where

$$\frac{\sqrt{(\lambda^2-4)}}{\mu} \equiv \sin\theta \equiv \frac{k}{\sqrt{(\lambda^2-4)}}.$$

Thus $\mu = \dfrac{\lambda^2-4}{k}$.

Putting $-\lambda \equiv X, \mu \equiv Y$, that is, relabelling the coordinates of P, we find locus of P is

$$\frac{X^2-4}{k} = Y,$$

i.e. $X^2 = 4+kY$.

C Coordinate geometry: Problems

Answers and hints will be found on pp. 66–9.

1 The point P (1, 1) lies on the ellipse $b^2x^2+a^2y^2 = a^2b^2$, where $a > b$. The tangent at P meets the x-axis at Q and the y-axis at R. O is the origin of coordinates. Find, in terms of a, the distances OQ and OR.

A circle is drawn on a major semi-axis of the ellipse as diameter to intersect the ellipse at L and M. Find, in terms of a, the distance OL (or OM). [L, J 1968, P II, Q 7]

2 Find the equation of the tangent to the hyperbola $b^2x^2 - a^2y^2 = a^2b^2$ at the point $(a \sec t, b \tan t)$ and show that the equation of the normal to the curve at this point is

$$ax \sin t + by = (a^2+b^2)\tan t.$$

Show that the product of the areas of the two triangles formed by each of these lines and the coordinate axes is independent of t.

Find the locus of the circumcentre of the triangle formed by the tangent and the coordinate axes.
[L, J 1968, P II, Q 8]

3 A circle is drawn to touch the straight line $y = x$ at the point (1, 1) and to pass through the point (4, 2). Find the equation of the circle and the equation of the other tangent which can be drawn to the circle from the origin. [L, S 1967, P I, Q 5]

4 Find the equation of the tangent to the parabola $y^2 = 4ax$ at the point $(at^2, 2at)$.

From a point (h, k) tangents are drawn to the parabola $y^2 = 4ax$. Show that the area of the triangle formed by the tangents and the chord of contact is

$$\frac{(k^2-4ah)^{\frac{3}{2}}}{2a}.$$
[L, S 1967, P I, Q 6]

5 Find the equations of the common tangents to the curves $(x-2)^2+(y+1)^2 = 4$ and $y = x^2-4x+11$.
[L, S 1967, P S, Q 5]

6 Prove that the locus of a point from which the tangents drawn to two given circles are equal is a straight line.

Define the limiting points of a system of coaxal circles and show that they are inverse points with respect to any circle of the system.

Prove that the chord of contact of the tangents drawn from any fixed point to a variable circle of the coaxal system passes through a fixed point, checking your result by considering the limiting points as degenerate circles of the system.
[L, J 1967, FM VI, Q 5]

7 A variable circle passes through the fixed point (a, 0) and is such that its intercept on the axis of x is always double its intercept on the axis of y. Show that the locus of the centre of the circle is the hyperbola

$$x^2-4y^2+6ax-3a^2 = 0.$$
[O, S 1967, PM II, Q 7]

8 Prove that the equation of the chord of the parabola $y^2 = 4ax$ of which (x', y') is the midpoint is

$$yy'-2ax = y'^2-2ax'.$$

If P is any point on the focal chord $y = x-a$, prove that the chord bisected at P touches the fixed parabola

$$(y+2a)^2 = 8a(x-a).$$
[C, S 1967, P III, Q 14]

9 If the line $x \sin\theta - y \cos\theta = c$ is a tangent to the ellipse

$$\frac{x^2}{a^2}+\frac{y^2}{b^2} = 1,$$

prove that

$$c^2 = a^2\sin^2\theta + b^2\cos^2\theta.$$

The point P $((a+b)\cos\theta, (a+b)\sin\theta)$ is joined to the point Q $(a\cos\theta, b\sin\theta)$ on the ellipse. Prove that (i) PQ is normal to the ellipse at Q, (ii) PQ is equal in length to the perpendicular from P to either of the two tangents of gradient $\tan\theta$.

Deduce that the locus of centres of circles which touch a given ellipse and also touch a variable pair of parallel tangents to the ellipse is a circle, and state its radius.
[C, S 1966, P III, Q 15]

10 The normal at the point P $(ap^2, 2ap)$ of the parabola $y^2 = 4ax$ meets the parabola again at Q, and the tangents at P and Q meet at T. Prove that

$$PT^2 = \frac{4a^2(1+p^2)^3}{p^2}.$$

Prove that, if the length of PT is a minimum, then $p = \pm 1/\sqrt{2}$. [O & C, S 1967, M & HM I, Q 8]

11 Two ellipses, E and E′, have the property that the foci of each ellipse are the ends of the minor axis of the other. Prove that

(i) the major axes of E and E′ are of equal length;
(ii) if the eccentricities of E and E′ are e and e', then $e^2+e'^2 = 1$;
(iii) the common tangents of E and E′ make angles $\sin^{-1}e$ with the major axis of E. [O & C, S 1967, M & HM V Sp, Q 4]

12 The equations of two lines are $y = m_1x+c_1$, $y = m_2x+c_2$, where $m_1 > m_2 > 0$. Prove that the acute angle between the lines is

$$\tan^{-1}\left[\frac{m_1-m_2}{1+m_1m_2}\right].$$

The perpendicular drawn from the point H (1, 3) meets the line PQ, whose equation is $3x+4y = 10$, at the point N. The line HN is produced to K such that HN = NK and through K a line KJ is drawn with positive gradient m to meet PQ at J such that angle KJN = 45°. Calculate

(i) the coordinates of K,
(ii) the value of m. [AEB, S 1968, P & A II, Q 3]

13 Prove that the equation of the normal to the parabola $y^2 = 4ax$ at the point $(at^2, 2at)$ is $tx+y = 2at+at^3$.

A line is drawn through the point $(-2a, 0)$ to meet the parabola at P and Q. Prove that the normals at P and Q meet on the parabola. [AEB, N 1967, P & AI, Q 4]

14 Show that, if a, b, c are positive constants satisfying the relation $ab = 2c^2$, then the ellipse

$$\frac{x^2}{a^2}+\frac{y^2}{b^2} = 1$$

touches the hyperbola $xy = c^2$.

A point $Q_1(x_1, y_1)$ moves on the line $y = x\tan\alpha$ and another point $Q_2(x_2, y_2)$ moves on the line $y = -x\tan\alpha$. Express the coordinates of the midpoint P of Q_1Q_2 in terms of x_1, x_2 and α. Show that, if Q_1 and Q_2 move in such a way that the length Q_1Q_2 remains equal to a constant $2k$, then the locus of P is an ellipse.

Prove that, whatever the value of a may be, the locus touches the hyperbola $2xy = k^2$. [JMB, S 1967, P I, Q 6]

15

(*a*) Show that the gradient of the line joining the points $(ct_1, c/t_1)$, $(ct_2, c/t_2)$ on the hyperbola $xy = c^2$ is $-1/(t_1 t_2)$. The points A, B, C lie on this hyperbola. The line through A perpendicular to BC meets the line through B perpendicular to AC at R. Prove that R lies on the hyperbola.

(b) Show that the line $y = mx+c$ is a tangent to the ellipse $x^2/a^2+y^2/b^2 = 1$ if $c^2 = a^2m^2+b^2$. Hence obtain the quadratic equation satisfied by the gradient of a tangent to the ellipse from a given point (x_1, y_1). Find the equation of the locus of the point if the two tangents from the point to the ellipse are at right angles. [JMB, S 1967, P II, Q 9]

16 C is the circle whose equation is $x^2+y^2 = a^2$. Two circles C_1 and C_2 both meet C orthogonally and also meet one another orthogonally. Prove that the centre of C_1 lies on the common chord of C and C_2.

Find the equation of the circle which meets $x^2+y^2 = 1$ and $(x-1)^2+(y-2)^2 = 4$ orthogonally, and which passes through the point (4, 5). [W, S 1969, P II, Q 5]

17 ABCD is a quadrilateral whose vertices are the points (3, 1), (2, 5), (−2, 2) and (−1, −1) respectively. Find

(a) the perimeter of the quadrilateral,
(b) its area,
(c) the angle between AC and BD,
(d) the ratio in which BD divides AC. [W, S 1969, P II, Q 8]

18 Obtain the equation of the chord whose extremities are the points P and Q, with parameters u and v, on the parabola $(at^2, 2at)$. Find the condition that PQ shall pass through the point $(a, 0)$ and show that the tangents at P and Q then intersect at right angles.

Find the equation of the locus of the midpoints of all such chords and identify this curve. [W, S 1968, P II, Q 3]

19 The normal at a variable point P of an ellipse meets the major axis at Q and the minor axis at R. Show that the locus of the midpoint of QR is an ellipse with the same eccentricity as the given ellipse. [W, S 1968, P S, Q 8]

D Further properties of triangles: Theory summary

1 The centre O of the circle through the vertices A, B, C of $\triangle$ABC is the meet of the perpendicular bisectors of the sides of the triangle.

$$\text{Circumradius,}\quad R = \frac{a}{2\sin A} = \frac{abc}{4\triangle}.$$

2 The centres I, I_1, I_2, I_3 of the circles which touch the sides of $\triangle$ABC are found by bisecting the angles of the triangle, internally and externally.

$$\text{Inradius,}\quad r = \frac{\triangle}{s} = 4R\sin\tfrac{1}{2}A\sin\tfrac{1}{2}B\sin\tfrac{1}{2}C = (s-a)\tan\tfrac{1}{2}A$$

and the exradius of the circle opposite A,

$$r_1 = \frac{\triangle}{s-a} = 4R\sin\tfrac{1}{2}A\cos\tfrac{1}{2}B\cos\tfrac{1}{2}C = s\tan\tfrac{1}{2}A,$$

with similar expressions for the other exradii r_2, r_3. Note, $s = \frac{1}{2}(a+b+c)$.

3

The perpendiculars AD, BE, CF from the vertices of $\triangle$ABC to the opposite sides meet at H, the orthocentre. $\triangle$DEF is the pedal triangle; the point H is its incentre.

$$\text{EF} = a\cos A = R\sin 2A,$$
$$\text{AH} = 2R\cos A,$$
$$\text{DH} = 2R\cos B\cos C.$$

4 The nine-point circle passes through the midpoints X, Y, Z of BC, CA, AB and passes also through D, E, F and through the midpoints of HA, HB, HC.

The centre N is the midpoint of OH, and the radius of the nine-point circle is $\frac{1}{2}R$.

$\triangle$ABC is the pedal triangle of $\triangle I_1I_2I_3$ and of $\triangle II_2I_3$. Circumradius of each of these triangles is $2R$.

5 The medians of $\triangle$ABC concur at a point G, the centroid. GX = $\frac{1}{3}$AX. G is on OH with OG = $\frac{1}{3}$OH. OGNH is a straight line, known as the Euler line. OG:GN:NH = 2:1:3.

6 For $\triangle$ABC,

$$\text{OI}^2 = R^2-2Rr,\qquad \text{OI}_1^2 = R^2+2Rr_1,$$
$$\text{OH}^2 = R^2(1-8\cos A\cos B\cos C),$$
$$\text{IH}^2 = 2r^2-4R^2\cos A\cos B\cos C,$$
$$\text{I}_1\text{H}^2 = 2r_1^2-4R^2\cos A\cos B\cos C,$$
$$\text{IN} = \tfrac{1}{2}R-r,\qquad \text{I}_1\text{N} = \tfrac{1}{2}R+r_1.$$

Feuerbach's theorem (derived from last pair of results): the nine-point circle touches the in-circle and the ex-circles.

E Further properties of triangles: Illustrative worked problems

1 A triangle ABC is obtuse angled at B. From a point P on the minor arc AB of the circumcircle of the triangle, perpendiculars PE and PF are drawn to the sides AC and AB respectively and the triangle AEF is similar to the triangle ABC. Prove that AP is perpendicular to BC.

If, however, P is on the major arc AC, show that the triangle AEF cannot be similar to the triangle ABC.

[L, S 1967, FM VI, Q 3]

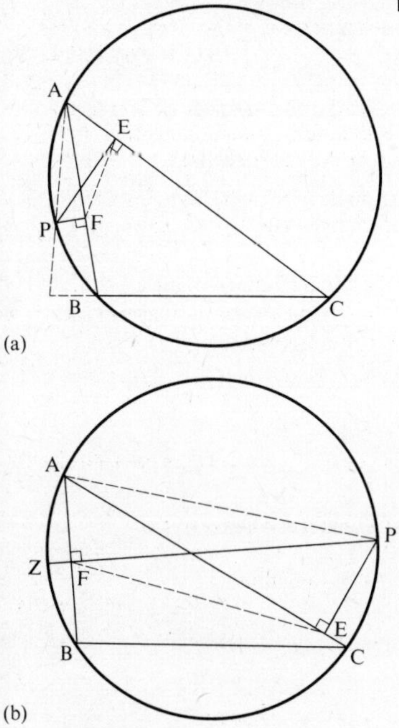

Figure 36

Solution. In Figure 36 (a), since $\triangle AEF$ is similar to $\triangle ABC$, it follows that $A\hat{F}E = \hat{C}$.

But APFE is a cyclic quadrilateral, since $A\hat{E}P = A\hat{F}P = 90°$,

hence $A\hat{P}E = A\hat{F}E = \hat{C}$

and $90° - P\hat{A}E = \hat{C}$,

and hence AP meets BC at right angles.

In Figure 36 (b), assuming $\triangle AFE$ is similar to $\triangle ABC$ again, when P is on the major arc, then we must have

$A\hat{E}F = \hat{C}$ (i.e. $FE||BC$).

But $A\hat{E}F = A\hat{P}F$ and $A\hat{P}F = A\hat{P}Z$ on arc $AZ < AB$

so that $A\hat{E}F < \hat{C}$,

so EF is not parallel to BC, that is, the triangles are not similar.

2 PQR is a triangle in which O is the circumcentre, H the orthocentre, G the centroid and M the midpoint of QR. Prove that $PH = 2OM$ and that O, G, H are collinear.

ABCD is a cyclic quadrilateral and H_1, H_2, H_3, H_4 are the orthocentres of the triangles ABD, CBD, ABC, ADC respectively. Prove that

$$H_1H_2 . H_3H_4 = AB.CD + AD.BC. \quad \text{[L, J 1967, FM V, Q 4]}$$

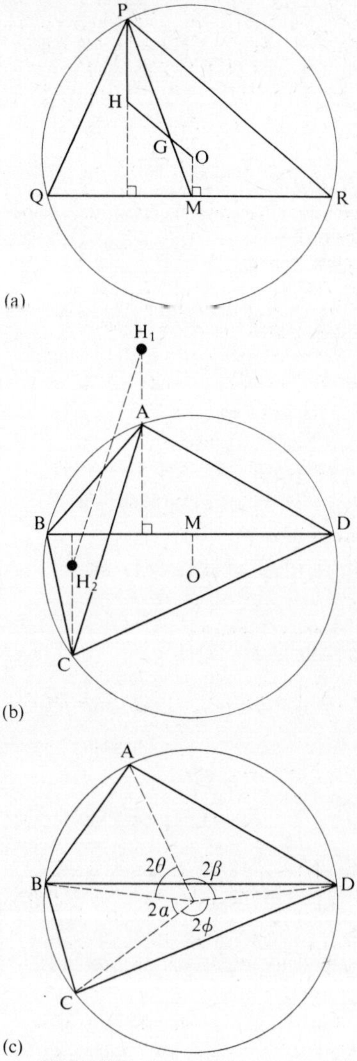

Figure 37

Solution. Compare $\triangle PHG$ with $\triangle MOG$ in Figure 37(a). We know

$$\frac{PG}{MG} = \frac{2}{1}$$

and $PH = 2R\cos A = 2OM$, therefore also

$$\frac{PH}{MO} = \frac{2}{1}.$$

In addition $H\hat{P}G = O\hat{M}G$, since OM and PH are both perpendicular to QR.

Hence the triangles are similar, HGO is a straight line and $PH/MO = 2$. In Figure 37(b), $AH_1 = CH_2 (= 2OM)$ and AH_1 is parallel to CH_2, hence ACH_2H_1 is a parallelogram and so $H_1H_2 = AC$. Similarly $H_3H_4 = BD$. It remains to prove, then, that

$$AC.BD = AB.CD+AD.BC$$

for the last part; refer to Figure 37(c).

The last statement is in fact Ptolemy's theorem, and can be proved trigonometrically thus:
With the notation of Figure 37(c),

$$\begin{aligned} AB.CD+AD.BC &= R^2(\sin\theta\sin\phi+\sin\alpha\sin\beta) \\ &= \tfrac{1}{2}R^2\{\cos(\theta-\phi)-\cos(\theta+\phi)+ \\ &\quad +\cos(\alpha-\beta)-\cos(\alpha+\beta)\} \\ &= \tfrac{1}{2}R^2\{\cos(\theta-\phi)+\cos(\alpha-\beta)\}, \end{aligned}$$

as $\alpha+\beta = 180°-(\theta+\phi)$.

Also
$$\begin{aligned} AC.BD &= R^2\sin(\theta+\alpha)\sin(\phi+\alpha) \\ &= \tfrac{1}{2}R^2\{\cos(\theta-\phi)-\cos(\theta+\phi+2\alpha)\} \\ &= \tfrac{1}{2}R^2\{\cos(\theta-\phi)-\cos(180°+\alpha-\beta)\}, \end{aligned}$$

which is the same result as above.

Hence finally $H_1H_2.H_3H_4 = AB.CD+AD.BC$.

F Further properties of triangles: Problems

Answers and hints will be found on p. 69.

1 In a triangle ABC, AD, BE, CF are altitudes, R is the circumradius and Δ is the area. Prove that the triangle DEF has

(i) circumradius $= \frac{1}{2}R$,
(ii) perimeter $= 4R\sin A\sin B\sin C$,
(iii) area $= 2\Delta\cos A\cos B\cos C$.

[L, J 1968, FM VI, Q 3]

2 The feet of the perpendiculars from the vertices A, B and C of a triangle to the opposite sides are P, Q and R respectively, and the point of intersection of these perpendiculars is O. The incentres of the triangles ABC, AQR, BRP, CPQ are I, X, Y, Z respectively. With the usual notation for triangle ABC, prove that:

(a) triangle AQR is similar to triangle ABC and the ratio of the corresponding sides is $\cos A$;
(b) AXI, BYI and CZI are straight lines and
$IX = (1-\cos A)IA = 2r\sin\frac{1}{2}A$;
(c) triangles IXY, IBA are similar;
(d) angle $XZY = 90°-\frac{1}{2}C$ and $XY = 2r\cos\frac{1}{2}C$;
(e) the circumradius of triangle XYZ is r. [O, S 1966, P II, Q 2]

3 In triangle ABC, $AB = AC = p$, the midpoint of BC is L and $AL = h$. Prove that the radius of the nine-point circle of the triangle is $p^2/4h$.

H, in CB produced, is the point of contact with CB produced of one of the escribed circles of this triangle ABC and HK is a diameter of this escribed circle. KL meets this escribed circle at M.

Prove that:

(a) $KH = 2h$;
(b) $LH = p$;
(c) triangles KHM and HLM are similar to triangle KLH and hence
$$\frac{KM}{ML} = \frac{4h^2}{p^2};$$
(d) this escribed circle touches the nine-point circle of triangle ABC at M. [O, S 1966, P II, Q 4]

4 By using the properties of the nine-point circle of triangle $I_1I_2I_3$, where I_1, I_2, I_3 denote, as usual, the centres of the escribed circles of triangle ABC, prove that:

(a) The circumcentre of triangle ABC is the midpoint of the line joining the incentre of triangle ABC to the circumcentre of triangle $I_1I_2I_3$;
(b) The diameter, $2R$, of the circumcircle of triangle ABC is equal to the radius of the circumcircle of triangle $I_1I_2I_3$;
(c) The sides of triangle $I_1I_2I_3$ are equal to
$4R\cos\frac{1}{2}A$, $4R\cos\frac{1}{2}B$, $4R\cos\frac{1}{2}C$;
(d) The area of triangle $I_1I_2I_3$ is
$8R^2\cos\frac{1}{2}A\cos\frac{1}{2}B\cos\frac{1}{2}C$. [O, S 1965, P II, Q 3]

G Coordinate geometry: Answers and hints

1 Equation of tangent RPQ is $b^2x+a^2y = a^2b^2$, hence Q is $(a^2, 0)$, R is $(0, b^2)$ and thus

$OQ = a^2$, $OR = b^2$.
But as (1, 1) lies on ellipse $b^2+a^2 = a^2b^2$,

$$OR = b^2 = \frac{a^2}{a^2-1}.$$

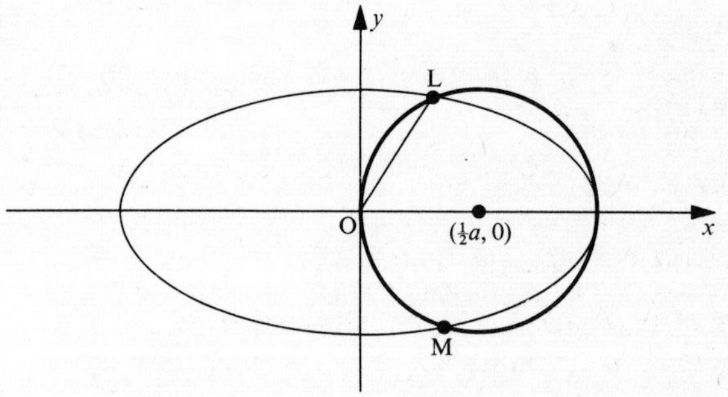

Figure 38

Equation of circle is $(x-\frac{1}{2}a)^2+y^2 = \frac{1}{4}a^2$,

i.e. $x^2+y^2-ax = 0$ **4.2**

Equation of ellipse is

$$\frac{1}{a^2-1}x^2+y^2 = \frac{a^2}{a^2-1} = 0. \qquad \textbf{4.3}$$

On eliminating y from equations **4.2** and **4.3** we obtain

$(a^2-2)x^2-a(a^2-1)x+a^2 = 0$

and solving we find the x-coordinate of L (and M). This gives

$$x = \frac{a}{a^2-2},$$

but $\quad OL^2 = OM^2 = x^2+y^2 = ax = \dfrac{a^2}{a^2-2},$

thus $\quad OL = OM = \dfrac{a}{\sqrt{(a^2-2)}}.$

2 Equation of tangent is

$(b \sec t)x-(a \tan t)y = ab.$

Equation of normal is

$$y-b \tan t = \frac{-1}{(b \sec t)/(a \tan t)}(x-a \sec t),$$

which on simplification reduces to quoted result.

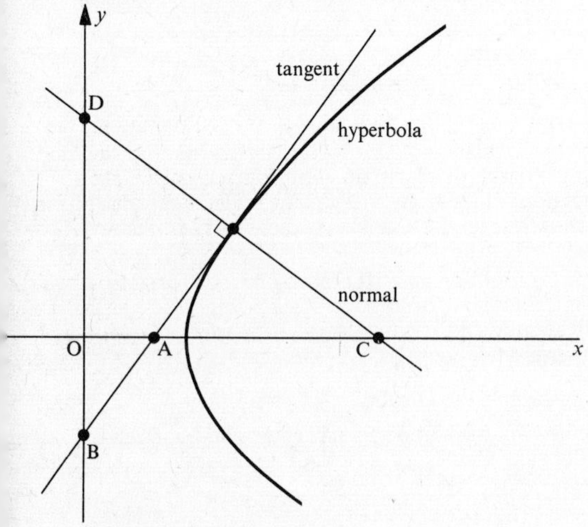

Figure 39

Area $\triangle OAB = \frac{1}{2}OA.OB = \frac{1}{2}(a \cos t)(b \cot t),$

area $\triangle OCD = \frac{1}{2}OC.OD$

$$= \frac{1}{2}\left[\frac{a^2+b^2}{a}\sec t\right]\left[\frac{a^2+b^2}{b}\tan t\right],$$

hence $\triangle OAB \times \triangle OCD$ is independent of t.
Circumcentre of $\triangle OAB$, $\quad (x, y) = (\frac{1}{2}OA, \frac{1}{2}OB)$
$\qquad = (\frac{1}{2}a \cos t, -\frac{1}{2}b \cot t)$

and, eliminating t, we obtain

$x^2b^2 = y^2(a^2-4x^2).$

3 Let centre of required circle be (x_0, y_0), then, from Figure 40,

$(1-x_0)^2+(1-y_0)^2 = (4-x_0)^2+(2-y_0)^2,$

which provides $\quad 3x_0+y_0 = 9.$ **4.4**

Also as centre lies on normal to $y = mx$, that is, on the line $y = -x+2$, we have

$y_0+x_0 = 2.$ **4.5**

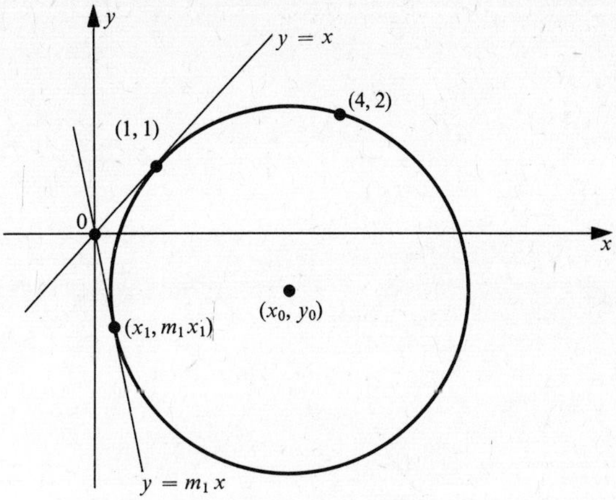

Figure 40

Solving equations **4.4** and **4.5** we obtain $(x_0, y_0) = (\frac{7}{2}, -\frac{3}{2})$ and hence equation of circle is

$(x-\frac{7}{2})^2+(y+\frac{3}{2})^2 = (1-\frac{7}{2})^2+(1+\frac{3}{2})^2,$

i.e. $\quad x^2+y^2-7x+3y+2 = 0.$ **4.6**

Let equation of other tangent be $y = m_1 x$. Hence substituting for y in equation **4.6** and using the condition that the quadratic should have equal roots, we obtain

$$m^2-42m+41 = 0,$$

i.e. $\quad (m-1)(m-41) = 0.$

Thus other tangent is $y = 41x$.

4 Equation of tangent is

$ty = x+at^2.$ **4.7**

Let points where tangents from (h, k) touch the parabola be characterized by t_1, t_2. Then area of triangle

$$\Delta = \frac{1}{2}\begin{vmatrix} h & k & 1 \\ at_1^2 & 2at_1 & 1 \\ at_2^2 & 2at_2 & 1 \end{vmatrix}$$

$$= \frac{1}{2}a^2\left[2\frac{h}{a}(t_1-t_2)-\frac{k}{a}(t_1^2-t_2^2)+2t_1t_2(t_1+t_2)\right],$$

and by using the fact that t_1, t_2 satisfy equation **4.7** when $x = h$, $y = k$, the quoted result follows.

5 Let $y = mx+c$ be equation of common tangent. Then substituting $y = mx+c$ in the curve equations and using the condition for equal roots, we obtain

$c = -1-2m\pm\sqrt{(m^2+1)},$

$(4+m)^2 = 4(11-c) = 4\{12+2m\mp\sqrt{(m^2+1)}\}.$ **4.8**

Solving equation **4.8** we obtain $m^2 = 120$ or $m^2 = 8$.
Thus equations of common tangents are

$y = \pm 2\sqrt{2}x+5\mp 4\sqrt{2}, \qquad y = \pm 2\sqrt{30}x-23\mp 4\sqrt{30}.$

6

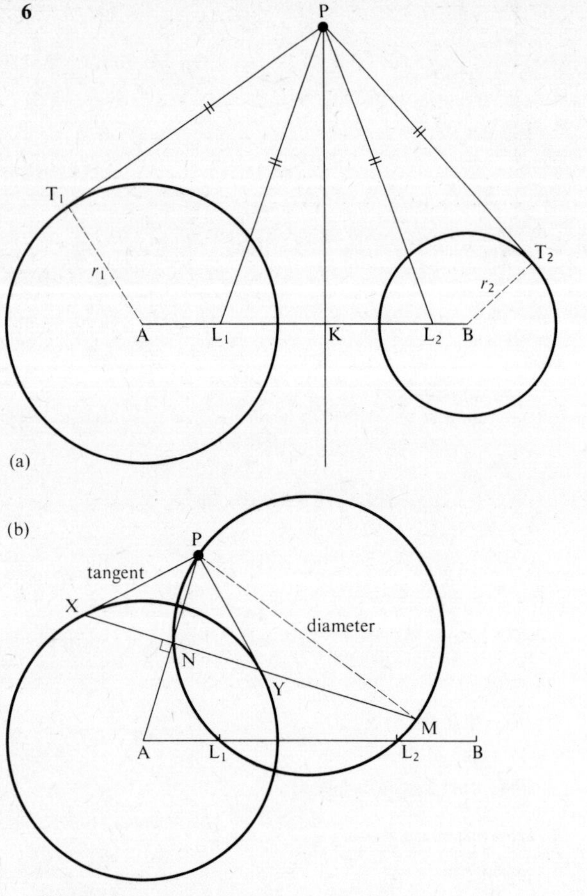

Figure 41

See Figure 41(a). Prove that $AK^2 - KB^2$ is constant by showing that $AK^2 - KB^2 = AP^2 - PB^2 = r_1^2 - r_2^2$. Hence deduce that K is a fixed point, and hence that locus of P is a straight line. Limiting points L_1, L_2 of coaxal system are on AB and

$$PL_1 = PL_2 = PT_1 = PT_2 = \ldots.$$

$$AL_1 . AL_2 = AT_1^2 = r_1^2,$$

whence L_1, L_2 are inverse with respect to any circle of the system.

See Figure 41(b). The circle PL_1L_2 passes through N, since $AN . AP = AL_1 . AL_2 = r_1^2$. If M is the opposite end of diameter through P, the chord of contact of tangents from P passes through M (a fixed point), whence result.

In special cases when radii of circles tend to zero, the chords of contact are L_1M and L_2M and both pass through M.

7 Let centre of variable circle be $O(x, y)$, and intercept $AB = t$. Then show $x = a - t$ or $t = a - x$ and use

$$OA^2 = \tfrac{1}{4}t^2 + x^2 = OD^2 = y^2 + t^2.$$

By eliminating t, quoted result follows.

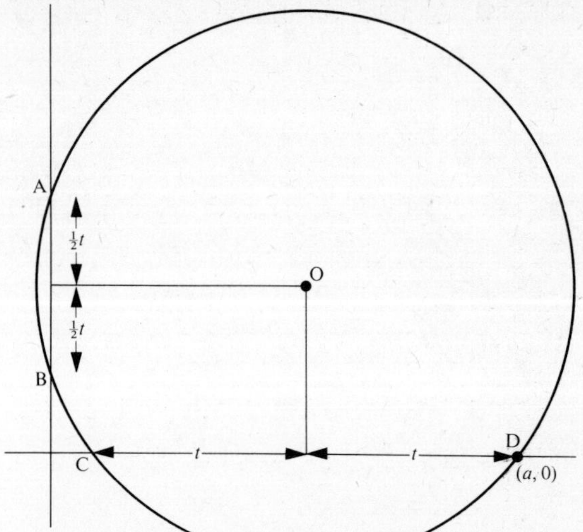

Figure 42

8 The endpoints of the chord having $M(x', y')$ as midpoint are, say, $P(x'+h, y'+mh)$, $Q(x'-h, y'-mh)$, where m is the gradient of chord. Each pair of coordinates satisfies the equation of the parabola; obtain two equations and subtract to find $m = 2a/y'$. This makes the equation of the chord as stated.

In the problem, take $M(a+\lambda, \lambda)$ on $y = x - a$ (λ, parameter) M is the midpoint of chord $\lambda y - 2ax = \lambda^2 - 2a(a+\lambda)$. For a neighbouring point $N(a+\lambda+d\lambda, \lambda+d\lambda)$ on $y = x - a$, the corresponding chord is

$$(\lambda + d\lambda)y - 2ax = (\lambda + d\lambda)^2 - 2a(a + \lambda + d\lambda).$$

Deduce for intersection point of chords that, as $d\lambda \to 0$,

$$y = 2\lambda - 2a \quad \text{and so} \quad x = a + \frac{\lambda^2}{2a}.$$

Eliminate λ from these two equations to obtain final result.

9 Tangent at (x', y') is

$$\frac{x'x}{a^2} + \frac{y'y}{b^2} = 1.$$

If this is identical with $x \sin\theta - y \cos\theta = c$, then

$$\frac{\sin\theta}{x'/a^2} = -\frac{\cos\theta}{y'/b^2} = c.$$

Deduce, using that (x', y') lies on the ellipse, that

$$c^2 = a^2\sin^2\theta + b^2\cos^2\theta.$$

Gradient $PQ = \dfrac{a}{b}\tan\theta$, tangent at Q has gradient, $-\dfrac{b}{a}\cot\theta$, whence next result. $PQ = \sqrt{(a^2\sin^2\theta + b^2\cos^2\theta)}$. The two tangents of gradient $\tan\theta$ are

$$x\sin\theta - y\cos\theta = \sqrt{(a^2\sin^2\theta + b^2\cos^2\theta)}.$$

R.H.S. is the perpendicular distance from P to either tangent, which is equal to PQ. Hence P is centre of circle of radius PQ which touches ellipse at Q, and the tangent at Q and the parallel tangent. As θ varies, locus of P is $x^2+y^2 = (a+b)^2$, i.e. a circle radius $a+b$.

10 Tangent at P is $x-py+ap^2 = 0$; solve with tangent at Q to find the coordinates of T. Use the fact that gradient of PQ,

$$\frac{2}{p+q} = -p \text{ to show T}\left(-a(p^2+2), -\frac{2a}{p}\right).$$ Deduce the result for PT^2. Differentiate PT^2 w.r.t. p^2 to show PT^2, and so PT, is a minimum when $p^2 = \frac{1}{2}$.

11 Use $b' = ae = \sqrt{\{a'^2(1-e'^2)\}}$ and $b = a'e' = \sqrt{\{a^2(1-e^2)\}}$. From these prove (i) $a = a'$, and result (ii).

The line $y = mx+c$ is tangent to E, $\frac{x^2}{a^2}+\frac{y^2}{b^2} = 1$ provided $c^2 = a^2m^2+b^2$; and it is tangent to E′: $\frac{x^2}{b'^2}+\frac{y^2}{a^2} = 1$ provided $c^2 = b'^2m^2+a^2$. Deduce that $m^2(1-e^2) = e^2$, and so

$$m = \frac{\pm e}{\sqrt{(1-e^2)}} = \pm\tan\alpha, \quad \text{say.}$$

Then $\sin\alpha = e$.

12 The first part is straightforward.
Let K have coordinates (a, b), then N has coordinates $(\frac{1}{2}(a+1), \frac{1}{2}(b+3))$. N lies on PQ, hence show $3a+4b = 5$. Also HK $\perp$ PQ, hence show $4a-3b = -5$. Then K is $(-\frac{1}{5}, \frac{7}{5})$. Gradient HK $= \frac{4}{3} = m_1$; angle between KJ and HK $= 45°$. Use the formula in the first part and show $m_2 = m = \frac{1}{7}$.

13 To obtain the equation of the normal is standard bookwork. If $P(ap^2, 2ap)$, $Q(aq^2, 2aq)$, P, Q and the given point are collinear, so that $\frac{p}{p^2+2} = \frac{q}{q^2+2}$. Deduce that $pq = 2$. The normals at P and Q meet at $(a(4+p^2+q^2), -2a(p+q))$, on using this relation. It is easy to show this point lies on $y^2 = 4ax$.

14 $P\left(\frac{x_1+x_2}{2}, \frac{x_1-x_2}{2}\tan\alpha\right)$.

Locus of P is $\frac{x^2}{(k\cot\alpha)^2}+\frac{y^2}{(k\tan\alpha)^2} = 1$.

15
(b) Quadratic equation satisfied by gradient m of tangent to the ellipse from (x_1, y_1) is $(mx_1-y_1)^2 = b^2+a^2m^2$. If the two tangents from (x_1, y_1) are perpendicular, equation of locus is $x_1^2+y_1^2 = a^2+b^2$.

16 The data gives $c_1 = c_2 = a^2$, and $2g_1g_2+2f_1f_2 = c_1+c_2$. Deduce that $g_1g_2+f_1f_2 = a^2$. The common chord of C_1 and C_2 is $g_2x+f_2y+a^2 = 0$, and it follows that the centre $(-g_1, -f_1)$ of C_1 lies on this chord. Verify the given circles meet orthogonally. Show that $g_2+2f_2 = -1$, and, because C_2 passes through (4, 5), that $4g_2+5f_2 = -21$, hence f_2 and g_2, and the equation of C_2 as $x^2+y^2-\frac{74}{3}x+\frac{34}{3}y+1 = 0$.

17
(a) Perimeter $= \sqrt{17}+5+\sqrt{10}+2\sqrt{5}$.
(b) Area $= 16\frac{1}{2}$.
(c) Angle between diagonals $= \tan^{-1}\frac{11}{30}$.
(d) The ratio of division is AP : PC $= 6:5$.

18 The equation of the chord PQ is $(u+v)y-2x = 2auv$. It passes through the focus $(a, 0)$ if $uv = -1$. The tangent at P has gradient $1/u$: at Q it has gradient $1/v$. The condition makes the tangents intersect at right angles. The midpoint of PQ has coordinates $(\frac{1}{2}a(u^2+v^2), a(u+v))$. The locus of midpoints is the parabola $y^2 = 2a(x-a)$.

19 Write down the equation of the normal PQR: determine the coordinates of Q and R. The midpoint of QR is

$$\left(\frac{\cos\theta}{2a}(a^2-b^2), -\frac{\sin\theta}{2b}(a^2-b^2)\right)$$

Its locus as θ varies is the ellipse

$$\left[\frac{2ax}{a^2-b^2}\right]^2+\left[\frac{2by}{a^2-b^2}\right]^2 = 1,$$

the semi-major axis of which is $(a^2-b^2)/2b$, and the semi-minor axis is $(a^2-b^2)/2a$. Hence show that the eccentricity is the square root of $1-(b^2/a^2) = e^2$.

H Further properties of a triangle: Answers and hints

1
(i) Note: circle through DEF is nine-point circle, which passes through midpoints of the sides of $\triangle$ABC.
(ii) Show EF $= a\cos A = R\sin 2A$, then perimeter of $\triangle$DEF is $R(\sin 2A+\sin 2B+\sin 2C)$. Factorize this using the fact that $A+B+C = 180°$.
(iii) $E\hat{D}F = 180°-2A$; then area

$$\triangle DEF = \tfrac{1}{2}DE.DF\sin(180°-2A),$$

which equals $bc\cos B\cos C\cos A$, whence result on using $\Delta = \frac{1}{2}bc\sin A$.

2 (a–e) must be done in order.
(a) Consider angles, noting BCQR is cyclic.
(b) Note IA $= r/\sin\frac{1}{2}A$.
(c) Prove IX/IB $=$ IY/IA $= 2\sin\frac{1}{2}A\sin\frac{1}{2}B$.
(d) $X\hat{Z}Y = \frac{1}{2}(A+B)$. To find XY consider $\triangle$XIY.
(e) Use sine rule for $\triangle$XYZ.

3 Use that radius of nine-point circle $= \frac{1}{2}R$, where $2R = p/\sin C$.
(a) If O is centre of escribed circle, show that OA is parallel to HBC first.
(b) LH $=$ OA $=$ OH/sin $O\hat{A}B$.
(c) Clear that the triangles are right angled and similar. Show KM/ML $= 2h\cot\theta/p$, where $\theta = H\hat{K}L$.
(d) Draw LN′ perpendicular to HL meeting OM produced at N′. OM/MN′ $= 4h^2/p^2$ from (c). Deduce MN′ $=$ radius of nine-point circle. Also MN′ $=$ N′L means N′ $\equiv$ N, and nine-point circle touches escribed circle at M.

4
(a) Circumcircle of ABC $\equiv$ nine-point circle of $I_1I_2I_3$. Let centre of this circle be N. If O′ is circumcentre of $I_1I_2I_3$ we know N is on O′I and O′N $=$ NI.
(b) Radius of nine-point circle of $I_1I_2I_3$, i.e. $R = \frac{1}{2}\times$ radius of circumcircle of $I_1I_2I_3$, whence result.
(c) Use sine rule for $\triangle I_1I_2I_3$, noting, e.g., $\hat{I}_3 = \frac{1}{2}(A+B)$, and result (b).
(d) Use $\triangle = \frac{1}{2}I_1I_3.I_2I_3\sin\hat{I}_3$.

Chapter Five
Further Geometry

A Further geometry: Theory summary

1 *Translation of axes*

(a) *Parallel translation of axes.* The original coordinates (x, y) are related to the new (X, Y) by

$$X = x-h, \qquad Y = y-k, \quad \text{where OO}' = (h, k).$$

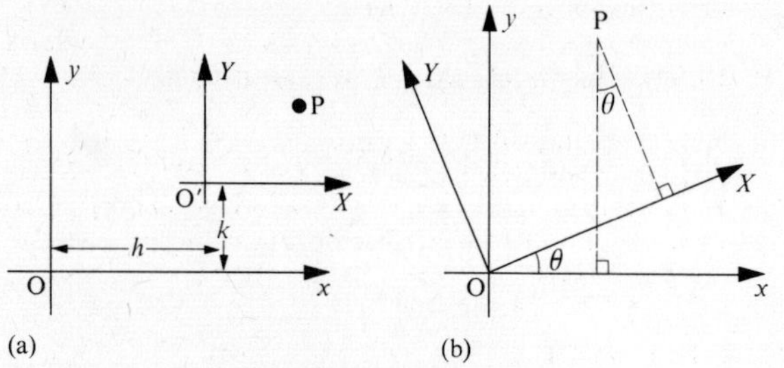

Figure 43 (a) Parallel translation of axes.
(b) Rotation of axes through angle θ

(b) *Rotation through angle θ.* The coordinates (x, y) are related to (X, Y) by

$$\begin{bmatrix} X \\ Y \end{bmatrix} = \begin{bmatrix} \cos\theta & \sin\theta \\ -\sin\theta & \cos\theta \end{bmatrix} \begin{bmatrix} x \\ y \end{bmatrix}.$$

2 *General equation of the second degree*

$$S \equiv ax^2+2hxy+by^2+2gx+2fy+c = 0 \quad \text{is:}$$

(a) a circle, if $a = b$ and $h = 0$;

(b) a pair of straight lines, if

$$\Delta \equiv \begin{vmatrix} a & h & g \\ h & b & f \\ g & f & c \end{vmatrix} = 0,$$

i.e. if $af^2+bg^2+ch^2 = 2fgh+abc$;

(c) a parabola, if $h^2 = ab$;

(d) an ellipse, if $0 < h^2 < ab$;

(e) a hyperbola, if $h^2 > ab$ and $\Delta \neq 0$ (Δ as in b);

(f) a rectangular hyperbola, if $a+b = 0, \Delta \neq 0$;

(g) a perpendicular line pair, if $a+b = 0$ and $\Delta = 0$.

3 *Pencils of conics*

If $S = 0$ and $S' = 0$ are any two conics, the equation $S+\lambda S' = 0$ represents a system of conics passing through the points of intersection of S and S'.

For example:

(a) If four points A, B, C, D are determined by the intersections of the lines $L_1 = 0, L_2 = 0, L_3 = 0, L_4 = 0$, then the equation of the general conic through A, B, C, D is

$$L_1 L_2+\lambda L_3 L_4 = 0.$$

(b) If A, B, C, D are the points of intersection of $S = 0$ with two lines $L_1 = 0$ and $L_2 = 0$, then the equation of a general conic through A, B, C, D is

$$S+\lambda L_1 L_2 = 0.$$

(c) If $L_1 = 0, L_2 = 0$ and $L_3 = 0$ are three straight lines, the equation

$$L_1 L_2+\lambda L_3^2 = 0$$

is a system of conics touching L_1 and L_2 at their intersection points with L_3.

4 *Cross-ratio, harmonic ranges and pencils*

(a) *Definition.* If a transversal meets a pencil of lines O{ACBD} at A, C, B, D, the ratio

$$\frac{\text{AC.BD}}{\text{AD.BC}} = \frac{\sin \text{AOC} \sin \text{BOD}}{\sin \text{AOD} \sin \text{BOC}} = \{\text{ABCD}\}$$

is the cross-ratio of the point range {ACBD}.

(b) {ABCD} is harmonic if the cross-ratio {ACBD} $= -1$.
The concurrent lines OA, OB, OC, OD then form a harmonic pencil. If one transversal is cut by the four concurrent lines in a harmonic range, then every transversal will be cut harmonically by the four lines.
If {ACBD} = {ADBC}, then O{ACBD} is harmonic.

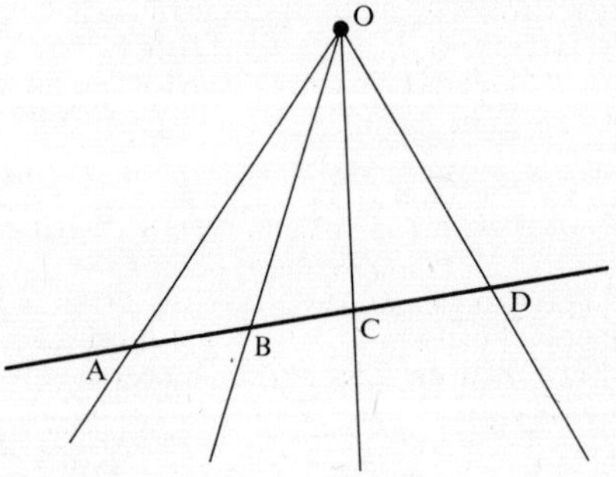

Figure 44

5 *Quadrilateral and quadrangle*

(a) *Complete quadrangle.* A complete quadrangle is formed by four vertices A, B, C, D no three of which are collinear. Their joins AB, CD; BC, AD; CA, BD form the six sides; the sides in each pair are opposite sides. The meets G, F, E of opposite sides are the diagonal points. △EFG is the diagonal-point triangle.

{ACEP} is harmonic; each set of four collinear points in the figure is harmonic.

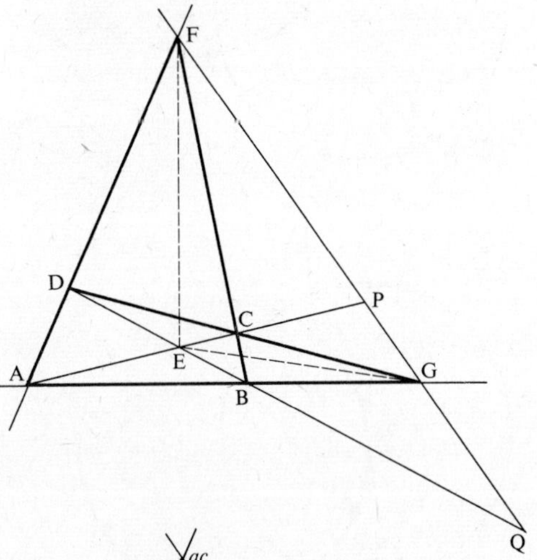

(a)

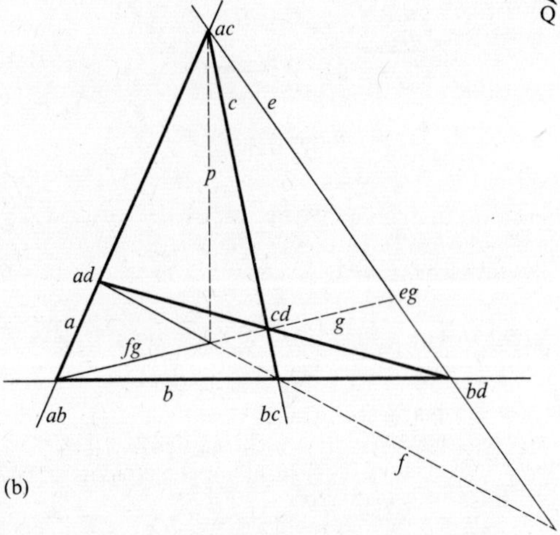

(b)

Figure 45 (a) Complete quadrangle. (b) Complete quadrilateral

(b) *Complete quadrilateral.* A complete quadrilateral is formed of four sides *a*, *b*, *c*, *d*, no two of which are concurrent. Their meets form the six vertices *ab*, *cd*; *bc*, *ad*; *ca*, *bd*; the vertices in each pair are opposite vertices.

The joins of opposite vertices form the three diagonal lines *g*, *f*, *e*. △*efg* is the diagonal-line trilateral.

{*acep*} is harmonic; each set of four concurrent lines in the figure is harmonic.

This figure is the dual of the previous one.

The midpoints of the three diagonal lines of a complete quadrilateral are collinear.

6 *Pole and polar*

(a) The polar of a point P (x', y') with respect to the general conic (in section A2) is

$$axx' + h(xy' + yx') + byy' + g(x + x') + f(y + y') + c = 0.$$

P is the pole of this straight line. For example, for the circle $x^2 + y^2 + 2gx + 2fy + c = 0$, the polar of P is $xx' + yy' + g(x + x') + f(y + y') + c = 0.$

If P lies on the conic it will give the tangent at the point. If P lies outside the conic, the polar is the join of the points of contact of the two tangents from (x', y') to the conic.

(b) *Properties.*

(i) If the polar of P passes through a point R, the polar of R passes through P.

(ii) If P is the pole of a line QR w.r.t. a circle C, then any line through P is cut harmonically by P, QR and C, i.e., {RP; HK} is harmonic. (See Figure 46b.)

(iii) If *p*, *r* are conjugate lines meeting at a point T outside a circle and if TX and TY are the tangents to the circle, then *p*, *r* are harmonically conjugate to TX, TY; i.e. T{PQ; *XY*} is harmonic.

(iv) If four points form a harmonic range, then their polars form a harmonic pencil.

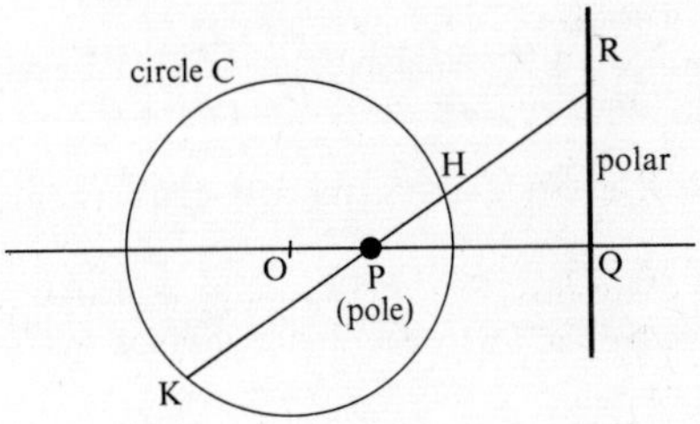

(a)

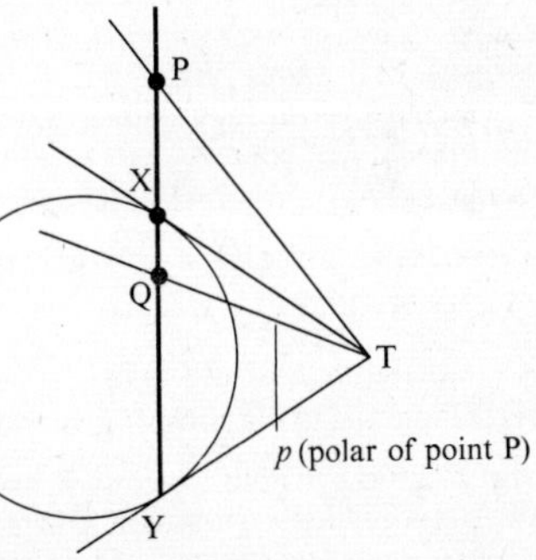

(b)

Figure 46

7 *Concurrence and collinearity*

(a) *Ceva's theorem.* If three concurrent lines are drawn through the vertices of △ABC to meet the opposite sides at D, E, F respectively, then

$$\frac{BD}{DC} \times \frac{CE}{EA} \times \frac{AF}{FB} = 1.$$

(b) *Menelaus' theorem.* If a transversal meets the sides BC, CA, AB of a triangle at D, E, F respectively, then

$$\frac{BD}{DC}\times\frac{CE}{EA}\times\frac{AF}{FB} = -1.$$

(c) *Desargues's theorem on triangles in perspective.* If the lines joining corresponding vertices of two triangles are concurrent, the corresponding sides intersect in collinear points; and conversely.

8 *Reflections, translations, and rotations in space*

Let $T_x = \begin{bmatrix} 1 & \lambda \\ 0 & 1 \end{bmatrix}$, $\quad T_y = \begin{bmatrix} 1 & 0 \\ \lambda & 1 \end{bmatrix}$;

$$E_O = \begin{bmatrix} \mu & 0 \\ 0 & \mu \end{bmatrix}; \quad R_\theta = \begin{bmatrix} \cos\theta & \sin\theta \\ -\sin\theta & \cos\theta \end{bmatrix};$$

$$S_x = \begin{bmatrix} \lambda & 0 \\ 0 & 1 \end{bmatrix}, \quad S_y = \begin{bmatrix} 1 & 0 \\ 0 & \lambda \end{bmatrix}.$$

Then T_x and T_y are operations of translation, E_O is an enlargement about the origin, S_x and S_y represent 'shears', R_θ is a matrix producing a clockwise rotation through an angle θ.

9 *Inversion in two dimensions*

(a) *Definition.* Two points P, P′ are mutually inverse w.r.t. a circle C having centre O and radius a if they are collinear with O and $OP.OP' = a^2$.

O is the centre of inversion and C is the circle of inversion.

(b) *Properties.*

(i) Each point on C is inverted into itself, and the points, other than O, which are inside C invert into points outside C, and vice versa.

(ii) A circle with O as centre inverts into a second circle with O as centre.

(iii) A circle which does not pass through the centre of inversion O inverts into a circle of the same type.

A circle C which passes through O inverts into a straight line parallel to the tangent to C at O.

A line through O inverts into itself.

(iv) Every circle inverts into a circle or a straight line.

(v) Tangent curves invert into tangent curves.

(vi) A circle other than the circle of inversion will invert into itself provided it is orthogonal to the circle of inversion.

(vii) Every inversion carries two points mutually inverse w.r.t. a circle into two points which are mutually inverse w.r.t. the transformed circle.

B Further geometry: Illustrative worked problems

1 P′ and Q′ are the inverse points of P and Q with respect to O, and k is the radius of inversion. Prove that

$$\frac{PQ}{P'Q'} = \frac{k^2}{OP'.OQ'}.$$

Using this property, invert the following theorem with respect to O: if the tangent at O to a circle intersects a chord RQ (produced) at P, then

$PQ.PR = PO^2$. [L, S 1966, FM VI, Q 5]

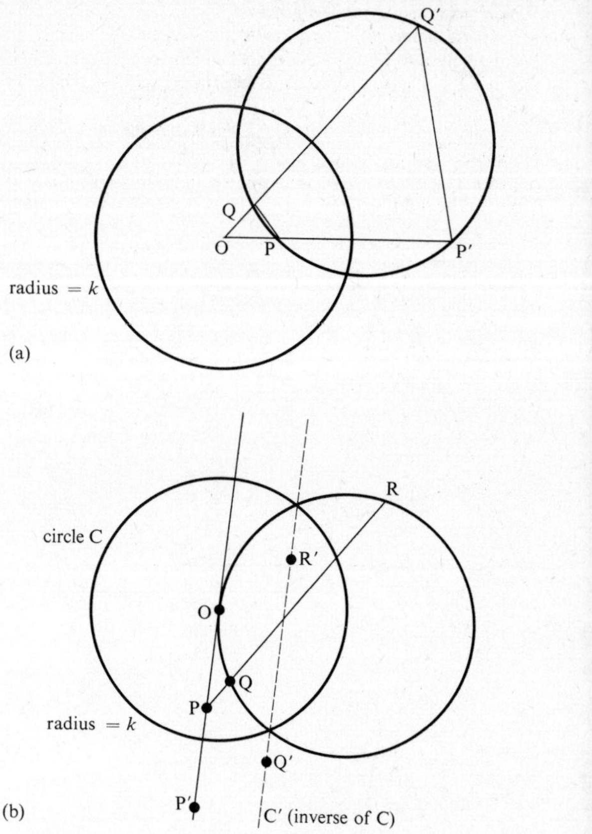

Figure 47

Solution. From the data $OP.OP' = OQ.OQ' = k^2$ (shown also in Figure 47a). This makes PQQ′P′ a cyclic quadrilateral, and also, therefore, △OP′Q′ is similar to △OQP. It follows that

$$\frac{PQ}{P'Q'} = \frac{OQ}{OP'} = \frac{OP}{OQ'} = \frac{k^2}{OP'.OQ'}.$$

In the problem, the inverse of the circle w.r.t. the point O through which it passes is the straight line C′ which is the common chord of C and the circle of inversion (radius k), see Figure 47(b). The theorem gives

$$\frac{PQ}{P'Q'} = \frac{k^2}{OP'.OQ'} \quad\text{and}\quad \frac{PR}{P'R'} = \frac{k^2}{OP'.OR'}.$$

P′ is on OP, and Q′ and R′ are the points on C′ lying on OQ and OR respectively.

Since $PQ.PR = OP^2$, we have

$$\frac{OP^2}{P'Q'.P'R'} = \frac{k^4}{OP'^2.OQ'.OR'} = \frac{OP^2}{OQ'.OR'},$$

whence $P'Q'.P'R' = OQ'.OR'$.

2 P is the arbitrary point $(4\cos\phi, 4\sin\phi)$ on the circle $x^2+y^2 = 16$ and p is the polar of P with respect to the ellipse $x^2/25+y^2/16 = 1$.

(a) Show that the tangent at P to C meets p on the line $x = 0$.
(b) Show that, for any value of ϕ, p is always tangential to some fixed ellipse E.
(c) If p touches the ellipse E in (b) at the point Q, show that PQ is parallel to $y = 0$.

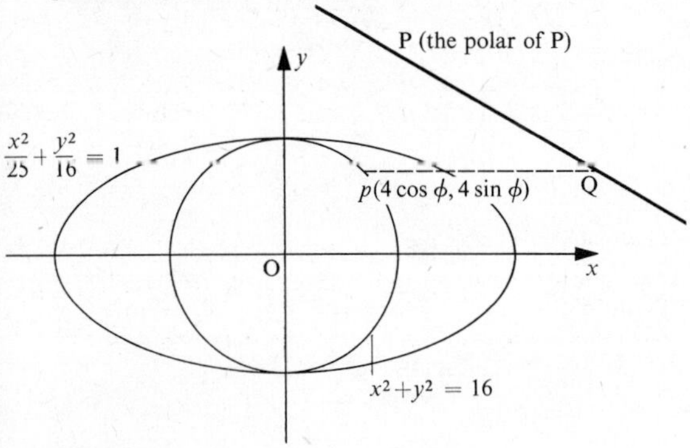

Figure 48

Solution. The polar of $P(4\cos\phi, 4\sin\phi)$ w.r.t. the ellipse $x^2/25+y^2/16 = 1$ is

$$p: \quad \frac{x.4\cos\phi}{25}+\frac{y.4\sin\phi}{16} = 1,$$

i.e. $$p: \quad \frac{16}{25}x\cos\phi+y\sin\phi = 4. \qquad \textbf{5.1}$$

The tangent to the circle at P is

$$4x\cos\phi+4y\sin\phi = 16,$$

i.e. $$x\cos\phi+y\sin\phi = 4. \qquad \textbf{5.2}$$

Lines **5.1** and **5.2** meet where

$$\frac{9x}{25}\cos\phi = 0,$$

i.e. where $x = 0$.

The polar p, i.e. line **5.1**, is tangent to the ellipse

$$\text{E:} \quad \frac{x^2}{(25/4)^2}+\frac{y^2}{16} = 1.$$

The point of contact of p with E is $Q\left(\frac{25}{4}\cos\phi, 4\sin\phi\right)$;

hence gradient of PQ = 0.

C Further geometry: Problems

Answers and hints will be found on pp. 74–5.

1 If the points P_i with coordinates (t_i, t_i^3), where $i = 1, 2, 3$, are collinear, prove that the tangents to the curve $y^2 = x^3$ at the points Q_i whose coordinates are (t_i^2, t_i^3) are concurrent.

If the points Q_i are collinear, prove that the tangents to the curve $y = x^3$ at the points P_i are concurrent.

Show that in the first case, either two of the points P_i are coincident or $\sum t_i = 0$, and obtain the corresponding conditions in the second case. [L, S 1967, FM VI, Q 4]

2
(i) The axes of coordinates Ox, Oy are rotated anticlockwise through an acute angle α to take up new positions $O\xi$, $O\eta$ respectively. Prove that if a point has coordinates (x, y) referred to the original axes and (ξ, η) referred to the new axes, then

$$x = \xi\cos\alpha-\eta\sin\alpha \quad \text{and} \quad y = \xi\sin\alpha+\eta\cos\alpha.$$

Using this transformation, reduce the conic

$$2x^2-5xy+2y^2 = 1$$

to the form $p\xi^2+q\eta^2 = 1$, determining α, p and q.

Hence state the nature of the conic, sketch it relative to the original axes and give its eccentricity.

(ii) Find the asymptote of

$$y^3 = x^2(3a-x)$$

and its form at the origin. Sketch the curve. [L, S 1966, FM VI, Q 8]

3 Prove that the angle of intersection of two curves is equal to the angle of intersection of the inverses of the curves.

A variable line AOB passes through a fixed point O and intersects two fixed lines CX and CY at A and B respectively. Prove that the circles AOC and BOC intersect at a constant angle

(a) by using elementary geometry,
(b) by inverting and proving the inverse theorem. [L, J 1966, FM VI, Q 5]

4 The sides AB, CD of a quadrangle ABCD meet at E; AC, BD meet at F; AD, BC meet at G. AB, FG meet at P; AC, EG meet at Q; BD, EG meet at R. Prove that

(i) APBE is a harmonic range,
(ii) AR, BQ, GP are concurrent. [C, S 1966, PS, Q 14]

5 Two triangles ABC, XYZ in different planes are in perspective, that is, AX, BY, CZ are concurrent. Prove that the points of intersection of the pairs of lines BC, YZ; CA, ZX; AB, XY lies on a line (the axis of perspective).

Prove that the axes of perspective for the four pairs of triangles

ABC, XYZ; XBC, AYZ; AYC, XBZ; ABZ, XYC

lie in a plane. [O & C, S 1966, V (P, S), Q 12]

6 Employ either a translation or a rotation of axes to determine the nature of the loci whose equations are

(a) $5x^2-6xy+5y^2 = 8$,
(b) $3x^2+7xy+4y^2-13x-15y+14 = 0$.

Sketch both loci on the same diagram, showing clearly their positions relative to the original coordinate axes and relative to one another. [W, S 1969, P II, Q 4]

7 Transform the equation

$$14x^2-4xy+11y^2-36x+48y+41 = 0$$

to rectangular axes through the point $(1, -2)$.

What does the resulting equation become if it be now referred to rectangular axes through the new origin inclined at an angle $-\tan^{-1}\frac{1}{2}$ to the original axes?

What is the nature of the corresponding curve? [W, S 1968, P II, Q 5]

D Further geometry: Answers and hints

1 If the P_i are collinear, then

$$\Delta = \begin{vmatrix} t_1 & t_1^3 & 1 \\ t_2 & t_2^3 & 1 \\ t_3 & t_3^3 & 1 \end{vmatrix} = 0,$$

giving $(t_3-t_2)(t_3-t_1)(t_2-t_1)(t_1+t_2+t_3) = 0$, and so $\sum t_i = 0$. The tangents at the Q_i to the curve $y^2 = x^3$ have equations $2y-3t_i x+t_i^3 = 0$ and will be concurrent for the same determinantal condition $\Delta = 0$ above.

If the Q_i are collinear, then

$$\Delta' = \begin{vmatrix} t_1^2 & t_1^3 & 1 \\ t_2^2 & t_2^3 & 1 \\ t_3^2 & t_3^3 & 1 \end{vmatrix} = 0,$$

giving $(t_3-t_2)(t_3-t_1)(t_2-t_1)(t_1t_2+t_2t_3+t_3t_1) = 0$, and so $\sum t_it_j = 0$. The tangents at P_i to $y = x^3$ are $y-3t_i^2x+2t_i^3 = 0$ and are concurrent for the condition $\Delta' = 0$.

In the first case, either $t_i = t_j$ or $\sum t_i = 0$. In the second case, either $t_i = t_j$ or $\sum t_it_j = 0$.

2

(i) The first part is standard theory. Under the transformation, the conic $2x^2-5xy+2y^2 = 1$ becomes $2\xi^2+2\eta^2-5\{(\xi^2-\eta^2)\sin\alpha\cos\alpha+\xi\eta\cos 2\alpha\} = 1$. Make $\alpha = \frac{1}{4}\pi$, to give the hyperbola

$$\frac{\xi^2}{2/9}-\frac{\eta^2}{2} = 1,$$

so $p = -\frac{1}{2}, q = \frac{9}{2}$. Hyperbola has $e = \sqrt{10}$.

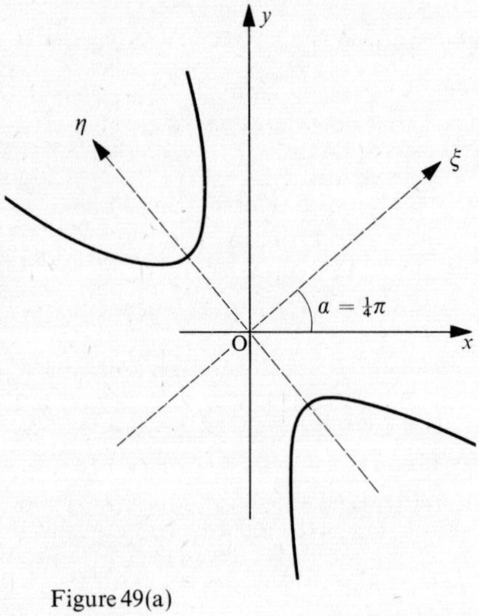

Figure 49(a)

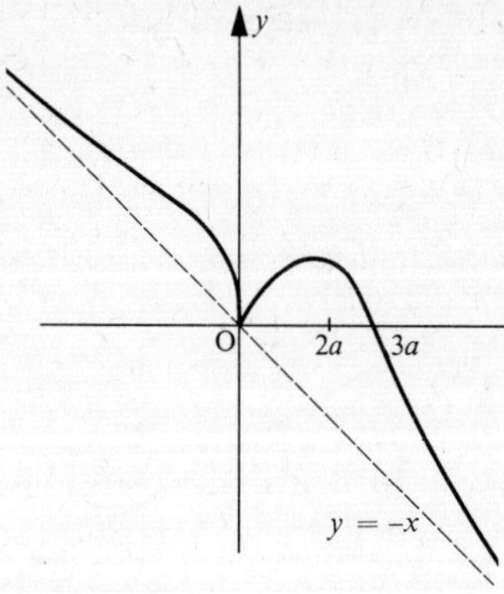

Figure 49(b)

(ii) Asymptote to curve is $y = -x$. Sketch of curve is as in Figure 49(b).
Near the origin, $y^3 = 3ax^2$, approximately.

(a)

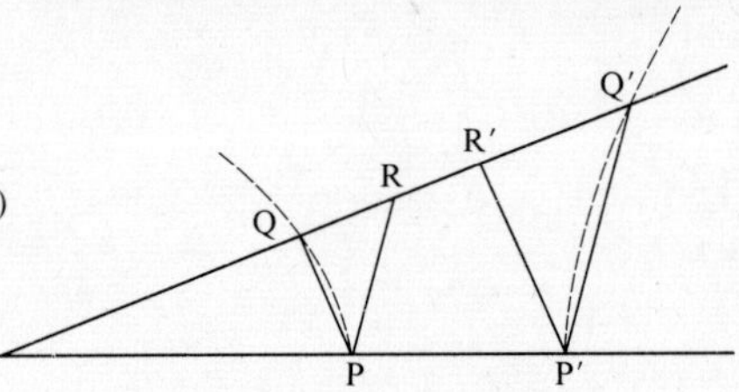

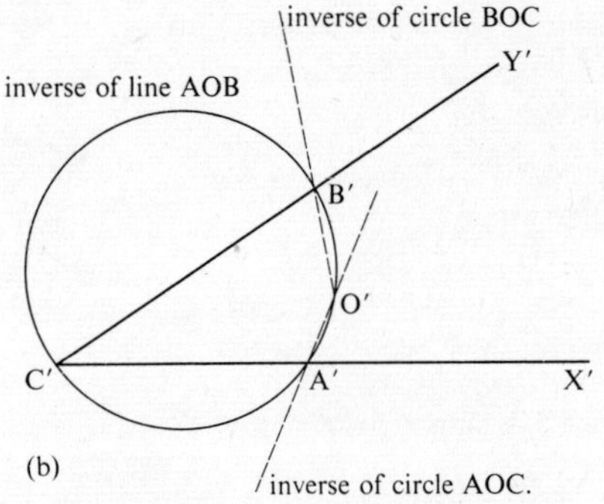

(b)

Figure 50

3

(a) See Figure 50(a). We have $k^2 = OP.OP' = OQ.OQ' = OR.OR'$. Hence $\triangle OQP$ is similar to $\triangle OP'Q'$ and $O\hat{Q}P = O\hat{P}'Q'$. Similarly, $O\hat{R}P = O\hat{P}'R'$. By subtraction, $Q\hat{P}R = R'\hat{P}'Q'$. Take the limit of this result for the answer.

(a) The angle between the tangents to the circles AOC, BOC at O (or C) is $B\hat{C}O+O\hat{C}A = B\hat{C}A$, a constant.

(b) The inverted figure is as shown in Figure 50(b). The circles invert to straight lines, and the angle between these inverses is $B'\hat{C}A'$, as in (a).

4

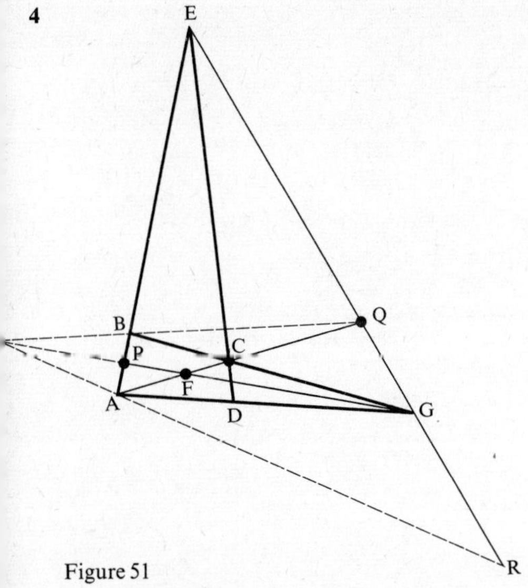

Figure 51

(i) To prove $\{APBE\} = -1$ consider Ceva's and Menelaus' theorems on $\triangle GAB$.

(ii) $\{APBE\} = F\{APBE\} = \{QGRE\} = \{EQGR\}$; and since both ranges $\{EBPA\}$ and $\{EQGR\}$ have the same cross-ratios and intersect at E it follows that QB, GP, RA are concurrent.

5 The first part is standard theory on Desargues's theorem. For the second part apply Desargues's theorem. Check that the axes of perspective of the given triangle-pairs are contained by the plane which passes through the six points:

$P_1(AB, XY)$, $P_2(AZ, XC)$, $P_3(BZ, YC)$, $P_4(AC, XZ)$, $P_5(BC, YZ)$, $P_6(AY, BX)$.

The axis of perspective for the first pair of triangles passes through P_1, P_4 and P_5; for the second pair, it passes through P_2, P_5 and P_6, etc.

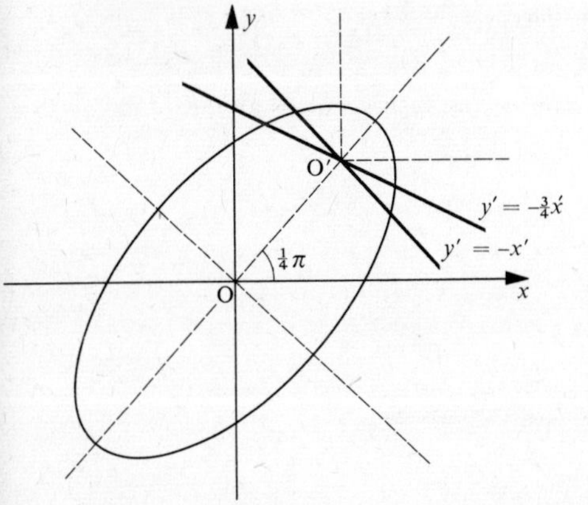

Figure 52

6

(a) Put $x = x'\cos\alpha - y'\sin\alpha, \qquad y = x'\sin\alpha + y'\cos\alpha$

into the given equation, yielding

$(5-6\sin\alpha\cos\alpha)x'^2 + (5+6\sin\alpha\cos\alpha)y'^2 - 6x'y'\cos 2\alpha = 8.$

Choose $\alpha = \frac{1}{4}\pi$. Then get $\frac{1}{4}x'^2 + y'^2 = 1$, an ellipse with $a = 2$ and $b = 1$ (see Figure 52).

(b) Put $x = x'+h$, $y = y'+k$. Show that on substituting $h = 1$ and $k = 1$, the x' and y' terms vanish to give $3x'^2 + 7x'y' + 4y'^2 = 0$, a pair of straight lines:

$3x' + 4y' = 0 \quad \text{and} \quad x' + y' = 0.$

These are shown in Figure 52 relative to the ellipse.

7 Substituting $x = 1+x'$, $y = -2+y'$ into the equation gives

$14x'^2 - 4x'y' + 11y'^2 = 25.$

Now rotate the axes, using

$$x' = \frac{2X}{\sqrt{5}} + \frac{Y}{\sqrt{5}}, \quad \text{and} \quad y' = -\frac{X}{\sqrt{5}} + \frac{2Y}{\sqrt{5}}.$$

Obtain the ellipse $3X^2 + 2Y^2 = 5.$

Chapter Six
Vector Analysis

A Vector analysis: Theory summary

1 *Definitions*

(a) A vector is a quantity having both magnitude and direction. It can be represented by a directed line segment. The magnitude of a vector **a** is denoted by $|\mathbf{a}|$ or a.

(b) A unit vector has a magnitude equal to 1. **i**, **j**, **k** represent the unit vectors in the positive directions along the x-, y- and z-axes.

(c) If λ is a scalar, $\lambda\mathbf{a}$ is a vector with a length (magnitude) $|\lambda|a$ and a direction the same as that of **a** when λ is positive and opposite if λ is negative.

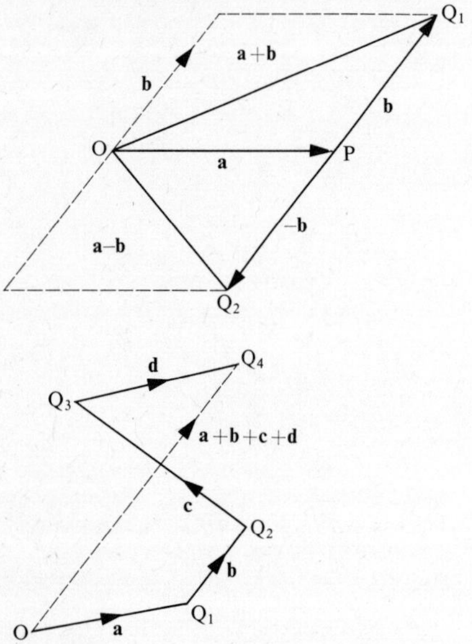

Figure 53

2 *Addition and subtraction of vectors*

(a) The sum $\mathbf{a}+\mathbf{b}$ of two vectors is performed by the triangle or parallelogram rule. In Figure 53(a), $\mathbf{a} = \overrightarrow{OP}$, $\mathbf{b} = \overrightarrow{PQ_1}$, $-\mathbf{b} = \overrightarrow{PQ_2}$, then $\mathbf{a}+\mathbf{b} = \overrightarrow{OQ_1}$, $\mathbf{a}-\mathbf{b} = \overrightarrow{OQ_2}$.

(b) For the addition of more than two vectors use the polygon rule, see Figure 53(b), where $\mathbf{a}+\mathbf{b}+\mathbf{c}+\mathbf{d} = \overrightarrow{OQ_4}$.

3 *Cartesian components of a vector*

(a) A vector **a** in the xy plane may be written in terms of its Cartesian components as

$$\mathbf{a} = a_x\mathbf{i}+a_y\mathbf{j} \quad \text{(see Figure 54a).}$$

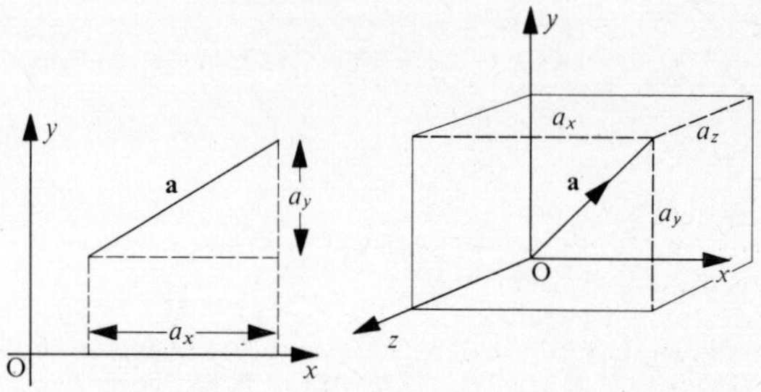

Figure 54

(b) A three-dimensional vector can similarly be written as

$$\mathbf{a} = a_x\mathbf{i}+a_y\mathbf{j}+a_z\mathbf{k} \quad \text{(see Figure 54b).}$$

(c) The magnitudes of the above vectors are, respectively,

$$|\mathbf{a}| = (a_x^2+a_y^2)^{\frac{1}{2}}, \qquad |\mathbf{a}| = (a_x^2+a_y^2+a_z^2)^{\frac{1}{2}}.$$

(d) If $\mathbf{a} = a_x\mathbf{i}+a_y\mathbf{j}+a_z\mathbf{k}$ and $\mathbf{b} = b_x\mathbf{i}+b_y\mathbf{j}+b_z\mathbf{k}$,
then $\mathbf{a}\pm\mathbf{b} = (a_x\pm b_x)\mathbf{i}+(a_y\pm b_y)\mathbf{j}+(a_z\pm b_z)\mathbf{k}$
and $\lambda\mathbf{a}+\mu\mathbf{b} = (\lambda a_x+\mu b_x)\mathbf{i}+(\lambda a_y+\mu b_y)\mathbf{j}+(\lambda a_z+\mu b_z)\mathbf{k}$.

4 *Position vector*

(a) If (x, y, z) are the coordinates of a point P, then

$$\overrightarrow{OP} = \mathbf{r} = x\mathbf{i}+y\mathbf{j}+z\mathbf{k}$$

is known as the position vector of P.

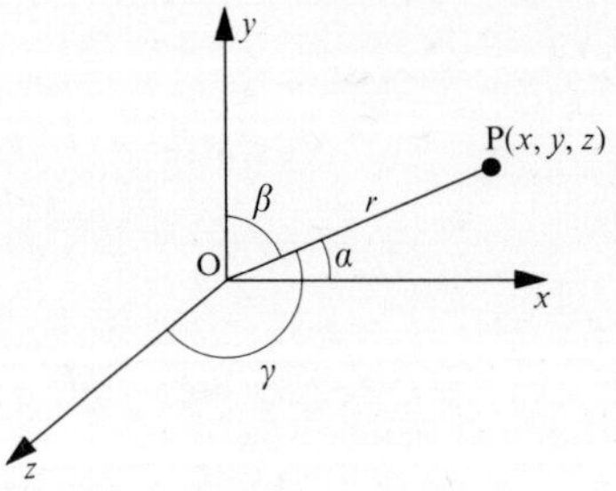

Figure 55

(b) If the angles which $\overrightarrow{OP}$ makes with the x-, y- and z-axes are respectively α, β and γ, then

$$x = r\cos\alpha, \qquad y = r\cos\beta, \qquad z = r\cos\gamma,$$

where $r = |\mathbf{r}| = (x^2+y^2+z^2)^{\frac{1}{2}}$.

$\cos\alpha$, $\cos\beta$, $\cos\gamma$ are known as the direction cosines of $\overrightarrow{OP}$.

5 *Scalar products*

(a) *Definition.* The scalar product of **a** and **b** is

$$\mathbf{a}.\mathbf{b} = ab\cos\theta,$$

where θ is the angle between **a** and **b**.

N.B. $\mathbf{i}.\mathbf{j} = \mathbf{j}.\mathbf{j} = \mathbf{k}.\mathbf{k} = 1\cos 0^\circ = 1$,

and $\mathbf{i}.\mathbf{j} = \mathbf{j}.\mathbf{k} = \mathbf{k}.\mathbf{i} = \cos 90^\circ = 0$.

(b) If **a** and **b** are expressed in terms of their Cartesian components,

$$\mathbf{a}.\mathbf{b} = a_x b_x + a_y b_y + a_z b_z.$$

As a special case, $\mathbf{a}.\mathbf{a} = a^2 = a_x^2 + a_y^2 + a_z^2$.

(c) *Distributive law.* $\mathbf{a}.(\mathbf{b}+\mathbf{c}) = \mathbf{a}.\mathbf{b}+\mathbf{a}.\mathbf{c}$.

6 *Vector products*

(a) *Definition.* The vector product of **a** into **b** is

$$\mathbf{a}\wedge\mathbf{b} = ab(\sin\theta)\mathbf{n},$$

where θ is the angle between **a** and **b** ($\leqslant 180^\circ$) and **n** is the unit vector perpendicular to **a** and **b** and directed so that a right-handed rotation about **n** through angle θ carries **a** to the direction of **b**.

$\mathbf{a}\wedge\mathbf{b} = ab(\sin\theta)\mathbf{n} = (\text{area parallelogram OACB})\mathbf{n}$

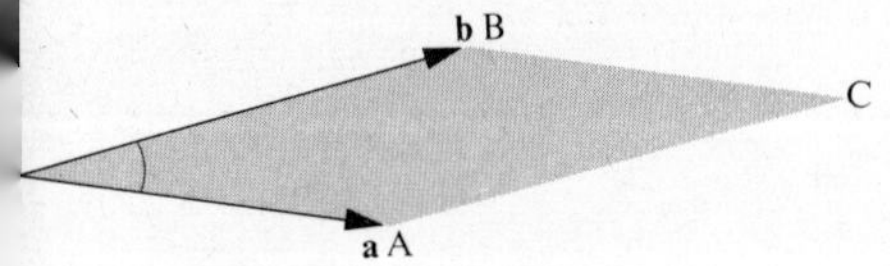

Figure 56

Note:

(i) $|\mathbf{a}\wedge\mathbf{b}|$ = area of parallelogram with sides **a**, **b**.
= 2 × area of triangle with sides **a**, **b**.

(ii) $\mathbf{b}\wedge\mathbf{a} = -\mathbf{a}\wedge\mathbf{b}$.

(iii) If **a** is parallel to **b**, $\mathbf{a}\wedge\mathbf{b} = \mathbf{0}$.

(iv) $\mathbf{i}\wedge\mathbf{i} = \mathbf{j}\wedge\mathbf{j} = \mathbf{k}\wedge\mathbf{k} = \mathbf{0}$;
$\mathbf{i}\wedge\mathbf{j} = \mathbf{k}, \quad \mathbf{j}\wedge\mathbf{k} = \mathbf{i}, \quad \mathbf{k}\wedge\mathbf{i} = \mathbf{j}$.

(b) If **a** and **b** are expressed in terms of their Cartesian components,

$$\mathbf{a}\wedge\mathbf{b} = (a_y b_z - a_z b_y)\mathbf{i} + (a_z b_x - a_x b_z)\mathbf{j} + (a_x b_y - a_y b_x)\mathbf{k}.$$

This can be remembered as a determinant:

$$\mathbf{a}\wedge\mathbf{b} = \begin{vmatrix} \mathbf{i} & \mathbf{j} & \mathbf{k} \\ a_x & a_y & a_z \\ b_x & b_y & b_z \end{vmatrix}.$$

(c) *Distributive law.* $\mathbf{a}\wedge(\mathbf{b}+\mathbf{c}) = \mathbf{a}\wedge\mathbf{b}+\mathbf{a}\wedge\mathbf{c}$.

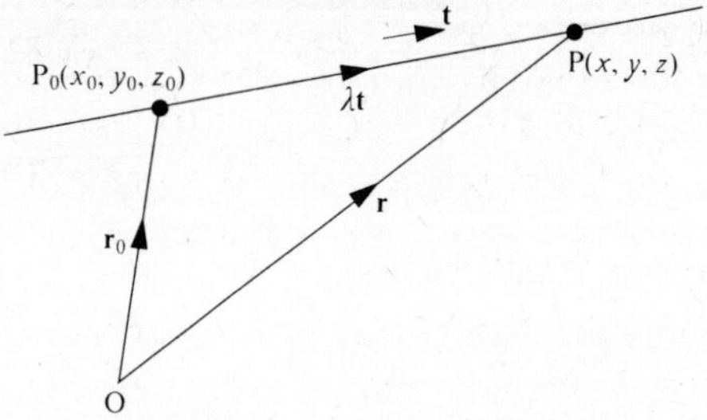

Figure 57

7 *The equation of a straight line*

(a) One vector form for the equation is

$$\mathbf{r} = \mathbf{r}_0 + \lambda\mathbf{t}, \qquad \textbf{6.1}$$

where $\mathbf{r}_0$ is the position vector of the point (x_0, y_0, z_0) through which the given line passes, and **t** is the unit vector in the direction of the line,

$\mathbf{t} = l\mathbf{i}+m\mathbf{j}+n\mathbf{k}$ (l, m, n are the direction cosines of **t**).

(b) The Cartesian equivalent of equation **6.1** is

$$(x-x_0)\mathbf{i}+(y-y_0)\mathbf{j}+(z-z_0)\mathbf{k} = \lambda(l\mathbf{i}+m\mathbf{j}+n\mathbf{k}),$$

hence, equating components,

$$\frac{x-x_0}{l} = \frac{y-y_0}{m} = \frac{z-z_0}{n}.$$

(c) Alternatively, equation **6.1** can be expressed as $(\mathbf{r}-\mathbf{r}_0)\wedge\mathbf{t} = 0$.

8 *The equation of a plane*

(a) The vector form of the equation is $(\mathbf{r}-\mathbf{r}_0).\mathbf{n} = 0$, where $\mathbf{r}_0$ is the position vector of the point (x_0, y_0, z_0) through which the given plane passes and **n** is the unit vector normal to the plane,

$\mathbf{n} = A\mathbf{i}+B\mathbf{j}+C\mathbf{k}$, say.

(b) In Cartesian form the equation becomes

$$A(x-x_0)+B(y-y_0)+C(z-z_0) = 0.$$

(c) Alternatively, the vector form of the equation of the plane can be expressed as $\mathbf{r}.\mathbf{n} = p$, where p is the perpendicular distance from the origin to the plane. This is so since $\mathbf{r}_0.\mathbf{n} = p$.

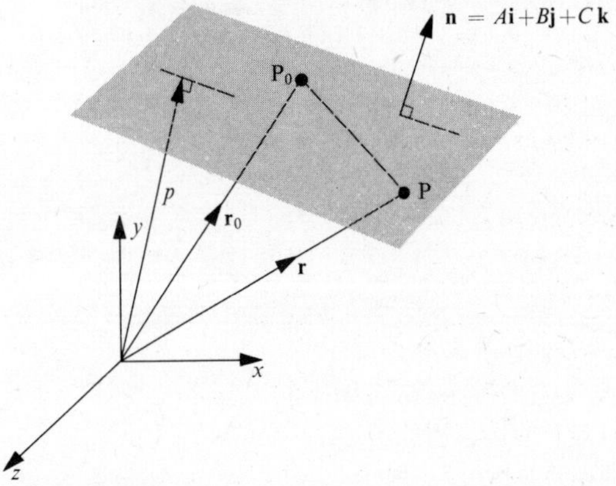

Figure 58

9 *Angles between planes and lines*

(a) If the unit vectors in the directions of two given straight lines are $\mathbf{t}_1$ and $\mathbf{t}_2$, the angle between the lines is

$$\cos^{-1}\mathbf{t}_1.\mathbf{t}_2 = \cos^{-1}(l_1 l_2 + m_1 m_2 + n_1 n_2).$$

(b) If the unit normals to two given planes are $\mathbf{n}_1$ and $\mathbf{n}_2$, the angle between the planes is

$$\cos^{-1}\mathbf{n}_1.\mathbf{n}_2 = \cos^{-1}(A_1 A_2 + B_1 B_2 + C_1 C_2).$$

(c) If a line having direction $\mathbf{t}$, meets a plane having a normal $\mathbf{n}$, the angle between the line and the plane is $\sin^{-1}\mathbf{t}.\mathbf{n}$.

10 *Shortest distances*

(a) The shortest distance from the point P, with position vector $\mathbf{r}'$, to the plane $\mathbf{r}.\mathbf{n} = p$ is $\pm(\mathbf{r}'.\mathbf{n} - p)$.

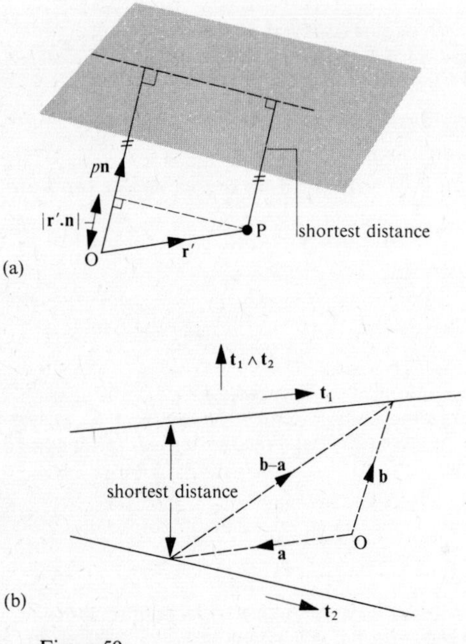

Figure 59

(b) The shortest distance between the non-intersecting lines $(\mathbf{r}-\mathbf{a})\wedge\mathbf{t}_1 = 0$ and $(\mathbf{r}-\mathbf{b})\wedge\mathbf{t}_2 = 0$, where $\mathbf{t}_1$ and $\mathbf{t}_2$ are unit vectors along the lines is

$$\frac{(\mathbf{b}-\mathbf{a}).\mathbf{t}_1\wedge\mathbf{t}_2}{|\mathbf{t}_1\wedge\mathbf{t}_2|}.$$

11 *Parametric curves*

The position vector to a point on a space curve may be expressed as $\mathbf{r} = x(t)\mathbf{i} + y(t)\mathbf{j} + z(t)\mathbf{k}$, where t is a parameter ranging over a set of values.

Examples. The equations of a circle and a parabola in the xy plane are, respectively,

$$\mathbf{r} = a(\cos t)\mathbf{i} + a(\sin t)\mathbf{j}, \quad (0 \leqslant t \leqslant 2\pi),$$

and $\mathbf{r} = at^2\mathbf{i} + 2at\mathbf{j}, \quad (-\infty < t < \infty)$.

The equation of a circular helix in three dimensions is

$$\mathbf{r} = a(\cos t)\mathbf{i} + a(\sin t)\mathbf{j} + bt\,\mathbf{k}.$$

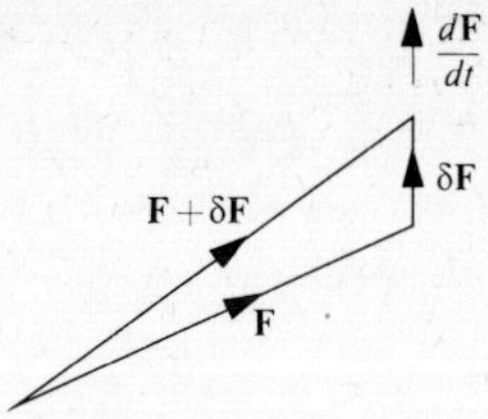

Figure 60

12 *Differentiation of vectors*

(a) *Definition.* $\dfrac{d\mathbf{F}}{dt} = \lim\limits_{\delta t\to 0}\dfrac{\mathbf{F}(t+\delta t) - \mathbf{F}(t)}{\delta t}$. (Vector $\mathbf{F}$ is a function of t.)

If $\mathbf{F} = u(t)\mathbf{i} + v(t)\mathbf{j} + w(t)\mathbf{k}$,

then $\dfrac{d\mathbf{F}}{dt} = \dfrac{du}{dt}\mathbf{i} + \dfrac{dv}{dt}\mathbf{j} + \dfrac{dw}{dt}\mathbf{k}$.

(b) If f is a scalar, and $\mathbf{F}$ is a vector function of t, then

$$\frac{d}{dt}(f\mathbf{F}) = \frac{df}{dt}\mathbf{F} + f\frac{d\mathbf{F}}{dt}.$$

(c) If $F = \mathbf{u}(t).\mathbf{v}(t)$,

then $\dfrac{dF}{dt} = \dfrac{d\mathbf{u}}{dt}.\mathbf{v}(t) + \mathbf{u}(t).\dfrac{d\mathbf{v}}{dt}$.

Note:

(i) If $\mathbf{v} = \mathbf{u}$, then

$$\frac{dF}{dt} = 2\mathbf{u}.\frac{d\mathbf{u}}{dt} = \frac{d}{dt}(u^2).$$

(ii) Additionally, if $\mathbf{u}$ is of constant magnitude,

$$\frac{dF}{dt} = 0;$$

since $\dfrac{d\mathbf{u}}{dt}$ is perpendicular to $\mathbf{u}$.

(d) If $\mathbf{F} = \mathbf{u}(t)\wedge\mathbf{v}(t)$,

then $\dfrac{d\mathbf{F}}{dt} = \dfrac{d\mathbf{u}}{dt}\wedge\mathbf{v} + \mathbf{u}\wedge\dfrac{d\mathbf{v}}{dt}$.

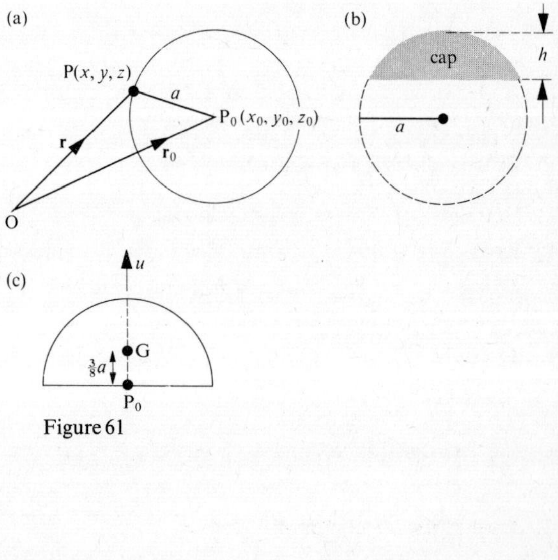

Figure 61

13 *Mensuration and properties of the sphere, cone and tetrahedron*

(a) *Sphere: Zone and cap of sphere.*

(i) The vector equation of a sphere, with centre at $\mathbf{r}_0$, and radius a is

$$(\mathbf{r}-\mathbf{r}_0)^2 = a^2.$$

The Cartesian form of this is

$$(x-x_0)^2+(y-y_0)^2+(z-z_0)^2 = a^2.$$

(ii) Volume of sphere is $\frac{4}{3}\pi a^3$.
Surface area of sphere is $4\pi a^2$.

(iii) Volume of a cap of height is $\pi h^2(a-\frac{1}{3}h)$.
Surface area of cap, or zone, of height h is $2\pi ah$.

(iv) Position vector of centroid of a uniform hemisphere (radius a), relative to centre P_0 of plane face is $(3a/8)\mathbf{u}$, where $\mathbf{u}$ is the unit vector along the axis of symmetry.

(b) *Cone: Frustrum of cone.*

(a) (b)

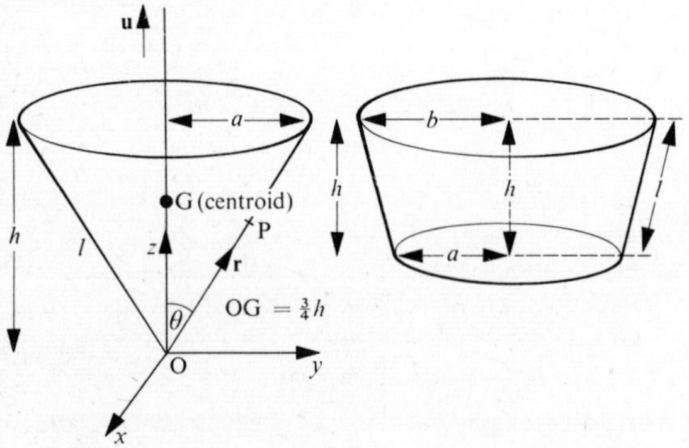

Figure 62

(i) The vector equation of the cone (shown in Figure 62a) is

$$(\mathbf{r}.\mathbf{u})^2 = \cot^2\theta\{\mathbf{u}\wedge(\mathbf{r}\wedge\mathbf{u})\}^2,$$

where θ is the semi-vertical angle of the cone and $\mathbf{u}$ is the unit vector along the axis of cone.

With the z-axis in the direction of $\mathbf{u}$, the Cartesian equation is

$$x^2+y^2 = z^2\tan^2\theta.$$

(ii) Volume of cone is $\frac{1}{3}\pi a^2 h$.
Curved surface area of cone is πal, $l = (a^2+h^2)^{\frac{1}{2}}$.

(iii) Position of centroid of uniform cone relative to O is $(3h/4)\mathbf{u}$.

(iv) Volume of frustrum (Figure 62b) is $\frac{1}{3}\pi h(a^2+ab+b^2)$.
Area of curved surface of frustum is $\pi l(a+b)$,
$l = \{h^2+(b-a)^2\}^{\frac{1}{2}}$.

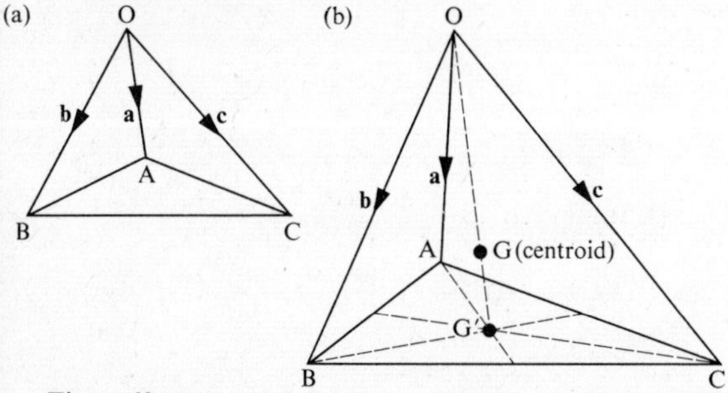

Figure 63

(c) *Tetrahedron.*

(i) Volume of tetrahedron is $\frac{1}{6}\mathbf{a}.(\mathbf{b}\wedge\mathbf{c})$.

(ii) Position vector of centroid of uniform solid tetrahedron relative to O is $\frac{1}{4}(\mathbf{a}+\mathbf{b}+\mathbf{c})$.

The centroid is at the point of intersection of the four lines joining the vertices of the tetrahedron to the centroids of the opposite faces.

B Vector analysis: Illustrative worked problems

1

(a) Form the scalar and vector products of the vectors

$$\mathbf{a} = (\cos\theta)\mathbf{i}+(\sin\theta)\mathbf{j}, \qquad \mathbf{b} = (\cos\phi)\mathbf{i}+(\sin\phi)\mathbf{j}$$

and hence show that:

$$\cos(\theta-\phi) = \cos\theta\cos\phi+\sin\theta\sin\phi,$$
$$\sin(\theta-\phi) = \sin\theta\cos\phi-\cos\theta\sin\phi.$$

(ii) The position vector of the foci of an ellipse are $\mathbf{c}$ and $-\mathbf{c}$ and the length of the major axis is $2a$. Show that the equation of the ellipse can be written in the form of

$$a^4-a^2(r^2+c^2)+(\mathbf{r}.\mathbf{c})^2 = 0$$

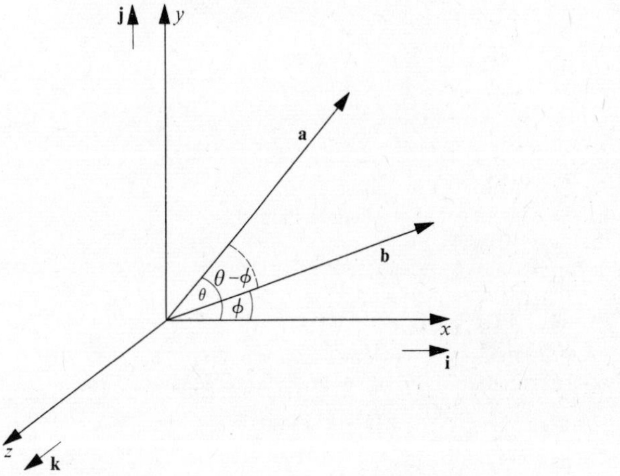

Figure 64

Solution.

(a)
$$\begin{aligned}\mathbf{a}.\mathbf{b} &= \{(\cos\theta)\mathbf{i}+(\sin\theta)\mathbf{j}\}.\{(\cos\phi)\mathbf{i}+(\sin\phi)\mathbf{j}\}\\ &= \cos\theta(\cos\phi)\mathbf{i}.\mathbf{i}+\sin\theta(\sin\phi)\mathbf{j}.\mathbf{j}+\\ &\quad+(\sin\theta\cos\phi+\cos\theta\sin\phi)\mathbf{i}.\mathbf{j}\\ &= \cos\theta\cos\phi+\sin\theta\sin\phi,\end{aligned}$$

as $\mathbf{i}.\mathbf{i} = \mathbf{j}.\mathbf{j} = 1$ and $\mathbf{i}.\mathbf{j} = \mathbf{j}.\mathbf{i} = 0$.

Also
$$\begin{aligned}\mathbf{a}.\mathbf{b} &= ab\cos(\theta-\phi)\\ &= \cos(\theta-\phi),\\ a &= (\mathbf{a}.\mathbf{a})^{\frac{1}{2}} = (\cos^2\theta+\sin^2\theta)^{\frac{1}{2}}\\ &= 1\end{aligned}$$

and
$$\begin{aligned}b &= (\mathbf{b}.\mathbf{b})^{\frac{1}{2}} = (\cos^2\phi+\sin^2\phi)^{\frac{1}{2}}\\ &= 1.\end{aligned}$$

Hence we have $\cos(\theta-\phi) = \cos\theta\cos\phi+\sin\theta\sin\phi$.

$$\mathbf{a}\wedge\mathbf{b} = \begin{vmatrix} \mathbf{i} & \mathbf{j} & \mathbf{k} \\ \cos\theta & \sin\theta & 0 \\ \cos\phi & \sin\phi & 0 \end{vmatrix}$$
$$= -(\sin\theta\cos\phi-\cos\theta\sin\phi)\mathbf{k},$$

but $\mathbf{a}\wedge\mathbf{b} = -\{ab\sin(\theta-\phi)\}\mathbf{k}$
$$= -\sin(\theta-\phi)\mathbf{k};$$

hence $\sin(\theta-\phi) = \sin\theta\cos\phi-\cos\theta\sin\phi$.

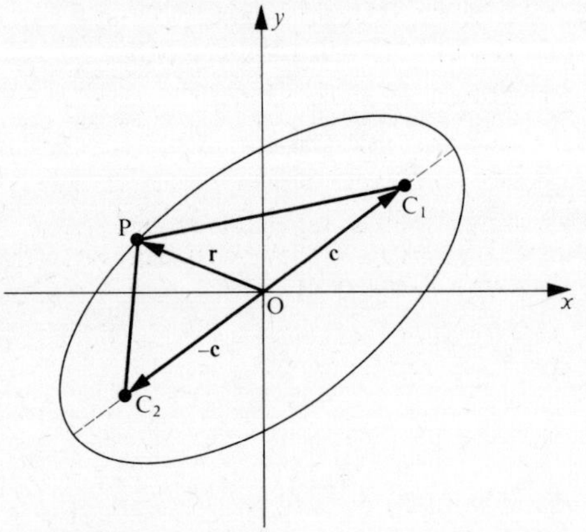

Figure 65

(b) By definition, for any point on the ellipse,

$$|\mathbf{PC}_1|+|\mathbf{PC}_2| = 2a, \qquad \textbf{6.2}$$

but $\mathbf{PC}_1 = (\mathbf{c}-\mathbf{r})$, $\quad \mathrm{PC}_1^2 = \mathbf{PC}_1.\mathbf{PC}_1 = c^2+r^2-2\mathbf{c}.\mathbf{r}$;
and $\mathbf{PC}_2 = -(\mathbf{c}+\mathbf{r})$, $\quad \mathrm{PC}_2^2 = \mathbf{PC}_2.\mathbf{PC}_2 = c^2+r^2+2\mathbf{c}.\mathbf{r}$;

$\mathbf{r} = \mathbf{OP}$ being the position vector of point P. Therefore, on substituting for $|\mathbf{PC}_1|$ and $|\mathbf{PC}_2|$ in equation **6.2**, we obtain

$$\sqrt{(c^2+r^2-2\mathbf{c}.\mathbf{r})}+\sqrt{(c^2+r^2+2\mathbf{c}.\mathbf{r})} = 2a$$

and, squaring both sides,

$$2(c^2+r^2)+2\sqrt{[(c^2+r^2-2\mathbf{c}.\mathbf{r})(c^2+r^2+2\mathbf{c}.\mathbf{r})]} = 4a^2$$
or $$\sqrt{\{(c^2+r^2)^2-4(\mathbf{c}.\mathbf{r})^2\}} = \{2a^2-(c^2+r^2)\}. \qquad \textbf{6.3}$$

On squaring again and simplifying, equation **6.3** produces the quoted form of $a^4-a^2(c^2+r^2)+(\mathbf{r}.\mathbf{c})^2 = 0$.

2 Prove that, if **a**, **b**, **c** are three coplanar vectors no two of which are parallel, then there are non-zero numbers λ, μ, ν such that

$$\lambda\mathbf{a}+\mu\mathbf{b}+\nu\mathbf{c} = \mathbf{0}.$$

The vectors **a**, **b**, **c** are the position vectors of the vertices A, B, C of a triangle referred to an origin O. The line AO meets BC at L; M and N are similarly defined. Show that the position vector of L is

$$\frac{\mu\mathbf{b}+\nu\mathbf{c}}{\mu+\nu}.$$

Hence prove Ceva's theorem that,

$$\frac{\mathrm{BL}}{\mathrm{LC}}\cdot\frac{\mathrm{CM}}{\mathrm{MA}}\cdot\frac{\mathrm{AN}}{\mathrm{NB}} = 1. \qquad \text{[O \& C, S 1967, M \& HM III, Q 7]}$$

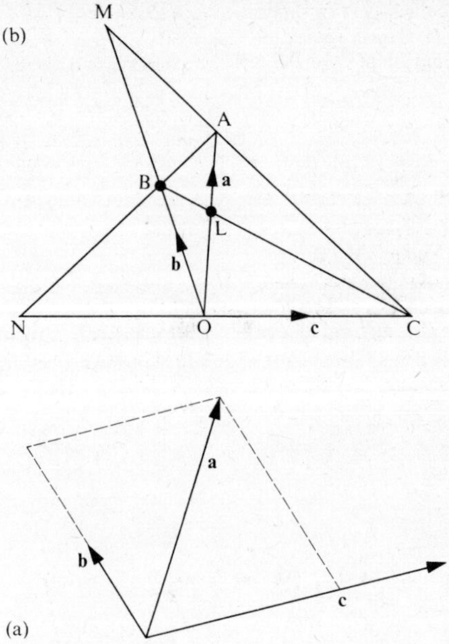

Figure 66

Solution. By completing the parallelogram shown in Figure 66(a) it is clear that we can express one of the vectors, here **a**, as the sum of scalar multiples of the other two,

i.e. $\mathbf{a} = l\mathbf{b}+m\mathbf{c} \quad (l \neq 0, m \neq 0)$.

Let $l = -\mu/\lambda$ and $m = -\nu/\lambda$, where $\lambda \neq 0, \mu \neq 0, \nu \neq 0$,

then $$\mathbf{a} = -\frac{\mu}{\lambda}\mathbf{b}-\frac{\nu}{\lambda}\mathbf{c},$$

so that $\lambda\mathbf{a}+\mu\mathbf{b}+\nu\mathbf{c} = \mathbf{0}$. **6.4**

In Figure 66(b), the position vector of L is $\mathbf{OL} = -\alpha\mathbf{a}$ say, since L lies on OA. By equation **6.4** we can write

$$\mathbf{OL} = \frac{\alpha}{\lambda}(\mu\mathbf{b}+\nu\mathbf{c}).$$

But L lies on BC so **OL** can also be expressed in the form $\mathbf{OB}+k\,\mathbf{BC}$ where k is some scalar,

$$\mathbf{OL} = \mathbf{b}+k(\mathbf{c}-\mathbf{b}) = (1-k)\mathbf{b}+k\mathbf{c}.$$

Comparing the two different forms:

$$1-k = \frac{\alpha}{\lambda}\mu \quad\text{and}\quad k = \frac{\alpha}{\lambda}\nu,$$

so that $\frac{\alpha}{\lambda}(\mu+\nu) = 1$ and $\frac{\alpha}{\lambda} = \frac{1}{\mu+\nu}$,

and hence $$\mathbf{OL} = \frac{\mu\mathbf{b}+\nu\mathbf{c}}{\mu+\nu}.$$

$$\mathbf{BL} = \mathbf{OL}-\mathbf{b} = \frac{\mu\mathbf{b}+\nu\mathbf{c}}{\mu+\nu}-\mathbf{b} = \frac{\nu(\mathbf{c}-\mathbf{b})}{\mu+\nu}$$

and $$\mathbf{LC} = \mathbf{c}-\mathbf{OL} = \mathbf{c}-\frac{\mu\mathbf{b}+\nu\mathbf{c}}{\mu+\nu} = \frac{\mu(\mathbf{c}-\mathbf{b})}{\mu+\nu},$$

whence $$\frac{\mathrm{BL}}{\mathrm{LC}} = \frac{\nu}{\mu}.$$

Similarly, CM/MA $= -\lambda/\nu$ and AN/NB $= -\nu/\mu$ can be shown, and

$$\frac{BL}{LC}\cdot\frac{CM}{MA}\cdot\frac{AN}{NB} = \frac{\nu}{\mu}\left(-\frac{\lambda}{\nu}\right)\left(-\frac{\nu}{\mu}\right) = 1.$$

C Vector analysis: Problems

Answers and hints will be found on pp. 82–3.

1 Given that **i** and **j** are perpendicular unit vectors, prove that the scalar product of the vectors $x_1\mathbf{i}+y_1\mathbf{j}$ and $x_2\mathbf{i}+y_2\mathbf{j}$ is $x_1x_2+y_1y_2$. What is the condition that these two vectors should be at right angles?

If, at a certain instant, these two vectors are the position vectors, relative to the origin, of points A_1 and A_2 which are moving in the plane of **i** and **j**, and if the velocities of A_1 and A_2 are, as usual, $\dot{x}_1\mathbf{i}+\dot{y}_1\mathbf{j}$ and $\dot{x}_2\mathbf{i}+\dot{y}_2\mathbf{j}$, write down an expression for the length A_1A_2 in terms of x_1, x_2, y_1 and y_2, and show that this length is constant if

$$(x_1-x_2)(\dot{x}_1-\dot{x}_2)+(y_1-y_2)(\dot{y}_1-\dot{y}_2) = 0.$$

Write down also, in terms of **i**, **j** and the *x*s and *y*s, the position vector of A_2 relative to A_1 and the velocity of A_2 relative to A_1, and verify that the condition that these two vectors shall be at right angles is the same as the condition that the length A_1A_2 shall be constant. [O, S 1968, M II, Q 4]

2 Define **a.b** and **a − b**, where **a** and **b** are vectors, and prove that, if **c** is a third vector in the plane of **a** and **b**, then

$$\mathbf{c}.(\mathbf{a}-\mathbf{b}) = \mathbf{c}.\mathbf{a}-\mathbf{c}.\mathbf{b}.$$

A, B, C and D are four points in a plane and O is a fixed point of the plane. If **a**, **b**, **c**, **d** are the position vectors relative to O of A, B, C, D respectively, give geometric interpretations of

(i) $(\mathbf{a}-\mathbf{b}).(\mathbf{a}-\mathbf{b}) = (\mathbf{c}-\mathbf{d}).(\mathbf{c}-\mathbf{d})$,
(ii) $\mathbf{a}-\mathbf{b} = k(\mathbf{c}-\mathbf{d})$, where k is a scalar,
(iii) $(\mathbf{a}-\mathbf{b}).(\mathbf{c}-\mathbf{d}) = 0$.

If $(\mathbf{a}-\mathbf{b}).(\mathbf{c}-\mathbf{d}) = 0$ and $(\mathbf{b}-\mathbf{c}).(\mathbf{a}-\mathbf{d}) = 0$; prove that $(\mathbf{c}-\mathbf{a}).(\mathbf{b}-\mathbf{d}) = 0$ and give a geometrical interpretation of this result. What is the interpretation of the result obtained by replacing **a**, **b**, **c**, **d** by $\dot{\mathbf{a}}$, $\dot{\mathbf{b}}$, $\dot{\mathbf{c}}$, $\dot{\mathbf{d}}$ respectively, where $\dot{\mathbf{a}}$ denotes, as usual, the velocity of A? [O, S 1967, M II, Q 4]

3 A, B, C, D and A′, B′, C′, D′ are two opposite faces of a rectangular block both perpendicular to the edges AA′, BB′, CC′, DD′. AB $= a$, AD $= b$, AA′ $= c$. The two tetrahedra ABCB′ and ADCD′ are cut away from the block. Prove that the angle θ between the two slant faces left on the block is given by

$$\cos\theta = \frac{b^2c^2+c^2a^2-a^2b^2}{b^2c^2+c^2a^2+a^2b^2}.$$

[O & C, S 1967, M & HM III, Q 2]

4 The position vectors of two points A and B, with respect to an origin O, are **a** and **b**. Show (from first principles) that the position vector of any point on AB is of the form $\alpha\mathbf{a}+\beta\mathbf{b}$, where $\alpha+\beta = 1$.

A point C, not on AB, has position vector **c**, and a point D has position vector $\alpha\mathbf{a}+\beta\mathbf{b}+\gamma\mathbf{c}$. If the plane ABC does not pass through O, find the condition on α, β, γ that D should lie in the plane ABC. [O & C, S 1966, M & HM IV, Q 1]

5

(a) Find the scalar product of **a** and **b** when

(i) $\mathbf{a} = \mathbf{i}+2\mathbf{j}-\mathbf{k}$, $\mathbf{b} = 2\mathbf{i}+3\mathbf{j}+\mathbf{k}$;
(i) $\mathbf{a} = \mathbf{i}-3\mathbf{j}+\mathbf{k}$, $\mathbf{b} = 2\mathbf{i}+\mathbf{j}+\mathbf{k}$.

(b) Find the angles and sides of the triangle whose vertices have position vectors $\mathbf{i}+\mathbf{j}-\mathbf{k}$, $2\mathbf{i}-\mathbf{j}+\mathbf{k}$, $-\mathbf{i}+\mathbf{j}+\mathbf{k}$.

6

(a) Prove that the shortest distance from the point **a** to the line joining the points **b** and **c** is given by

$$\frac{|\mathbf{a}\wedge\mathbf{b}+\mathbf{b}\wedge\mathbf{c}+\mathbf{c}\wedge\mathbf{a}|}{|\mathbf{b}-\mathbf{c}|}.$$

(b) If **r** is the position vector of a point and $\dot{\mathbf{r}} = d\mathbf{r}/dt$ its time derivative, interpret the equations

(i) $\mathbf{r}.\dot{\mathbf{r}} = 0$. (i) $\mathbf{r}\wedge\dot{\mathbf{r}} = \mathbf{0}$.

7

(a) Find the equation of the plane through the points A(1, 0, 0), B(0, 1, 0) and C(0, 0, 2). If O is the origin (0, 0, 0) and P is the point (1, 1, 2), find the cosine of the angle between OP and the normal to the plane ABC, and show that the distance between the feet of the perpendiculars from O and P to the plane ABC is $\sqrt{2}$.

(b) The two planes

$$2x-7y+5z+1 = 0 \quad\text{and}\quad x+4y-3z = 0$$

intersect in the line AB. Write down the general equation for a plane through AB. Obtain the equation of the plane through AB which is parallel to the line CD whose equation is

$$\frac{x}{-1} = \frac{y-1}{3} = \frac{z-3}{13}.$$

Hence find the shortest distance between the lines AB and CD. [N, S 1967, P II, Q 11]

8

(a) In a parallelogram ABCD, X is the midpoint of AB and the line DX cuts the diagonal AC at P. Writing $\overrightarrow{AB} = \mathbf{a}$, $\overrightarrow{AD} = \mathbf{b}$, $\overrightarrow{AP} = \lambda\overrightarrow{AC}$ and $\overrightarrow{DP} = \mu\overrightarrow{DX}$, express $\overrightarrow{AP}$ (i) in terms of λ, **a** and **b**, (ii) in terms of μ, **a** and **b**. Deduce that P is a point of trisection of both AC and DX.

(b) Define the scalar product **a.b** and the vector product $\mathbf{a}\wedge\mathbf{b}$ of two vectors **a** and **b**.

The points P, Q, R have coordinates (1, 1, 1), (1, 3, 2), (2, 1, 3) respectively, referred to rectangular axes O*xyz*. Calculate the products $\overrightarrow{PQ}.\overrightarrow{PR}$, $\overrightarrow{PQ}\wedge\overrightarrow{PR}$ and deduce the values of the cosine of the angle QPR and the area of the triangle PQR. [N, S 1967, FM I, Q 6]

9 Two planes, Π_1 and Π_2, have equations

$$2x+y+z = 1 \quad\text{and}\quad 3x+y-z = 2,$$

respectively. Prove that the plane Π_3 which is perpendicular to Π_1 and contains the line of intersection of Π_1 and Π_2 has the equation $x-2z = 1$.

Points P and Q lie on the planes Π_1 and Π_3, respectively, and the line PQ is perpendicular to Π_2. If the coordinates of P are (−2, 4, 1), find the coordinates of Q. Determine the angle between the line PQ and the perpendicular from P to the line of intersection of the three planes. [N, S 1967, FM II, Q 5]

D Vector analysis: Answers and hints

1

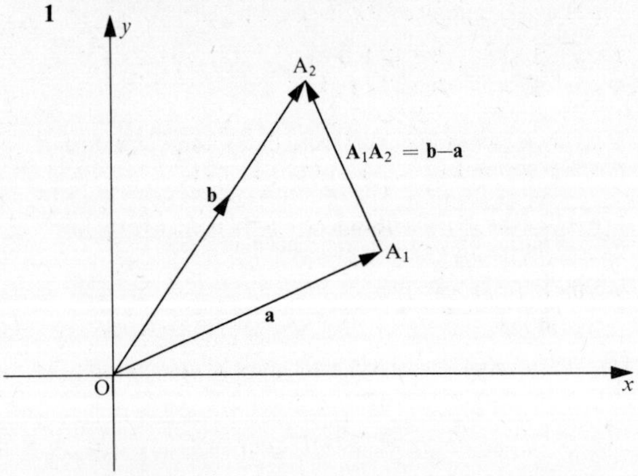

Figure 67

The condition the vectors are at 90° is $x_1 x_2 + y_1 y_2 = 0$.

$$\mathbf{A_1 A_2} = (x_2 - x_1)\mathbf{i} + (y_2 - y_1)\mathbf{j},$$
$$|\mathbf{A_1 A_2}|^2 = (x_2 - x_1)^2 + (y_2 - y_1)^2.$$

$|\mathbf{A_1 A_2}|$ is constant if

$$\frac{d}{dt}|\mathbf{A_1 A_2}|^2 = 0.$$

This condition leads to quoted result.

Position vector of A_2 relative to A_1 is

$$\mathbf{A_1 A_2} = (x_2 - x_1)\mathbf{i} + (y_2 - y_1)\mathbf{j}.$$

Velocity of A_2 relative to A_1 is

$$\frac{d}{dt}\mathbf{A_1 A_2} = (\dot{x}_2 - \dot{x}_1)\mathbf{i} + (\dot{y}_2 - \dot{y}_1)\mathbf{j}$$

and if $\mathbf{A_1 A_2} \perp \dot{\mathbf{A_1 A_2}}$, $\mathbf{A_1 A_2} . \dot{\mathbf{A_1 A_2}} = 0$, this leads to same condition as $|\mathbf{A_1 A_2}|^2 = \text{constant}$.

2

(i) $|\mathbf{AB}| = |\mathbf{CD}|$.

(ii) $\mathbf{AB} \| \mathbf{CD}$, $\quad \mathbf{AB} = k\,\mathbf{CD}$.

(iii) $\mathbf{AB} \perp \mathbf{CD}$.

If $(\mathbf{a}-\mathbf{b}).(\mathbf{c}-\mathbf{d}) = 0$ and $(\mathbf{b}-\mathbf{c}).(\mathbf{a}-\mathbf{d}) = 0$,

$$\begin{aligned} \text{then}\quad & (\mathbf{a}-\mathbf{b}).(\mathbf{c}-\mathbf{d}) + (\mathbf{b}-\mathbf{c}).(\mathbf{a}-\mathbf{d}) \\ &= \mathbf{a.c} - \mathbf{a.d} - \mathbf{b.c} + \mathbf{b.d} + \mathbf{a.b} - \mathbf{b.d} - \mathbf{a.c} + \mathbf{c.d} \\ &= \mathbf{a}.(\mathbf{b}-\mathbf{d}) - \mathbf{c}.(\mathbf{b}-\mathbf{d}) \\ &= (\mathbf{a}-\mathbf{c}).(\mathbf{b}-\mathbf{d}) = 0. \end{aligned}$$

Thus $(\mathbf{c}-\mathbf{a}).(\mathbf{b}-\mathbf{d}) = 0$.

Interpretation. As

$\mathbf{AB} \perp \mathbf{CD}$, $\quad \mathbf{BC} \perp \mathbf{AD}$, $\quad \mathbf{AC} \perp \mathbf{BD}$,

point C is the orthocentre of △ABD.

Interpretation of $(\dot{\mathbf{c}} - \dot{\mathbf{a}}).(\dot{\mathbf{b}} - \dot{\mathbf{d}}) = 0$. The relative velocity of point C relative to A is perpendicular to the relative velocity of point B relative to D.

3 Choose origin at A, x-axis along AB, y-axis along AD and z-axis along AA′. Face AB′C has a normal with direction ratios

$$\left(\frac{1}{a}, \frac{-1}{b}, \frac{-1}{c}\right).$$

Hence the unit normal to face AB′C is

$$\mathbf{n}_1 = \frac{(1/a)\mathbf{i} - (1/b)\mathbf{j} - (1/c)\mathbf{k}}{\sqrt{(1/a^2 + 1/b^2 + 1/c^2)}}$$

and similarly the unit normal to face AD′C is

$$\mathbf{n}_2 = \frac{(1/a)\mathbf{i} - (1/b)\mathbf{j} + (1/c)\mathbf{k}}{\sqrt{(1/a^2 + 1/b^2 + 1/c^2)}}.$$

The formula $\cos\theta = \mathbf{n}_1 . \mathbf{n}_2$ now gives

$$\cos\theta = \frac{1/a^2 + 1/b^2 - 1/c^2}{1/a^2 + 1/b^2 + 1/c^2},$$

and the result follows.

4 The point P on AB has position vector $\mathbf{OP} = \mathbf{OA} + \lambda\mathbf{AB}$ for some λ, whence $\mathbf{OP} = \mathbf{a} + \lambda(\mathbf{b} - \mathbf{a}) = (1-\lambda)\mathbf{a} + \lambda\mathbf{b}$. Thus $\alpha = 1 - \lambda$, $\beta = \lambda$ and $\alpha + \beta = 1$ follows. This result is used in the second part: if D lies on the plane ABC, D will lie on some line CP, where P is on AB. Hence, by the above, the position vector of D is of the form

$$\mathbf{OD} = (1-\mu)\mathbf{OC} + \mu\mathbf{OP} = (1-\mu)\mathbf{c} + \mu[(1-\lambda)\mathbf{a} + \lambda\mathbf{b}].$$

Write this as $\alpha\mathbf{a} + \beta\mathbf{b} + \gamma\mathbf{c}$, and prove that $\alpha + \beta + \gamma = 1$.

5

(a)

(i) $2 + 6 - 1 = 7$.

(ii) $2 - 3 + 1 = 0$.

(b) Denote position vectors of vertices by **OA**, **OB**, **OC**, respectively.

$$\begin{aligned} \text{Then}\quad \mathbf{AB} &= \mathbf{OB} - \mathbf{OA} = \mathbf{i} - 2\mathbf{j} + 2\mathbf{k}, \\ |\mathbf{AB}| &= \sqrt{(1^2 + 2^2 + 2^2)} = 3; \\ \mathbf{BC} &= \mathbf{OC} - \mathbf{OB} = -3\mathbf{i} + 2\mathbf{j}, \\ |\mathbf{BC}| &= \sqrt{(3^2 + 2^2)} = \sqrt{13}. \\ \mathbf{CA} &= \mathbf{OA} - \mathbf{OC} = 2\mathbf{i} - 2\mathbf{k}, \\ |\mathbf{CA}| &= \sqrt{(2^2 + 2^2)} = 2\sqrt{2}. \end{aligned}$$

$$\begin{aligned} \text{Also}\quad \mathbf{AB.AC} &= -(2-4) = 3.2\sqrt{2}\cos A, \\ A &= \cos^{-1}\frac{\sqrt{2}}{6}; \\ \mathbf{BA.BC} &= 7 = 3\sqrt{13}\cos B, \\ B &= \cos^{-1}\frac{7}{3\sqrt{13}}; \\ \mathbf{CA.CB} &= 6 = 2\sqrt{26}\cos C, \\ C &= \cos^{-1}\frac{3}{\sqrt{26}}. \end{aligned}$$

6

(a) Denote points defined by **a**, **b**, **c** by A, B, C and the perpendicular distance from A to line BC by p. Then

$$\text{Area} \quad \triangle ABC = \tfrac{1}{2}p\,.\,BC = \tfrac{1}{2}p|\mathbf{b}-\mathbf{c}|$$
$$= \tfrac{1}{2}AC\,.\,BC \sin C$$
$$= \tfrac{1}{2}|(\mathbf{a}-\mathbf{c})\wedge(\mathbf{b}-\mathbf{c})|$$

$$\text{and} \quad p = \frac{|(\mathbf{a}-\mathbf{c})\wedge(\mathbf{b}-\mathbf{c})|}{|\mathbf{b}-\mathbf{c}|}$$
$$= \frac{|\mathbf{a}\wedge\mathbf{b}+\mathbf{c}\wedge\mathbf{a}+\mathbf{b}\wedge\mathbf{c}|}{|\mathbf{b}-\mathbf{c}|}.$$

(a)

(i) $\mathbf{r}\,.\,\dot{\mathbf{r}} = 0$ indicates that velocity and position vectors are at 90° and motion is circular

$$\left(\text{since} \quad \mathbf{r}\,.\,\dot{\mathbf{r}} = \frac{1}{2}\frac{d}{dt}(\mathbf{r}\,.\,\mathbf{r}), \quad \text{i.e. } |\mathbf{r}| = \text{constant}\right).$$

(ii) $\mathbf{r}\wedge\mathbf{r} = \mathbf{0}$ indicates linear motion, $\dot{\mathbf{r}}$ parallel to $\mathbf{r}$.

7

(a) Equation of plane ABC is $x+y+\frac{1}{2}z = 1$. Unit normal is $\frac{2}{3}(\mathbf{i}+\mathbf{j}+\frac{1}{2}\mathbf{k})$; cosine of angle is

$$\frac{2}{\sqrt{6}}.$$

(b) General equation of plane through AB is

$$(2+\lambda)x+(4\lambda-7)y+(5-3\lambda)z+1 = 0.$$

Plane parallel to CD is $7x-2y+z+2 = 0$.

Shortest distance is $\dfrac{1}{\sqrt{6}}$.

8

(a)

(i) $\overrightarrow{AP} = \lambda(\mathbf{a}+\mathbf{b})$.

(ii) $\overrightarrow{AP} = \frac{1}{2}\mu\mathbf{a}+\mathbf{b}(1-\mu)$.

(b) $\overrightarrow{PQ}\,.\,\overrightarrow{PR} = 2, \qquad \overrightarrow{PQ}\wedge\overrightarrow{PR} = 4\mathbf{i}+\mathbf{j}-2\mathbf{k}.$

$\cos Q\hat{P}R = \frac{2}{5}, \qquad \text{area } \triangle QPR = \frac{1}{2}\sqrt{21}.$

9 Coordinates of Q are (1, 5, 1); required angle is $\cos^{-1}\frac{5}{11}$.

Chapter Seven
Differentiation and Integration

A Differentiation: Theory summary

1 *Differential coefficient or derivative*

The differential coefficient or the derivative of the function $y = f(x)$ is defined as

$$\frac{dy}{dx} = \lim_{\delta x \to 0} \frac{\delta g}{\delta x},$$

where $\delta y = f(x+\delta x) - f(x)$.

dy/dx measures the gradient of the tangent to the curve $y = f(x)$ at the point (x, y).

2 *Small increments*

If the increment δx is small enough,

$$\delta y \approx \frac{dy}{dx}\delta x,$$

that is, in Figure 68, δx is sufficiently small so that gradient of chord PQ $\approx$ gradient of tangent at P.

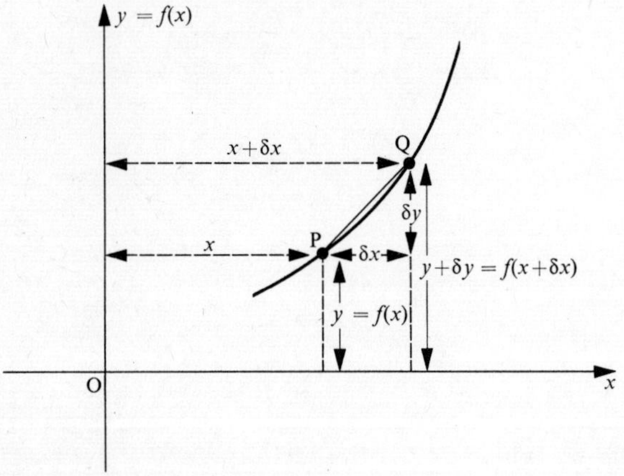

Figure 68

3 *Some important derivatives*

Function $y = f(x)$	*Derivative* $\frac{dy}{dx}$	
c	0	
x^n	nx^{n-1}	
$\log_e x$	$\frac{1}{x}$	
$\log_a x$	$\frac{1}{x \log_e a}$	
e^{px}	pe^{px}	
a^x	$a^x \log_e a$	
$\sin x$	$\cos x$	
$\cos x$	$-\sin x$	
$\tan x$	$\sec^2 x$	
$\operatorname{cosec} x$	$-\operatorname{cosec} x \cot x$	
$\sec x$	$\sec x \tan x$	
$\cot x$	$-\operatorname{cosec}^2 x$	
$\sin^{-1} x$	$\frac{1}{\sqrt{(1-x^2)}}$	
$\cos^{-1} x$	$\frac{-1}{\sqrt{(1-x^2)}}$	
$\tan^{-1} x$	$\frac{1}{1+x^2}$	
$\sinh x$	$\cosh x$	
$\cosh x$	$\sinh x$	
$\tanh x$	$\operatorname{sech}^2 x$	
$\operatorname{cosech} x$	$-\operatorname{cosech} x \coth x$	
$\operatorname{sech} x$	$-\operatorname{sech} x \tanh x$	
$\coth x$	$-\operatorname{cosech}^2 x$	
$\sinh^{-1} x$	$\frac{1}{\sqrt{(1+x^2)}}$	
$\cosh^{-1} x$	$\frac{1}{\sqrt{(x^2-1)}}$	$(x > 1)$
$\tanh^{-1} x$	$\frac{1}{1-x^2}$	$(\lvert x\rvert < 1)$

4 *Differentiation of a product*

If u, v and w are functions of x,

(a) $$\frac{d}{dx}(uv) = u\frac{dv}{dx} + v\frac{du}{dx};$$

(b) $$\frac{d}{dx}(uvw) = uv\frac{dw}{dx} + uw\frac{dv}{dx} + vw\frac{du}{dx}.$$

5 *Differentiation of a quotient*

$$\frac{d}{dx}\left[\frac{u}{v}\right] = \frac{1}{v^2}\left[v\frac{du}{dx} - u\frac{dv}{dx}\right].$$

6 *Differentiation of a function of a function; the 'chain rule'*

$$\frac{d}{dx}[f_1\{f_2(x)\}] = \frac{df_1}{df_2}\frac{df_2}{dx}.$$

7 *Implicit differentiation*
(That is, when y is not explicitly stated in the form $y = f(x)$).

Example. If

$$y^2+ax^3+bxy^4 = c,$$

then differentiating implicitly with respect to x yields

$$2y\frac{dy}{dx}+3ax^2+4bxy^3\frac{dy}{dx}+by^4 = 0.$$

8 *Logarithmic differentiation*

If $\quad y = u^a v^b w^c,$

then $\quad \log_e y = a \log_e u+b \log_e v+c \log_e w$

and thus $\quad \dfrac{1}{y}\dfrac{dy}{dx} = \dfrac{a}{u}\dfrac{du}{dx}+\dfrac{b}{v}\dfrac{dv}{dx}+\dfrac{c}{w}\dfrac{dw}{dx}.$

(Note: 4(b) is a special case of the above result when $a = b = c = 1$).

9 *Differentiation of parametric forms*
If y and x are expressed as functions of a parameter t, say, that is, $y = y(t)$, $x = x(t)$, then:

(a) $\dfrac{dx}{dt} = \dfrac{1}{dt/dx}.$

(b) $\dfrac{dy}{dx} = \dfrac{dy/dt}{dx/dt} \equiv \dfrac{\dot{y}}{\dot{x}}$, where the dot denotes differentiation w.r.t. t.

(c) $\dfrac{d^2y}{dx^2} = \dfrac{d}{dx}\left(\dfrac{\dot{y}}{\dot{x}}\right) = \dfrac{\dot{x}\ddot{y}-\dot{y}\ddot{x}}{\dot{x}^3}$, where $\ddot{y} = \dfrac{d^2y}{dt^2}$, $\ddot{x} = \dfrac{d^2x}{dt^2}$.

10 *Equations of tangent and normal at any point on a curve*

(a) *Tangent*. The equation of a tangent to the curve $y = f(x)$ at the point (x_1, y_1) is

$$y-y_1 = \left(\frac{dy}{dx}\right)_{x_1,\, y_1}(x-x_1).$$

(b) *Normal*. The equation of a normal to the curve $y = f(x)$ at the point (x_1, y_1) is

$$y-y_1 = \frac{-1}{(dy/dx)_{x_1,\, y_1}}(x-x_1).$$

11 *Maxima, minima and points of inflexion*

(a) y is a maximum at $x = a$ if

$$\frac{dy}{dx} = 0 \quad\text{and}\quad \frac{d^2y}{dx^2} < 0 \quad\text{at } x = a;$$

or equivalently,

$$\frac{dy}{dx} = 0 \quad\text{and}\quad \frac{dy}{dx}$$

changes from a positive to a negative value as x increases through the value a.

(b) y is a minimum at $x = a$ if

$$\frac{dy}{dx} = 0 \quad\text{and}\quad \frac{d^2y}{dx^2} > 0 \quad\text{at } x = a;$$

or equivalently

$$\frac{dy}{dx} = 0 \quad\text{and}\quad \frac{dy}{dx}$$

changes from a negative to a positive value as x increases through the value a.

(c) There is a point of inflexion at $x = a$ if

$$\frac{d^2y}{dx^2} = 0$$

and d^2y/dx^2 changes sign as x increases through the value a. Note that dy/dx may have any value.

(d) If $\dfrac{dy}{dx} = 0$ and $\dfrac{d^2y}{dx^2} = 0$ at $x = a$,

the point is either a maximum, minimum or point of inflexion. The point is a maximum if dy/dx changes from a positive to a negative value, a minimum if dy/dx changes from a negative to a positive value, or a point of inflexion if d^2y/dx^2 changes sign, as x increases through the value a.

12 *Curve sketching in Cartesian coordinates*
The following should normally be considered:

(a) Inspect curve equation to detect any symmetry, and any obvious change in origin.

(b) Determine whether asymptotes are present, that is, lines to which curve approaches as x and or y tends to $\pm\infty$.

(c) Construct a brief table of values to facilitate plotting over ranges of interest.

(d) Find points (if any) where the curve cuts the axes.

(e) Determine the gradient dy/dx and use it to determine regions where slope is positive and negative; and for points of maxima, minima and inflexion.

(f) Determine any limitation on possible ranges of x and/or y.

13 *Rates of change with respect to time*
Examples:

(a) Velocity, $v = \dfrac{ds}{dt}$

(rate of change of distance s w.r.t. t).

(b) Acceleration, $a = \dfrac{dv}{dt} = \dfrac{dv}{ds}\dfrac{ds}{dt} = v\dfrac{dv}{ds} = \dfrac{1}{2}\dfrac{d}{ds}(v^2).$

(c) Rate of change of a volume $V = \frac{4}{3}\pi r^3$ of a balloon, say, is

$$\frac{dV}{dt} = 4\pi r^2\frac{dr}{dt}.$$

B Differentiation: Illustrative worked problems

1

(i) If $y = \tan nx$, express both dy/dx and d^2y/dx^2 in terms of n and y only, and prove that

$$\frac{d^2y}{dx^2} = 2ny\frac{dy}{dx}.$$

(ii) Find the stationary values of the function $y = x^2e^{-x}$, and sketch its graph. [L, J 1968, P I, Q 8]

Solution.

(i) If $\quad y = \tan nx$,

then $\quad \dfrac{dy}{dx} = n\sec^2 nx = n(1+\tan^2 nx) = n(1+y^2)$. **7.1**

Differentiating expression **7.1** implicitly w.r.t. x yields

$$\frac{d^2y}{dx^2} = 2ny\frac{dy}{dx} = 2n^2y(1+y^2).$$

(ii) If $\quad y = x^2e^{-x}$,

then $\quad \dfrac{dy}{dx} = 2xe^{-x} - x^2e^{-x} = xe^{-x}(2-x)$.

Stationary values of y occur when $dy/dx = 0$, and since $e^{-x} > 0$ that is when $x = 0, x = 2$. The corresponding stationary values are $y = 0$ and $y = 4e^{-2}$.
Notes on graph sketch of $y = x^2e^{-x}$:

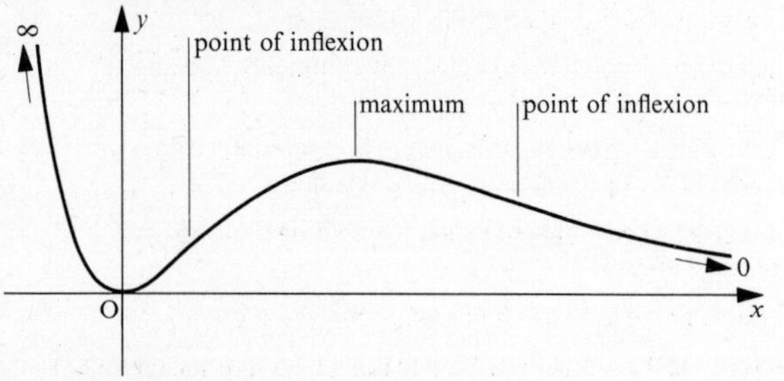

Figure 69 Sketch of curve $y = x^2e^{-x}$

(a) As $x \to +\infty, y \to 0$; as $x \to -\infty, y \to +\infty$.

(b) When $x = 0$, $dy/dx = 0$ and changes from a positive to a negative value as x increases through 0, hence point (0, 0) is a minimum.

When $x = 2$, $dy/dx = 0$ and changes from a negative to a positive value as x increases through 2, hence point $(2, 4e^{-2})$ is a maximum.

$$\frac{d^2y}{dx^2} = e^{-x}(2-2x) - e^{-x}(2x-x^2)$$
$$= e^{-x}(2-4x+x^2)$$
$$= 0 \quad \text{when } x = 2\pm\sqrt{2}.$$

d^2y/dx^2 also changes sign at $x = 2\pm\sqrt{2}$, hence points of inflexion occur at $x = 2-\sqrt{2}$ and at $x = 2+\sqrt{2}$.

2 A hollow right circular cone together with its base circumscribes a fixed sphere of radius a. Show that the area of the inner curved surface of the cone is $\pi a^2 \cot\theta(\cot\theta + \tan 2\theta)$, where 2θ is the angle between each generator and the base. Find the value of θ for which this area is a minimum and the value of this minimum area. [L, J 1968, P II, Q 9]

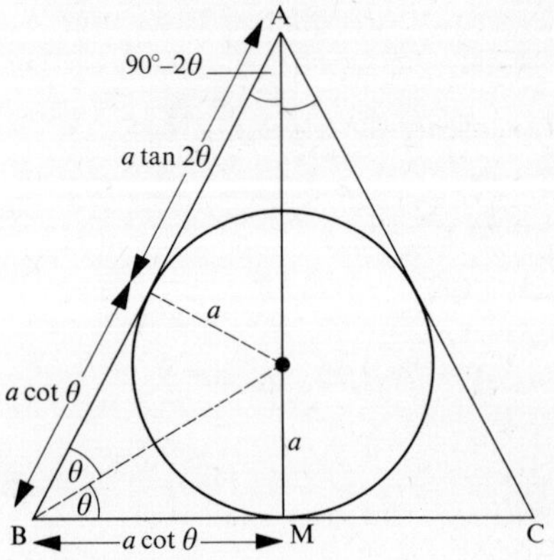

Figure 70

Solution:
A cross-section of the sphere and circumscribing cone is shown in Figure 70.

From this diagram it is seen that:

Slant length $\quad AB = a\cot\theta + a\tan 2\theta$;
Base radius $\quad BM = a\cot\theta$.
Hence $\quad$ Surface area of cone $= \pi \times AB \times BM$
$$= \pi a^2\cot\theta(\cot\theta + \tan 2\theta). \quad \textbf{7.2}$$

On substituting $t = \tan\theta$, and since

$\tan 2\theta = 2\tan\theta/(1-\tan^2\theta)$,

equation **7.2** becomes

$$S = \pi a^2\left[\frac{1}{t^2} + \frac{1}{t}\frac{2t}{1-t^2}\right] = \pi a^2\frac{1+t^2}{t^2(1-t^2)} \quad \textbf{7.3}$$

and $\quad \dfrac{dS}{d\theta} = \dfrac{dS}{dt}\dfrac{dt}{d\theta} = \dfrac{(t^2-t^4)2t-(1+t^2)(2t-4t^3)}{\{t^2(1-t^2)\}^2}(1+t^2)$

$$= \frac{2(1+t^2)}{t^3(1-t^2)^2}(t^4+2t^2-1) = 0$$

for a maximum or minimum.
Thus $\quad t^4+2t^2-1 = 0$,
hence $t^2 = \frac{1}{2}(-2\pm\sqrt{8}) = \sqrt{2}-1$
(positive alternative taken since t^2 cannot be negative)
and $t = +(\sqrt{2}-1)^{\frac{1}{2}}$.

Also $dS/d\theta$ changes from a negative to a positive value as t increases through the latter value. Thus the angle making S a minimum is

$$\theta = \tan^{-1}(\sqrt{2}-1)^{\frac{1}{2}},$$

whilst using equation **7.3**

$$S = \pi a^2\frac{\sqrt{2}}{(\sqrt{2}-1)(2-\sqrt{2})} = \frac{\pi a^2}{3-2\sqrt{2}}.$$

3

(i) Differentiate with respect to x

(a) $\left[\dfrac{x^2-1}{x^2+1}\right]^{\frac{1}{3}}$,

(b) $\log_e \sec\sqrt{(2x)}$,

expressing your results in the simplest form.

(ii) If $y = x^{-\frac{1}{2}}\sin x$, prove that

$$x^2\frac{d^2y}{dx^2}+x\frac{dy}{dx}+\left(x^2-\frac{1}{4}\right)y = 0. \qquad \text{[L, S 1967, P I, Q 7]}$$

Solution

(i)

(a) If $y = \left[\dfrac{x^2-1}{x^2+1}\right]^{\frac{1}{3}}$,

then $$\frac{dy}{dx} = \frac{1}{3}\left[\frac{x^2-1}{x^2+1}\right]^{-\frac{2}{3}}\frac{(x^2+1)2x-(x^2-1)2x}{(x^2+1)^2}$$

$$= \frac{4x}{3(x^2-1)^{\frac{2}{3}}(x^2+1)^{\frac{4}{3}}}$$

(b) If $y = \log_e \sec\sqrt{(2x)} = \log_e\{\cos\sqrt{(2x)}\}^{-1}$,

then $\dfrac{dy}{dx}$

$$= \cos\sqrt{(2x)}[-\{\cos\sqrt{(2x)}\}^{-2}]\{-\sin\sqrt{(2x)}\}\tfrac{1}{2}.(2x)^{-\frac{1}{2}}.2$$

$$= \frac{1}{\sqrt{(2x)}}\frac{\sin\sqrt{(2x)}}{\cos\sqrt{(2x)}} = \frac{1}{\sqrt{(2x)}}\tan\sqrt{(2x)}.$$

(ii) If $y = x^{-\frac{1}{2}}\sin x$,

then $$\frac{dy}{dx} = -\tfrac{1}{2}x^{-\frac{3}{2}}\sin x + x^{-\frac{1}{2}}\cos x$$

and $\dfrac{d^2y}{dx^2}$

$$= +\tfrac{3}{4}x^{-\frac{5}{2}}\sin x-\tfrac{1}{2}x^{-\frac{3}{2}}\cos x-\tfrac{1}{2}x^{-\frac{3}{2}}\cos x-x^{-\frac{1}{2}}\sin x$$
$$= x^{-\frac{1}{2}}\sin x(\tfrac{3}{4}x^{-2}-1)-x^{-\frac{3}{2}}\cos x.$$

Thus $x^2\dfrac{d^2y}{dx^2}+x\dfrac{dy}{dx}$

$$= x^{-\frac{1}{2}}\sin x(\tfrac{3}{4}-x^2)-x^{\frac{1}{2}}\cos x-\tfrac{1}{2}x^{-\frac{1}{2}}\sin x+x^{\frac{1}{2}}\cos x$$
$$= -(x^2-\tfrac{1}{4})x^{-\frac{1}{2}}\sin x = -(x^2-\tfrac{1}{4})y,$$

hence $$x^2\frac{d^2y}{dx^2}+x\frac{dy}{dx}+(x^2-\tfrac{1}{4})y = 0.$$

4 Within a given circle of radius a a fixed point A is distant h from the centre. A variable chord PQ moves in such a way that AP = AQ. Show that there are two positions of PQ for which the area of the triangle APQ is numerically a maximum and that, if Δ_1 and Δ_2 denote these maximum areas,

$$\Delta_1\Delta_2 = \tfrac{1}{4}a(a^2-h^2)^{\frac{3}{2}}. \qquad \text{[L, S 1966, PS, Q 8]}$$

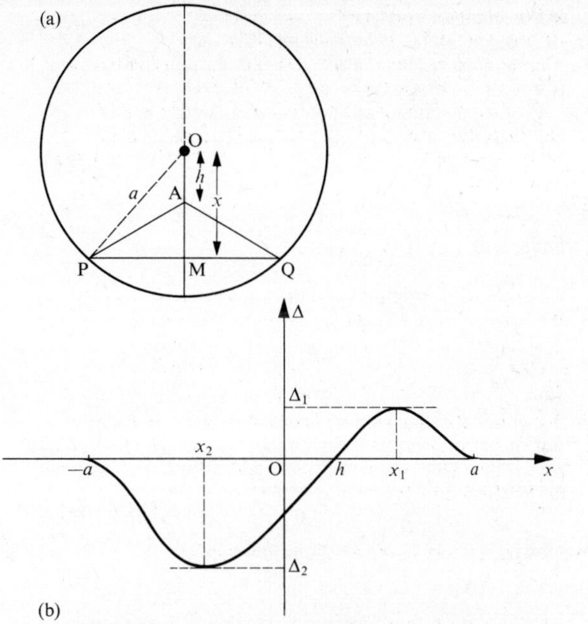

Figure 71

Solution. Since AP = PQ the variable chord PQ must move perpendicular to the diameter of the circle through A, see Figure 71(a). Let midpoint of PQ be M and OM = x, then area triangle APQ,

$$\Delta = \tfrac{1}{2}\text{AM}.\text{PQ} = (x-h)\sqrt{(a^2-x^2)},$$

and $$\frac{d\Delta}{dx} = \frac{-x(x-h)}{\sqrt{(a^2-x^2)}}+\sqrt{(a^2-x^2)}.$$

$d\Delta/dx = 0$ when $a^2-x^2 = x(x-h)$

i.e. if $2x^2-hx-a^2 = 0.$ **7.4**

The solutions of equation **7.4** are

$x = x_1 = \tfrac{1}{4}\{h+\sqrt{(h^2+8a^2)}\}$ (position of PQ above centre O),

$x = x_2 = -\tfrac{1}{4}\{\sqrt{(h^2+8a^2)}-h\}$ (position of PQ below centre O).

These values refer to maximum Δ since the curve of Δ against x passes through zero at $x = h$ and at $x = \pm a$, as shown in Figure 71(b).

The corresponding areas are

$$\Delta_1 = (x_1-h)\sqrt{(a^2-x_1^2)} \quad\text{and}\quad \Delta_2 = (h-x_2)\sqrt{(a^2-x_2^2)}.$$

Thus $$\Delta_1\Delta_2 = (x_1-h)(h-x_2)\sqrt{\{(a^2-x_1^2)(a^2-x_2^2)\}}$$
$$= (x_1-h)(h-x_2)\sqrt{\{\tfrac{1}{2}(a^2-x_1h)\tfrac{1}{2}(a^2-x_2h)\}}$$

(using equation **7.4**)

$$= \{(x_1+x_2)h-h^2-x_1x_2\}\tfrac{1}{2}\sqrt{\{a^4-(x_1+x_2)ha^2+x_1x_2h\}},$$

but, using equation **7.4** again,

$$x_1x_2 = -\tfrac{1}{2}a^2 \quad\text{and}\quad x_1+x_2 = \tfrac{1}{2}h.$$

Hence $$\Delta_1\Delta_2 = (\tfrac{1}{2}h^2-h^2+\tfrac{1}{2}a^2)\tfrac{1}{2}\sqrt{(a^4-\tfrac{1}{2}h^2a^2-\tfrac{1}{2}h^2a^2)}$$
$$= \tfrac{1}{4}a(a^2-h^2)^{\frac{3}{2}}.$$

C Differentiation: Problems

Answers and hints will be found on pp. 96–8.

1 A point moves along the x-axis so that its coordinate at time t is $x = te^{-\frac{1}{2}t^2}$, where t takes all positive and negative values.

Find the maximum and minimum values of x and of dx/dt and sketch the graphs of x plotted against t and of dx/dt plotted against t. [L, S 1967, P I, Q 8]

2

(i) Differentiate $\log_e \sqrt{\dfrac{1+2x^2}{1-2x^2}}$

with respect to x, simplifying your answer.

(ii) If $y = \sin(e^x + k)$, where k is a constant, show that

$$\frac{d^2y}{dx^2} - \frac{dy}{dx} + e^{2x}y = 0.$$

(iii) A frustum of a right circular cone has height h and ends of radii a and $2a$. From it is cut a right prism whose cross-section is a regular octagon and whose axis coincides with the axis of the frustum. Find the volume of the largest such prism. [L, J 1967, P I, Q 8]

3 Find the values of x for which the function

$$y = 2x^3 + 3x^2 - 12x + k,$$

where k is a constant, has maximum or minimum values.

If the minimum value is one-tenth of the maximum value, prove that $k = 10$.

The tangents to the curve $y = 2x^3 + 3x^2 - 12x + 10$ which are parallel to the x-axis cut the curve again at P and Q respectively. Find the coordinates of the point at which PQ again meets the curve. [L, S 1966, P I, Q 8]

4 A volume 76 cm^3 of metal is moulded into the shape of a box with a square base, rectangular sides and no lid, the base and sides being 1 cm thick. If the internal width of the box is x cm, show that its capacity is

$$\frac{x^2(72-4x-x^2)}{4(x+1)} \text{ cm}^3.$$

If the capacity is a maximum, show, by logarithmic differentiation, or otherwise, that

$$72 - 4x - x^2 = 2x(x+1)$$

and determine the maximum capacity. [L, J 1966, P I, Q 8]

5 A curve C is defined parametrically by the equations $x = X(t)$, $y = Y(t)$. The point Q with coordinates (p, q) is a given point not lying on the curve. P is the point on C nearest to Q. Show that t, the parameter of P, satisfies the equation

$$\{p - X(t)\}\frac{dX(t)}{dt} + \{q - Y(t)\}\frac{dY(t)}{dt} = 0.$$

Deduce that Q lies on the normal at P.

Find the coordinates of the point on the parabola

$$y^2 = 4ax$$

which is nearest to the point $(5a, 52a)$.

[C, S 1967, Sp O, Q 4]

6 Given that $y = \tan(k\tan^{-1}x)$, where $y = 0$ when $x = 0$, prove that

$$(1+x^2)\frac{dy}{dx} = k(1+y^2).$$

By differentiating this equation and using Maclaurin's series, show that

$$\tan(k\tan^{-1}x) = kx + \tfrac{1}{3}k(k^2-1)x^3 + \ldots .$$

[O & C, S 1967, M & HM II, Q 3]

7 Find the maximum and minimum values, and the values of x at which they occur, of the functions

(i) $2(x^2+1)^{\frac{1}{2}} - 3(x^2+4)^{\frac{1}{2}}$;

(ii) $5(x^2+1)^{\frac{1}{2}} - 3x$. [O & C, S 1966, M & HM II, Q 3]

8

(i) Differentiate with respect to x

(a) $\dfrac{2+x^2}{1-x^2}$,

(b) $\sin 2x \cos^2 x$.

(ii) If $y = (x^2-7)e^{-x}$, find the range of values of x for which d^2y/dx^2 is negative. [O & C, S 1967, M for Sc, I, Q 8]

9

(a) If $x = a(\theta - \sin\theta)$, $y = a(1-\cos\theta)$, determine the value of dy/dx in terms of θ. Prove that

$$y^2\frac{d^2y}{dx^2} = -a.$$

(b) In a triangle ABC the side $b = 4$ cm and $c = 5$ cm. The side a is increasing at the rate of $\frac{1}{2}$ cm s^{-1}. Determine in degrees s^{-1} the rate at which the angle A is increasing when $a = 6$ cm.

[AEB, N 1967, P & A II, Q 4]

10 Prove that the tangent at the point having parameter t on the curve $x = \frac{1}{2}t^2$, $y = \frac{1}{3}t^3$ is

$$6y - 6tx + t^3 = 0.$$

Show that this tangent intersects the curve again at the point having parameter $-\frac{1}{2}t$ and find the values of t for which the tangent is also normal to the curve.

Prove that the locus of the point of intersection of perpendicular tangents to the curve $x = \frac{1}{2}t^2$, $y = \frac{1}{3}t^3$ is the parabola

$$36y^2 = 6x - 1.$$

Verify that this parabola touches the curve at the feet of the normals which are also tangents to the curve.

[JMB, S 1967, PS, Q 5]

11 Prove that the maximum value of

$$\frac{e^{tx}}{1+e^x},$$

where t is a constant such that $0 < t < 1$, is $t^t(1-t)^{1-t}$.

If this maximum value is denoted by A, prove that $\frac{1}{2} \leqslant A < 1$. [JMB, S 1966, PS, Q 7]

12

(a) Find dy/dx if $y = x^n n^x$, where n is constant.

(b) Find dy/dx and d^2y/dx^2 at the point $(1, 2)$ on the curve

$$x^2 + 2xy - 3y^2 + 4x - y + 5 = 0.$$ [W, S 1969, P I, Q 8]

13 Find, in the simplest forms possible, the derivatives of the functions

(a) $\tan^{-1}(\tanh \frac{1}{2}x)$,

(b) $\tanh^{-1}\left[\dfrac{x+a}{1+ax}\right]$ (where a is a constant),

(c) $\log(x-1)-\dfrac{2x-1}{(x-1)^2}$. [W, S 1968, P I, Q 10]

14

(a) If $x^4+y^4 = a^4$, find d^4y/dx^4 when $x = 0$ and $y = a$.

(b) Find the nth derivative of $(x^3-6x^2+11x-6)^{-1}$ in a simple form. [W, S 1968, PS, Q 4]

D Integration: Theory summary

1 *Indefinite and definite integrals*

(a) One interpretation of the symbol $\int$ is that it denotes the reverse process of differentiation.

Thus if $\dfrac{d}{dx}\{y(x)\} = f(x)$,

then $y(x) = \int f(x)\,dx + C.$

$\int f(x)\,dx$ is known as an indefinite integral of $f(x)$ and C as an arbitrary constant of integration. Further,

$$\int_a^b f(x)\,dx = y(b)-y(a) \equiv [y(x)]_a^b$$

denotes the definite integral of $f(x)$ between the limits of a and b, and contains no arbitrary constant.

(b) The following are a few properties of definite integrals:

$$\int_a^b f(x)\,dx = -\int_b^a f(x)\,dx,$$

$$\int_a^b f(x)\,dx = \int_a^c f(x)\,dx + \int_c^b f(x)\,dx,$$

$$\int_a^b f(x)\,dx = \int_a^b f(t)\,dt, \quad \text{where } t \text{ is any variable.}$$

2 *Integration as the limit of a sum*

If $a = x_0 < x_1 < x_2 < \ldots < x_{r-1} < x_r < \ldots < x_n = b$,

$\delta x_r = x_r - x_{r-1}$ and $x_{r-1} \leqslant t_r \leqslant x_r$

for $r = 1, 2, 3, \ldots, n$,

then, defining $S \equiv f(t_1)\,\delta x_1 + f(t_2)\,\delta x_2 + \ldots + f(t_r)\,\delta x_r + \ldots + f(t_n)\,\delta x_n$,

we have $\lim\limits_{\substack{n\to\infty \\ \text{all } \delta x_r \to 0}} S = \int_a^b f(x)\,dx.$

This is the interpretation of the symbol $\int$ as a summation.

3 *Table of standard integrals*

Function $f(x)$	*Integral* $\int f(x)\,dx$
x^n	$\dfrac{x^{n+1}}{n+1}$ $(n \neq -1)$
$\dfrac{1}{x}$	$\log_e x$
e^{px}	$\dfrac{1}{p}e^{px}$
a^x	$\dfrac{a^x}{\log_e a}$
$\sin x$	$-\cos x$
$\cos x$	$\sin x$
$\tan x$	$\log_e \sec x$
$\operatorname{cosec} x$	$\log_e \tan \frac{1}{2}x = \log_e(\operatorname{cosec} x - \cot x)$
$\sec x$	$\log_e \tan(\frac{1}{4}\pi + \frac{1}{2}x) = \log_e(\sec x + \tan x)$
$\cot x$	$\log_e \sin x$
$\sec^2 x$	$\tan x$
$\operatorname{cosec}^2 x$	$-\cot x$
$\sec x \tan x$	$\sec x$
$\operatorname{cosec} x \cot x$	$-\operatorname{cosec}^2 x$
$\dfrac{1}{a^2-x^2}$	$\dfrac{1}{2a}\log_e\dfrac{a+x}{a-x} = \dfrac{1}{a}\tanh^{-1}\dfrac{x}{a}$ $(x^2 < a^2)$
$\dfrac{1}{a^2+x^2}$	$\dfrac{1}{a}\tan^{-1}\dfrac{x}{a}$
$\dfrac{1}{\sqrt{(a^2-x^2)}}$	$\sin^{-1}\dfrac{x}{a}$ or $-\cos^{-1}\dfrac{x}{a}$ $(x^2 < a^2)$
$\dfrac{1}{x\sqrt{(x^2-a^2)}}$	$\dfrac{1}{a}\sec^{-1}\dfrac{x}{a}$ or $-\operatorname{cosec}^{-1}\dfrac{x}{a}$ $(x^2 > a^2)$
$\sinh x$	$\cosh x$
$\cosh x$	$\sinh x$
$\tanh x$	$\log_e \cosh x$
$\operatorname{cosech} x$	$\log_e \tanh \frac{1}{2}x$
$\operatorname{sech} x$	$2\tan^{-1} e^x$
$\coth x$	$\log_e \sinh x$
$\operatorname{sech}^2 x$	$\tanh x$
$\operatorname{cosech}^2 x$	$-\coth x$
$\operatorname{sech} x \tanh x$	$-\operatorname{sech} x$
$\operatorname{cosech} x \coth x$	$-\operatorname{cosech} x$
$\dfrac{1}{\sqrt{(a^2+x^2)}}$	$\sinh^{-1}\dfrac{x}{a} = \log_e\left[\dfrac{x}{a}+\sqrt{\left(\dfrac{x^2}{a^2}+1\right)}\right]$
$\dfrac{1}{\sqrt{(x^2-a^2)}}$	$\cosh^{-1}\dfrac{x}{a} = \log_e\left[\dfrac{x}{a}+\sqrt{\left(\dfrac{x^2}{a^2}-1\right)}\right]$ $(x^2 > a^2)$
$\dfrac{1}{x\sqrt{(a^2-x^2)}}$	$\dfrac{1}{a}\operatorname{sech}^{-1}\dfrac{x}{a}$ $(x^2 < a^2)$
$\dfrac{1}{x\sqrt{(x^2+a^2)}}$	$\dfrac{1}{a}\operatorname{cosech}^{-1}\dfrac{x}{a}$
$\dfrac{1}{x^2-a^2}$	$-\dfrac{1}{a}\coth^{-1}\dfrac{x}{a}$ $(x^2 > a^2)$

4

(a) $\int \frac{f'(x)}{f(x)} dx = \log_e f(x)$, where $f'(x) \equiv \frac{d}{dx}\{f(x)\}$.

(b) $\int \{f(x)\}^n f'(x)\, dx = \frac{\{f(x)\}^{n+1}}{n+1}$ for all $n \neq -1$.

(c) A special case of (b), with $n = -\frac{1}{2}$ is

$$\int \frac{f'(x)}{\sqrt{\{f(x)\}}} dx = 2\sqrt{\{f(x)\}}.$$

5 *Integration by means of partial fractions*
If the function

$$f(x) = \frac{a_0 + a_1 x + a_2 x^2 + \ldots + a_n x^n}{b_0 + b_1 x + b_2 x^2 + \ldots + b_m x^m}$$

is to be integrated, factorize the denominator into linear or quadratic factors and express $f(x)$ in practical fractions. Note, if the degree of the numerator, $n > m$, then first reduce $f(x)$ to a polynomial and a proper fraction.

Example.

If $f(x) = \frac{2x^5 + x^4 - 8x^2 - 12x + 4}{x^3 - x^2 - 4}$,

then on dividing and converting to partial fractions,

$$f(x) = 2x^2 + 3x + \frac{1}{x-2} + \frac{2x+3}{x^2+x+2}$$

and $\int f(x)\, dx$

$$= \int \left[2x^2 + 3x + \frac{1}{x-2} + \frac{2x+1}{x^2+x+2} + \frac{2}{(x+\frac{1}{2})^2 + \frac{7}{4}} \right] dx$$

$$= \tfrac{2}{3}x^3 + \tfrac{3}{2}x^2 + \log_e(x-2) + \log_e(x^2+x+2) + {}$$

$$+ 2 \times \frac{2}{\sqrt{7}} \tan^{-1} \frac{2x+1}{\sqrt{7}} + C.$$

6 *Integration by substitution*
Certain integrals can more easily be determined by the introduction of a new variable to reduce the integral to a known form.

(a) *General method.* If $y = \int f(x)\, dx$, then on substituting a new variable u

such that $x = \phi(u)$,

$f(x) = f\{\phi(u)\} \equiv F(u)$, say,

and $dx = \frac{d\phi}{du} du \equiv \phi'(u)\, du$,

then $y = \int F(u)\, \phi'(u)\, du$.

(b) *Table of some trial substitutions.*

Type of integrand $f(x)$	*Suggested substitutions*
$\sqrt{(a^2 - x^2)}$ occurring in $f(x)$	$a^2 - x^2 = u^2$ or $x = a \sin\theta$ or $x = a \tanh\theta$
$\sqrt{(a^2 + x^2)}$ occurring in $f(x)$	$a^2 + x^2 = u^2$ or $x = a \tan\theta$ or $x = a \sinh\theta$
$\sqrt{(x^2 - a^2)}$ occurring in $f(x)$	$x^2 - a^2 = u^2$ or $x = a \sec\theta$ or $x = a \cosh\theta$
$\sqrt{(ax+b)}$ occurring in $f(x)$	$\sqrt{(ax+b)} = u$
$\sqrt{\{(x-a)(x-b)\}}$ occurring in $f(x)$	$\sqrt{(x-b)} = u\sqrt{(x-a)}$
$(a-x)(x-b)$ occurring in $f(x)$	$x = a\cos^2\theta + b\sin^2\theta$
$f(x) = x\, g(x^2)$	$x^2 = u$
$f(x) = g\{x\sqrt{(ax^2+bx+c)}\}$	$x = \frac{1}{u}$
$f(x) = \cos^m x \sin^n x$	if m odd, $\sin x = u$; if n odd, $\cos x = u$; if m and n both even express $f($ as a sum of sines or cosines of multiples of x
other trigonometrical functions	$\tan\frac{1}{2}x = t$ or $\tan x = u$

7 *Integration by parts*

$$\int u \frac{dv}{dx} dx = uv - \int v \frac{du}{dx} dx$$

or $\int u\, dv = uv - \int v\, du$.

8

(a) The area bounded by the curve $y = f(x)$, the x-axis and the ordinates at $x = a$, $x = b$ (see Figure 72a) is

$$\int_a^b y\, dx.$$

(b) The area bounded by the curve $y = f(x)$, the y-axis and the lines $y = c$, $y = d$ (see Figure 72b) is

$$\int_c^d x\, dy$$

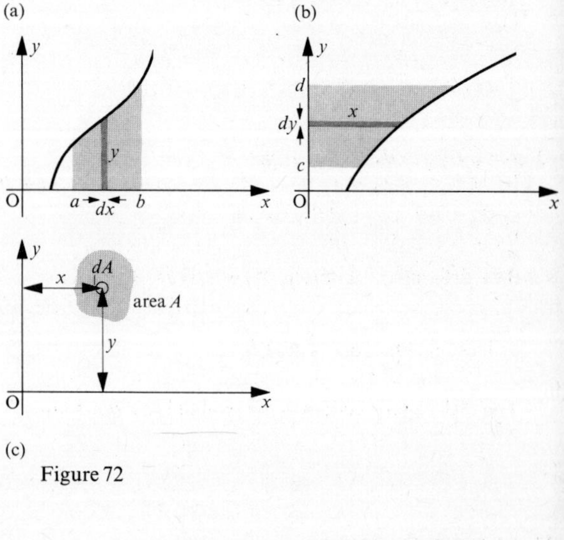

Figure 72

9

(a) The volume generated by revolving the area of Figure 72(a) through 360° about the x-axis is

$$\int_a^b \pi y^2\,dx.$$

(b) The volume generated by revolving the area of Figure 72(b) through 360° about the y-axis is

$$\int_c^d \pi x^2\,dy.$$

10 *Centroid (centre of gravity)*

(a) The coordinates $(\bar{x}, \bar{y})$ of the centroid of a closed area A (see Figure 72c) are given by

$$\bar{x} = \frac{\int x\,dA}{A}, \qquad \bar{y} = \frac{\int y\,dA}{A}.$$

Example. The coordinates of the centroid of the area of Figure 72 (a) are given by

$$\bar{x} = \frac{\int_a^b xy\,dx}{\int_a^b y\,dx}, \qquad \bar{y} = \frac{\frac{1}{2}\int_a^b y^2\,dx}{\int_a^b y\,dx}.$$

(b) The coordinates $(\bar{x}, \bar{y}, \bar{z})$ of the centroid of a volume V are given by

$$\bar{x} = \frac{\int_V x\,dV}{V}, \qquad \bar{y} = \frac{\int_V y\,dV}{V}, \qquad \bar{z} = \frac{\int_V z\,dV}{V}.$$

(a) The mean value of $f(x)$, from $x = a$ to $x = b$, is

$$\overline{f(x)} = \frac{1}{b-a}\int_c^b f(x)\,dx.$$

(b) The root-mean-square value of $f(x)$, from $x = a$ to $x = b$, is

$$f(x)_{\text{r.m.s.}} = \left[\frac{1}{b-a}\int_a^b \{f(x)\}^2\,dx\right]^{\frac{1}{2}}.$$

E Integration: Illustrative worked problems

1 Evaluate

(i) $$\int_0^{\frac{1}{2}\pi} (\sin\theta + \cos 2\theta)^2\,d\theta,$$

(ii) $$\int_2^3 \frac{(3-x)\,dx}{1+\sqrt{(x-2)}},$$

(iii) $$\int_0^{\frac{1}{2}\pi} x\sin x\cos x\,dx.$$ [L, J 1968, P II, Q 10]

Solution.

(i) $$\int_0^{\frac{1}{2}\pi} (\sin\theta + \cos 2\theta)^2\,d\theta$$

$$= \int_0^{\frac{1}{2}\pi} (\sin^2\theta + 2\sin\theta\cos 2\theta + \cos^2 2\theta)\,d\theta$$

$$= \int_0^{\frac{1}{2}\pi} \{\tfrac{1}{2}(1-\cos 2\theta) + \sin 3\theta - \sin\theta + \tfrac{1}{2}(\cos 4\theta + 1)\}\,d\theta$$

$$= \left[\tfrac{1}{2}\theta - \tfrac{1}{4}\sin 2\theta - \tfrac{1}{3}\cos 3\theta + \cos\theta + \tfrac{1}{8}\sin 4\theta + \tfrac{1}{2}\theta\right]_0^{\frac{1}{2}\pi}$$

$$= \tfrac{1}{2}\pi - \tfrac{2}{3}.$$

(ii) Let $x-2 = u^2$, then $dx = 2u\,du$ and

$$\int_2^3 \frac{(3-x)\,dx}{1+\sqrt{(x-2)}} = \int_0^1 \frac{(1-u^2)2u\,du}{1+u}$$

$$= \int_0^1 2u(1-u)\,du$$

$$= 2\left[\tfrac{1}{2}u^2 - \tfrac{1}{3}u^3\right]_0^1 = \tfrac{1}{3}.$$

(iii) $$I = \int_0^{\frac{1}{2}\pi} x\sin x\cos x\,dx = \tfrac{1}{2}\int_0^{\frac{1}{2}\pi} x\sin 2x\,dx,$$

and on integrating by parts (with $u = x$; $dv/dx = \sin 2x$, $v = -\frac{1}{2}\cos 2x$):

$$I = \tfrac{1}{2}\left[x(-\tfrac{1}{2}\cos 2x)\right]_0^{\frac{1}{2}\pi} - \tfrac{1}{2}\int_0^{\frac{1}{2}\pi} -\tfrac{1}{2}\cos 2x\,dx$$

$$= 0 + \tfrac{1}{4}\left[\tfrac{1}{2}\sin 2x\right]_0^{\frac{1}{2}\pi} = \tfrac{1}{8}.$$

2 O is a point on the axis of symmetry of a disc of radius a and it is at a distance x from the disc. Prove that the mean distance of points of the disc from O is

$$\frac{2}{3a^2}\{(a^2+x^2)^{\frac{3}{2}} - x^3\}.$$

A circular cone of semi-angle α has height h. Prove that the mean value of the distances of points inside the cone from the vertex of the cone is

$\frac{1}{2}h(\sec^3\alpha - 1)\cot^2\alpha.$ [O & C, S 1967, M & HM III, Q 10]

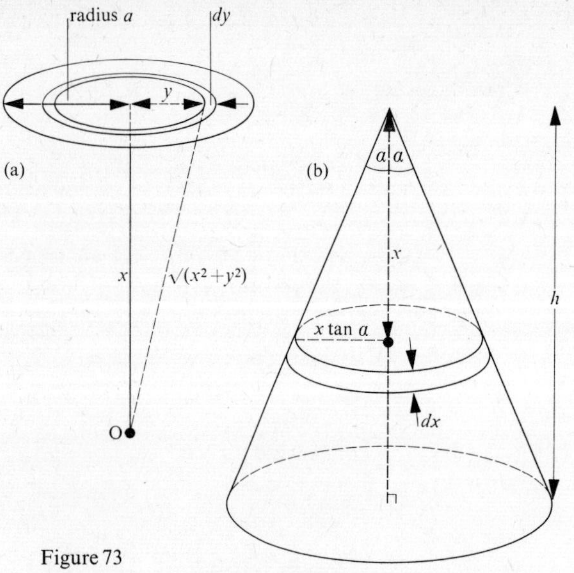

Figure 73

Solution.

Disc. See Figure 73(a). The points of the disc on a ring of thickness dy, at a radius y, are at a distance $\sqrt{(x^2+y^2)}$ from the point O on the axis and they are a fraction $2\pi y\, dy/\pi a^2$ of the total number of points of the whole disc.

Hence mean value of the distance of the points of the disc from O is

$$\int_0^a \sqrt{(x^2+y^2)}\,\frac{2\pi y}{\pi a^2}\,dy = \frac{2}{a^2}\int_0^a y\sqrt{(x^2+y^2)}\,dy$$

$$= \frac{2}{3a^2}\left[(x^2+y^2)^{3/2}\right]_{y=0}^{y=a}$$

$$= \frac{2}{3a^2}\{(a^2+x^2)^{3/2}-x^3\}.$$

Cone. See Figure 73(b). The required mean value for the cone can be derived using the result obtained for the disc above. Thus, for the elemental disc shown, $a = x \tan \alpha$, and hence for the elemental disc, the mean value of the distance from O equals

$$\frac{2}{3x^2 \tan^2\alpha}\,x^3(\sec^3\alpha-1) = \tfrac{2}{3}x \cot^2\alpha(\sec^3\alpha-1).$$

But the points on the disc, of thickness (depth) dx, are a fraction

$$\frac{\pi(x \tan \alpha)^2\,dx}{\frac{1}{3}\pi(h \tan \alpha)^2 h} = \frac{3}{h^3}x^2\,dx$$

of the number in the whole cone.

Hence, the mean value of the distance of the points of the cone from O is

$$\int_0^h \frac{2}{3}x\cot^2\alpha(\sec^3\alpha-1)\frac{3}{h^3}x^2\,dx = \frac{2}{h^3}\cot^2\alpha(\sec^3\alpha-1)\int_0^h x^3\,dx$$

$$= \tfrac{1}{2}h\cot^2\alpha(\sec^3\alpha-1).$$

3 If O is the origin and A is the point $(\frac{1}{3}, \frac{1}{4})$ on the curve $y = \cos^2 \pi x$, calculate the area bounded by the y-axis, the straight line OA and the arc of the curve joining the points (0, 1) and $(\frac{1}{3}, \frac{1}{4})$.

Find also the volume of the solid of revolution formed by rotating this area about the x-axis. [L, S 1967, P I, Q 10]

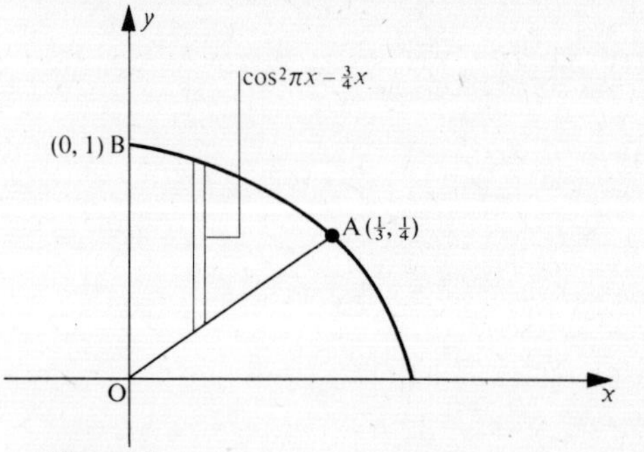

Figure 74

Solution.

$$\text{Area OBA} = \int_0^{\frac{1}{3}} (\cos^2 \pi x - \tfrac{3}{4}x)\,dx$$

$$= \int_0^{\frac{1}{3}} \{\tfrac{1}{2}(1+\cos 2\pi x)-\tfrac{3}{4}x\}\,dx$$

$$= \left[\frac{1}{2}x+\frac{1}{4\pi}\sin 2\pi x-\frac{3}{8}x^2\right]_0^{\frac{1}{3}} = \frac{\sqrt{3}}{8\pi}+\frac{1}{8}.$$

Volume v_1 generated by rotating area OBAC about the x-axis is

$$v_1 = \int_0^{\frac{1}{3}} \pi(\cos^2\pi x)^2\,dx = \pi\int_0^{\frac{1}{3}} \cos^4 \pi x\,dx,$$

but
$$\cos^4\pi x = \{\tfrac{1}{2}(1+\cos 2\pi x)\}^2$$
$$= \tfrac{1}{4}(\cos^2 2\pi x+2\cos 2\pi x+1)$$
$$= \tfrac{1}{8}(1+\cos 4\pi x)+\tfrac{1}{2}\cos 2\pi x+\tfrac{1}{4},$$

Thus
$$v_1 = \pi\left[\frac{1}{32\pi}\sin 4\pi x+\frac{1}{8}x+\frac{1}{4\pi}\sin 2\pi x+\frac{1}{4}x\right]_0^{\frac{1}{3}}$$

$$= \pi\left[\frac{1}{32\pi}\left(-\frac{\sqrt{3}}{2}\right)+\frac{1}{4\pi}\frac{\sqrt{3}}{2}+\frac{1}{8}\right] = \frac{1}{8}\pi+\frac{7\sqrt{3}}{64}.$$

Volume v_2 of cone generated by revolution of area OAC is

$$v_2 = \frac{1}{3}\pi\left(\frac{1}{4}\right)^2\frac{1}{3} = \frac{\pi}{144}.$$

Hence, required volume of revolution of area OBA is

$$v_1 - v_2 = \frac{17\pi}{144}+\frac{7\sqrt{3}}{64}.$$

4

(i) Find $\displaystyle\int \frac{(4x+1)\,dx}{\sqrt{(35+4x-4x^2)}}.$

(ii) Evaluate $\displaystyle\int_0^{\frac{1}{4}\pi} \frac{dx}{9\cos^2 x - \sin^2 x}$. [L, S 1967, P S, Q 9]

Solution.

(i)
$$\int \frac{(4x+1)\,dx}{\sqrt{(35+4x-4x^2)}}$$
$$= -\frac{1}{2}\int \frac{(-8x+4)\,dx}{\sqrt{(35+4x-4x^2)}} + 3\int \frac{dx}{\sqrt{(35+4x-4x^2)}}$$
$$= -\sqrt{(35+4x-4x^2)} + \frac{3}{2}\int \frac{dx}{\sqrt{\{36/4-(x-\frac{1}{2})^2\}}}$$
$$= -\sqrt{(35+4x-4x^2)} + \tfrac{3}{2}\sin^{-1}\tfrac{1}{3}(x-\tfrac{1}{2}) + c.$$

(ii)
$$I = \int_0^{\frac{1}{4}\pi} \frac{dx}{9\cos^2 x - \sin^2 x} = \int_0^{\frac{1}{4}\pi} \frac{\sec^2 x\,dx}{9-\tan^2 x}.$$

Then on letting $t = \tan x$, $dt = \sec^2 x\,dx$

$$I = \int_0^1 \frac{dt}{3^2-t^2} = \frac{1}{2\times 3}\left[\log_e \frac{3+t}{3-t}\right]_0^1$$
$$= \tfrac{1}{6}\log_e 2.$$

F Integration: problems

Answers and hints will be found on pp. 98–101.

1

(i) If $a > 1$ and
$$\int_1^a \frac{x^4-1}{x^3}\,dx = \frac{9}{8},$$
find a.

(ii) If n is a positive integer, find in terms of n the three possible values of
$$\int_{\frac{1}{2}\pi}^{\pi} \cos nx\,dx.$$

(iii) Evaluate
$$\int_0^{\frac{1}{4}\pi} \frac{2\cos x - \sin x}{2\sin x + \cos x}\,dx,$$
correct to three decimal places. [L, J 1968, P I, Q 9]

2 Evaluate, correct to three decimal places,

(a) $\displaystyle\int_0^1 \frac{2x^2\,dx}{2x+1}$,

(b) $\displaystyle\int_0^{\frac{1}{3}\pi} \sin^2 x\cos^2 x\,dx$,

(c) $\displaystyle\int_e^{e^2} \frac{dx}{x\log_e x}$. [L, S 1967, P I, Q 9]

3 Sketch the curve $x = \tan\theta + 1$, $y = \sec\theta$ for values of x between 1 and 4.

Find the centroid of the area bounded by the two branches of this curve and the lines $x = 1$ and $x = 4$.

[L, S 1967, P, Q 10]

4

(i) If $f(x)$ and $\phi(x)$ are differentiable functions of x such that $f(0) = 0$, $\phi(0) = 1$ and $\phi^2(x) \neq f^2(x)$ for any value of x in the range $0 \leqslant x \leqslant a$, prove that
$$\int_0^a \frac{\phi' f - \phi f'}{\phi^2 - f^2}\,dx = \tfrac{1}{2}\log\frac{\phi(a)-f(a)}{\phi(a)+f(a)},$$
where accents denote differentiation with respect to x.

(ii) Prove that
$$\int_0^{\frac{1}{2}\pi} \log\sin x\,dx = \int_0^{\frac{1}{2}\pi} \log\cos x\,dx = -\tfrac{1}{2}\pi\log 2.$$

[L, S 1967, FM V, Q 5]

5 A segment of a circle of radius r is cut off by a chord AB of length $2r\sin\alpha$ where $\alpha < \frac{1}{2}\pi$, and M is the midpoint of the arc AB. A solid gold wedding ring has the shape formed by the rotation of this segment about an axis in its plane, parallel to AB and distant $a + r\cos\alpha$ from it, M being distant $a + r$ from the axis. If the weight of the ring is equal to that of four golden spheres of diameter AB, find a in terms of r and α, and show that the ratio of the total surface area of the ring to that of the spheres is
$$\frac{\sin\alpha\,(\alpha+\sin\alpha)}{2\alpha-\sin 2\alpha} + \frac{1}{4}\cot\frac{\alpha}{2}.$$
[L, S 1967, FM VI, Q 9]

6 The bowl of a wine glass is shaped by the rotation of the portion of the curve $y = \log_e x$ between $x = 1/e$ and $x = e$, about the y-axis, through 2π radians, and the bottom of the bowl is a circular glass disc of radius $1/e$.
Find:

(a) the volume of wine it can hold,

(b) the height above the bottom of the bowl of the centroid of the wine in a full glass,

(c) the height above the bottom of the bowl of the surface of the wine when the glass is half full. [L, S 1966, P II, Q 10]

7 The area in the first quadrant bounded by the ellipse $\frac{1}{8}x^2 + \frac{1}{2}y^2 = 1$, the hyperbola $\frac{1}{2}x^2 - y^2 = 1$ and the x-axis is rotated through four right angles about the x-axis. Prove that the volume of the solid thus formed is $\frac{2}{3}\pi(5\sqrt{2}-6)$.

[O, S 1967, M & PM I, Q 11]

8 By using the change of variable from t to u where $(a-t)(a-u) = a^2-1$, or otherwise, prove that, when $a > 1$,
$$\int_{-1}^{1} \frac{1}{(1-t^2)^{\frac{1}{2}}}\,\frac{dt}{a-t} = \frac{\pi}{(a^2-1)^{\frac{1}{2}}}.$$

Hence show that

$$\int_{-1}^{1} (1-t^2)^{\frac{1}{2}} \frac{dt}{a-t} = \pi\{a-(a^2-1)^{\frac{1}{2}}\}$$

and deduce that, when $0 \leqslant x < 1$,

$$\int_0^{\pi} \frac{\sin^2\theta}{1-2x\cos\theta+x^2}\,d\theta = \tfrac{1}{2}\pi.$$ [O, S 1967, PM S, Q 8]

9 The curves $y = 3\sin x$, $y = 4\cos x$ $(0 \leqslant x \leqslant \frac{1}{2}\pi)$ intersect at the point A, and meet the axis of x at the origin O and the point B$(\frac{1}{2}\pi, 0)$ respectively. Prove that the area enclosed by the arcs OA, AB and the line OB is 2 square units.

If N is the foot of the perpendicular from A to the axis of x, find by integration the volume obtained when the area enclosed by AN, NB and the arc AB is completely rotated about the x-axis, giving the answer correct to two significant figures. [C, S 1967, II, Q 9]

10

(i) Prove that

$$\int_0^{\frac{1}{3}} x^2 e^{-3x}\,dx = \frac{2e-5}{27e}.$$

(ii) Find

$$\int \frac{7\cos^2 x+2\sin^2 x}{9\cos^2 x+4\sin^2 x}\,dx.$$ [C, S 1966, P III, Q 7]

11 By using the substitution $x = \tan\theta$ evaluate the integral

$$\int_0^1 \frac{1-x^2}{(1+x^2)^2}\,dx$$

and prove that

$$\int_0^{\frac{1}{4}\pi} \frac{1}{\sin^2\theta-2\sin\theta\cos\theta+2\cos^2\theta}\,d\theta = \tfrac{1}{4}\pi.$$

[O & C, S 1967, M & HM II, Q 6]

12 Integrate the following functions w.r.t. x:

(i) $\dfrac{x^3}{x^2+1}$,

(ii) $\dfrac{\cos x}{\sqrt{(\sin x)}}$,

(iii) $\sqrt{(2ax-x^2)}$.

By means of the substitution $\cos\theta = u$ or otherwise, evaluate:

$$\int_0^{\frac{1}{2}\pi} \frac{\sin^3\theta\,d\theta}{(1+\cos\theta)^2}.$$ [O & C, S 1966, M for Sc I, Q 8]

13 Sketch the curve whose equation is given by $x = 2\sin t$, $y = 2\tan t$ for values of t such that $0 \leqslant t < \frac{1}{2}\pi$.
Calculate

(i) the area bounded by the curve, the x-axis and the line $x = \sqrt{3}$,

(ii) the average ordinate in this area.

[AEB, N 1967, P & II, Q 5]

14

(a) Evaluate the following definite integrals.

(i) $\displaystyle\int_1^2 x^2 \log_e x\,dx.$

(ii) $\displaystyle\int_0^1 (2x^2+1)(2x^3+3x+4)^{\frac{1}{2}}\,dx.$

(b) Find the coordinates of the centroid of the region defined by the inequalities $x > 0$, $y > 0$ and $y < 1-x^2$.

[JMB, S 1967, P I, Q 9]

15

(a) Let $y = \sin x+3c\sin 3x$,

where c is a constant. Let h and k be the mean values of y and y^2, respectively, over the interval $0 \leqslant x \leqslant \pi$.
Prove that

$$h = \frac{2(1+c)}{\pi}, \qquad k = \tfrac{1}{2}(1+9c^2).$$

Verify that $k > h^2$ for all values of c.

(b) If a is positive and small compared with 1, show that

$$\tan^{-1}\frac{1}{a} \simeq \tfrac{1}{2}\pi - a.$$

Hence, or otherwise, show that

$$\frac{\int_0^a dx/(a^2+x^2)}{\int_0^a dx/(a^2+x^2)} \simeq \frac{1}{2}+\frac{a}{\pi}.$$ [JMB, S 1966, M Sp, Q 4]

16 Find $\displaystyle\int \frac{2x^3+5x^2+2x+4}{(x-1)^2(9x^2+4)}\,dx$

and $\displaystyle\int_{\sin^{-1}(3/4)}^{\sin^{-1}(3\sqrt{3}/4)} \frac{\cos\theta\,d\theta}{\sqrt{(5+4\cos^2\theta)}}.$ [W, S 1969, P I, Q 10]

17 Find the area bounded by the x-axis and the curve $y = (2+x^2)^{-1}$.

Find the volume generated when the area bounded by the curve and the lines $y = \frac{1}{4}$ and $y = \frac{1}{2}$ is rotated through half a revolution about the y-axis. [W, S 1968, P II, Q 7]

18 Integrate with respect to x

(a) $\log\left(1+\dfrac{a^2}{x^2}\right)$, where a is constant,

(b) $\sqrt{(x^2+x+1)}$. [W, S 1968, PS, Q 3]

G Mixed integration and differentiation: Problems

Answers and hints will be found on pp. 101–3.

1 Find the maximum and minimum values of the function

$$f(x) = \frac{(x-2)(x-3)}{(x-5)}.$$

Find also the area bounded by the x-axis and the arc of the curve $y = f(x)$ joining the points (2, 0), (3, 0).

[L, J 1968, P I, Q 10]

2

(i) Show graphically that the equation $\sinh^{-1} x = \operatorname{sech}^{-1} x$ has only one real root. Prove that this root is $\{\tfrac{1}{2}(\sqrt{5}-1)\}^{\frac{1}{2}}$.

(ii) Prove that

$$\int_{4/5}^{1} \operatorname{sech}^{-1} x\, dx = 2\tan^{-1} 2 - \frac{\pi}{2} - \frac{4}{5}\log 2.$$

[L, J 1968, FM V, Q 3]

3 The tangent at the point $(a\cos^3\theta, a\sin^3\theta)$ to the curve $x^{\frac{2}{3}} + y^{\frac{2}{3}} = a^{\frac{2}{3}}$ cuts the x-axis at A and the y-axis at B. Find the equation and the length of AB.

If that portion of the curve which is in the first quadrant is rotated completely about the x-axis, find the centre of gravity of the solid formed.

[L, S 1967, P II, Q 9]

4

(i) If $x = \log(u+2)$ and $d^2y/du^2 = ye^{-2x}$, show that

$$\frac{d^2y}{dx^2} - \frac{dy}{dx} - y = 0.$$

(ii) Two solid spheres of radii R and r, having their centres at A and B are placed in a dark room. A point source of light, free to move in the line AB, illuminates the largest total area of the spheres when it is at a point C. Find the ratio AC : CB.

[L, S 1967, PS, Q 8]

5

(i) Show that, if $a^2 > b^2$, the function

$$\cos^{-1}\left[\frac{a\cos x + b}{a + b\cos x}\right]$$

increases with x at a rate which lies between

$$\left[\frac{a-b}{a+b}\right]^{\frac{1}{2}} \quad \text{and} \quad \left[\frac{a+b}{a-b}\right]^{\frac{1}{2}}.$$

(ii) Evaluate $\displaystyle\int_{-1}^{1} \frac{(x+1)(4x-1)}{(2-x)^2(x^2+3)}\,dx.$

[L, J 1967, P II, Q 8]

6 Show that the curves

$$xy^2 = a^2(2a-x), \qquad yx^2 = a^2(2a-y)$$

have one, and only one, real common point and that the curves touch at this point.

Sketch the curves on the same diagram.

Show that the area enclosed by the common tangent and either curve is $\tfrac{1}{2}(\pi-3)a^2$.

[L, S 1966, PS, Q 10]

7 Draw a rough sketch of the curve whose equation is

$$a^2y^2 = x^2(a^2-x^2).$$

Show that each of the lines $y = \pm\tfrac{1}{2}a$ touches the curve at two distinct points.

Find the area enclosed by one of the loops of the curve.

[O, S 1967, M & PM I, Q 9]

8 Show that, for the curve $y = x - \dfrac{3\sin x}{2+\cos x}$,

(i) $$\frac{dy}{dx} = \left[\frac{1-\cos x}{2+\cos x}\right]^2,$$

(ii) $$\frac{d^2y}{dx^2} = \frac{6\sin x(1-\cos x)}{(2+\cos x)^3}.$$

Show that the points of inflexion of this curve are evenly spaced along the line $y = x$, and that the tangents at these points of inflexion are parallel either to $y = 0$ or to $y = 4x$.

Sketch the curve and prove that, if A and B are any two consecutive points of inflexion, then the area between the straight line AB and the curve is $\log_e 27$.

[O, S 1966, P I, Q 6]

9 Given that

$$y = \sqrt{(px^2+2qx+r)} \quad \text{and} \quad z = \sqrt{(px^2+2qx+s)},$$

where p, q, r and s are constants, prove that

$$\frac{d}{dx}\log_e(y+z) = \frac{px+q}{yz},$$

and give a similar formula for the derivative of $\log_e(y-z)$.

Using your results,

(i) show that

$$\int_0^1 \frac{x\,dx}{\sqrt{\{(x^2+1)(x^2+4)\}}} - \log_e\frac{\sqrt{2}+\sqrt{5}}{3};$$

(ii) evaluate $\displaystyle\int_0^1 \frac{dx}{\sqrt{\{(x+4)(x+9)\}}}$;

(iii) find $\displaystyle\int \frac{dx}{\sqrt{(x^2+a^2)}}$,

where a is a constant.

[C, S 1966, Sp 0, Q 4]

10

(i) Find the Maclaurin expansion of $\log_e \cos x$ as far as the term in x^4.

Your answer should involve only even powers of x with negative coefficients. If terms in x and x^3 had been present, what property of the cosine function would indicate that a mistake had been made?

(ii) Evaluate

$$\int_0^1 xe^{-\frac{1}{2}x^2}\,dx, \quad \text{leaving your answer in terms of } e,$$

$$\int_{\frac{1}{2}\pi}^{\pi} \cos^3 x\,dx.$$

[O & C, S 1967, M for Sc, I, Q 10]

11

(a) Differentiate e^{x^2} with respect to x and hence determine $\int xe^{x^2}\,dx$. Evaluate

$$\int_0^1 x^3 e^{x^2}\,dx$$

(b) Use the substitution $x = \sin^2\theta$ to evaluate

$$\int_0^{\frac{1}{2}} \left[\frac{x}{1-x}\right] dx.$$

[AEB, S 1968, P & A II, Q 5]

12 Show that the curve

$$y = \frac{\log_e x}{x}$$

has a turning point at P $(e, 1/e)$. Determine the nature of this turning point. Sketch the graph of y for positive values of x.

The ordinates at P and at the point Q $(\sqrt{e}, 1/2\sqrt{e})$ meet the x-axis at M and N respectively. Show that the area of the triangle OQN equals half the area bounded by the curve, the x-axis and the ordinate PM. [AEB, N 1967,, P & AI, Q 5]

13

(a) Write down the derivative of $\sinh x$, and deduce that

$$\frac{d}{dx}\sinh^{-1} x = \frac{1}{\sqrt{(x^2+1)}}.$$

If $y = \cosh(a \sinh^{-1} x)$, where a is a constant, show that

$$(x^2+1)\frac{d^2y}{dx^2}+x\frac{dy}{dx}-a^2y = 0.$$

(b) The curve $x^2-y^2 = 1$ cuts the positive x-axis at Q, and the point P of the curve has coordinates $(\cosh t, \sinh t)$, where $t > 0$. If O is the origin, show that the area bounded by the arc PQ and the lines OP and OQ is $\frac{1}{2}t$.

[JMB, S 1967, FM I, Q 4]

H Differentiation: Answers and hints

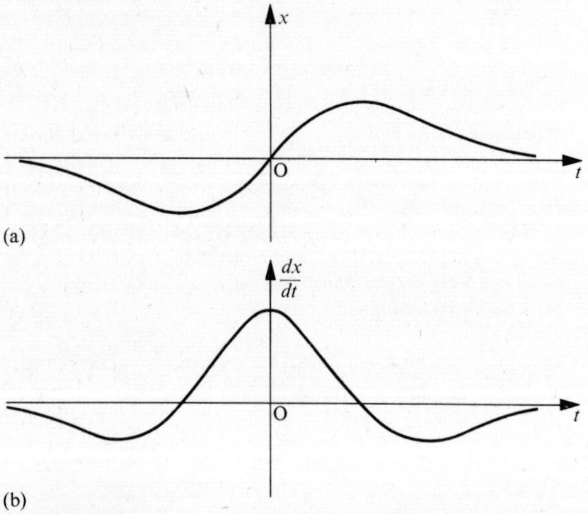

Figure 75

1 $x = e^{-\frac{1}{2}}$ (maximum), $x = -e^{-\frac{1}{2}}$ (minimum).

$\dfrac{dx}{dt} = 1$ (maximum), $\dfrac{dx}{dt} = -2e^{-\frac{3}{2}}$ (minimum).

2

(i) $\dfrac{4x}{1-4x^2}$.

(ii) $\dfrac{dy}{dx} = e^x\cos(e^x+k)$;

$$\frac{d^2y}{dx^2} = -e^{2x}\sin(e^x+k)+e^x\cos(e^x+k).$$

$$\frac{d^2y}{dx^2}-\frac{dy}{dx} = -e^{2x}\sin(e^x+k) = -e^{2x}y,$$

and hence $\dfrac{d^2y}{dx^2}-\dfrac{dy}{dx}+e^{2x}y = 0.$

(iii) Let height and radius of cylinder ABCD from which prism is cut be x and r respectively, see Figure 76.

Volume of prism $v = 8r^2 \sin\frac{1}{8}\pi(\sin\frac{3}{8}\pi)x$

and $x = \dfrac{h}{a}(2a-r)$.

$\dfrac{dv}{dr} = 0$ when $r = \dfrac{4}{3}a$,

thus $v_{max} = \dfrac{64}{27}\sqrt{2}\,a^2h.$

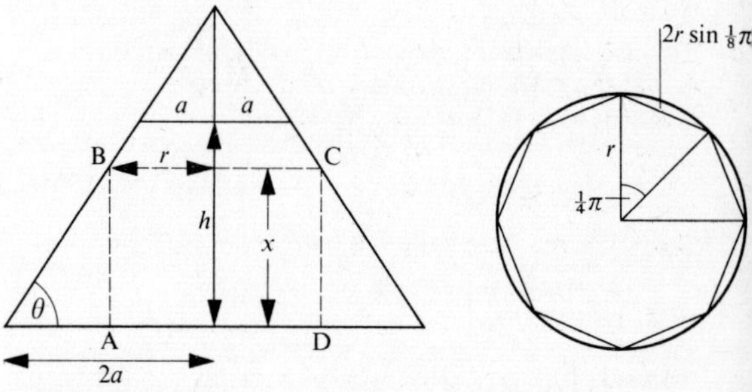

Figure 76

3 $x = 1$ (minimum), $x = -2$ (maximum).

$$\frac{y_{max}}{y_{min}} = \frac{k+20}{k-7} = 10,$$

$k = 10$.

Tangents parallel to x-axis (i.e. when $dy/dx = 0$), are $y = 3$, $y = 30$. $y = 3$ cuts curve again at $x = -\frac{7}{2}$, thus point P is $(-\frac{7}{2}, 3)$. $y = 30$ cuts curve again at $x = \frac{5}{2}$, thus point Q is $(\frac{5}{2}, 30)$. Equation of PQ is $4y = 18x+75$ and cuts curve again at $(-\frac{1}{2}, 16\frac{1}{2})$.

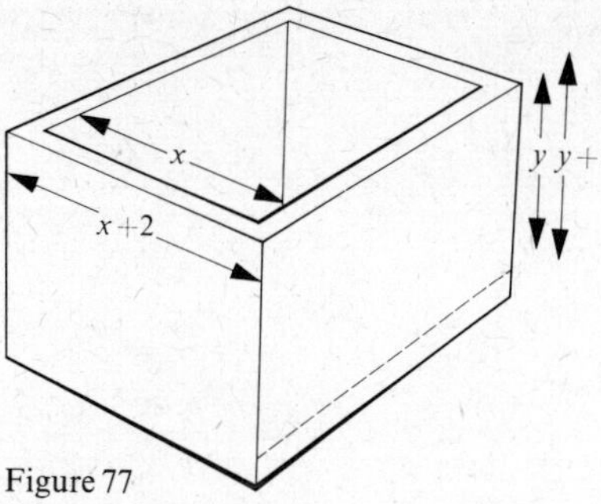

Figure 77

4 If internal height of box is y, then, on referring to Figure 77,

$76 = (x+2)^2(y+1) - x^2 y$,

$$y = \frac{76-(x+2)^2}{4\,(x+1)}.$$

Hence, internal capacity,

$$C = x^2 y = \frac{x^2(72-4x-x^2)}{4(x+1)}.$$

On logarithmic differentiation,

$$\frac{1}{C}\frac{dC}{dx} = \frac{2x}{x^2} - \frac{4}{4(x+1)} + \frac{-4-2x}{72-4x-x^2}$$

and $\dfrac{dC}{dx} = 0$ when $3x^2+6x-72 = 0$,

i.e. when $x = 4$, and so $C_{max} = 32\text{ cm}^3$.

5 The distance PQ is the square root of $\{p-X(t)\}^2 + \{q-Y(t)\}^2$. For a minimum, the derivative w.r.t. t is zero, whence the equation for the parameter t. It can be written

$$\frac{q-Y}{p-X} = \frac{dY}{dX}.$$

The l.h.s. is the gradient of PQ, the r.h.s. is the gradient of the tangent to C at P. This shows PQ is normal to C (at P). Now parametrize the parabola into $x = at^2$, $y = 2at$. Apply the first result to show that the parameter t of P satisfies $(5a-at^2)\,2at + (52a-2at)\,2a = 0$. Reduce to $t^3-3t-52 = 0$. Show that $t = 4$ is the only real root. This gives P$(16a, 8a)$.

6 $$\frac{dy}{dx} = \sec^2(k\tan^{-1}x)\frac{k}{1+x^2} = \frac{(1+y^2)k}{1+x^2}.$$

Multiply up to give the result quoted. Another differentiation gives $2xy' + (1+x^2)y'' = 2kyy'$, and a third, $2(y'+xy'') + 2xy'' + (1+x^2)y''' = 2k(yy''+y'^2)$. When $x = 0$, the three equations yield $y' = k$, $y'' = 0$ and $y''' = 2k^3 - 2k$. Substitute into the Maclaurin series to obtain the series for y.

7 Denote each of the expressions by y.

(i) Show that $y' = 0$ when $x = 0$ and when $x^2 = \frac{7}{5}$. The corresponding values of y are -4 and $-\sqrt{3}$.

(ii) Show that $y' = 0$ when $x = \pm\frac{3}{4}$. For $x = \frac{3}{4}$, $y = 4$; for $x = -\frac{3}{4}$, $y = \frac{17}{2}$.

8

(i)

(a) Use the quotient rule to show

$$\frac{d}{dx}\left[\frac{2+x^2}{1-x^2}\right] = \frac{6x}{(1-x^2)^2}.$$

(b) Use the product rule to show, after factorization, that

$$\frac{d}{dx}(\sin 2x\cos^2 x) = 2\cos x\cos 3x$$

(ii) Show that $y'' = e^{-x}(x+1)(x-5)$, and deduce that if $y'' < 0$, $-1 < x < 5$.

9

(a) $\dfrac{dx}{d\theta} = a(1-\cos\theta)$, $\qquad \dfrac{dy}{d\theta} = a\sin\theta$.

$$\frac{dy}{dx} = \frac{dy/d\theta}{dx/d\theta} = \frac{\sin\theta}{1-\cos\theta}.$$

$$\frac{d^2y}{dx^2} = \frac{d}{d\theta}\left[\frac{\sin\theta}{1-\cos\theta}\right]\frac{d\theta}{dx} = \frac{-1}{a(1-\cos\theta)^2} = \frac{-a}{y^2}.$$

(b) $a^2 = b^2 + c^2 - 2bc\cos A = 41 - 40\cos A$.

$$2a\frac{da}{dt} = 40\sin A\frac{dA}{dt},$$

$$\frac{dA}{dt} = \frac{a\,da/dt}{20\sin A}\text{ rad s}^{-1}.$$

Hence when

$a = 6$, $\cos A = \frac{1}{8}$, $\sin A = \dfrac{\sqrt{63}}{8}$ and

$$\frac{dA}{dt} = \frac{3}{20}\times\frac{8}{\sqrt{63}}\times\frac{180}{\pi} = 8{\cdot}62\text{ degrees}^{-1}.$$

10 Value of t for which tangent is also normal is $\pm\sqrt{2}$.

12

(a) Take logarithms,

i.e. $\log y = n\log x + x\log n$,

then differentiate

$$\frac{1}{y}\frac{dy}{dx} = \frac{n}{x} + \log n,$$

$$\frac{dy}{dx} = x^n n^x\left(\frac{n}{x} + \log n\right).$$

(b) Differentiate implicitly w.r.t. x,

$$2x + 2y + 2x\frac{dy}{dx} - 6y\frac{dy}{dx} + 4 - \frac{dy}{dx} = 0,\ \text{hence}\ \frac{dy}{dx} = \frac{10}{11}\text{ at } (1, 2)$$

Differentiate again,

$$2 + 2\frac{dy}{dx} + 2\left[\frac{dy}{dx} + x\frac{d^2y}{dx^2}\right] - 6\left[\frac{dy}{dx}\right]^2 - 6y\frac{d^2y}{dx^2} - \frac{d^2y}{dx^2} = 0$$

from which $\dfrac{d^2y}{dx^2} = \dfrac{82}{1331}$ at $x = 1$, $y = 2$ i.e. $(1, 2)$.

13

(a) Let $y = \tan^{-1}(\tanh \frac{1}{2}x)$, then $\tan y = \tanh \frac{1}{2}x$ and

$$(\sec^2 y)\frac{dy}{dx} = \frac{1}{2}\operatorname{sech}^2 \frac{1}{2}x,$$

$$\frac{dy}{dx} = \frac{\frac{1}{2}\operatorname{sech}^2 \frac{1}{2}x}{1+\tanh^2 \frac{1}{2}x} = \frac{1}{2}\cosh x.$$

(b) Let $$y = \tanh^{-1}\left[\frac{x+a}{1+ax}\right],$$

then $$\tanh y = \frac{x+a}{1+ax}$$

and $$(\operatorname{sech}^2 y)\frac{dy}{dx} = \frac{1+ax-a(x+a)}{(1+ax)^2},$$

from which $$\frac{dy}{dx} = \frac{1}{1-x^2}.$$

(c) $$\frac{d}{dx}\left[\log(x-1)-\frac{2x-1}{(x-1)^2}\right] = \frac{1}{x-1}-\frac{(x-1)^2 2-(2x-1)2(x-1)}{(x-1)^4}$$

$$= \frac{x^2+1}{(x-1)^3}.$$

14

(a) Differentiate equation four times

$$\frac{d^4 y}{dx^4} = \frac{-6}{a^3} \quad \text{at } (0, a)$$

(b) Express in terms of partial fractions,

i.e. $$\frac{1}{x^3-6x^2+11x-6} = \frac{\frac{1}{2}}{x-1}-\frac{1}{x-2}+\frac{\frac{1}{2}}{x-3},$$

then $$\frac{d^n}{dx^n}(x^3-6x^2+11x-6)^{-1}$$

$$= (-1)^n n!\left[\frac{1}{2(x-1)^{n+1}}-\frac{1}{(x-2)^{n+1}}+\frac{1}{2(x-3)^{n+1}}\right].$$

I Integration: Answers and hints

1

(i) $a = 2$

(ii) $-\frac{1}{n}$ (for $n = 1, 5, 9, \ldots$), $+\frac{1}{n}$ (for $n = 3, 7, \ldots$), 0 (n even)

(iii) $$[\log_e(2\sin x+\cos x)]_0^{\frac{1}{4}\pi} = \log_e \frac{3}{\sqrt{2}}$$

$$= 0{\cdot}752 \quad \text{(to three decimal places).}$$

2

(a) 0·275 (b) 0·158 (c) $\log_e 2 = 0{\cdot}693$
(use $\log_e x = u$ substitution)

3

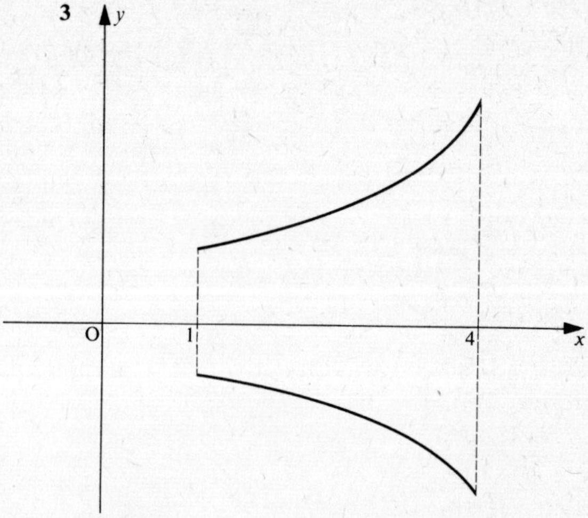

Figure 78

By symmetry, $\bar{y} = 0$;

$$\bar{x} = \frac{1}{A}\int_1^4 2xy\,dx,$$

where $$A = 2\int_1^4 y\,dx = 2\int_1^4 \sqrt{\{(x-1)^2+1\}}\,dx$$

and, on substituting $x-1 = \sinh\theta$,

$$A = \left[\sinh^{-1}(x-1)+(x-1)\sqrt{\{1+(x-1)^2\}}\right]_1^4 = 11{\cdot}31.$$

$$\int_1^4 2xy\,dx = 2\int (1+\sinh\theta)\cosh^2\theta\,d\theta$$

$$= [A+\tfrac{2}{3}\cosh^3\theta] = 31{\cdot}73.$$

Thus $\bar{x} = 31{\cdot}73/11{\cdot}31 = 2{\cdot}8$,

and centroid at $(\bar{x}, \bar{y}) = (2{\cdot}8, 0)$.

4

(i) $$\int_0^a \frac{\phi' f-\phi f'}{\phi^2-f^2}\,dx = \frac{1}{2}\int_0^a \left[\frac{\phi'-f'}{\phi-f}-\frac{\phi'+f'}{\phi+f}\right]dx$$

$$= \tfrac{1}{2}[\log(\phi-f)-\log(\phi+f)]_0^a$$

$$= \tfrac{1}{2}\log\left[\frac{\phi(a)-f(a)}{\phi(a)+f(a)}\right].$$

(ii) Let $x = \frac{1}{2}\pi - u$,

then $I = \int_0^{\frac{1}{2}\pi} \log \sin x \, dx = \int_{\frac{1}{2}\pi}^{0} -\log \sin(\frac{1}{2}\pi - u) \, du$

$= \int_0^{\frac{1}{2}\pi} \log \cos u \, du.$

$2I = \int_0^{\frac{1}{2}\pi} \log(\sin x \cos x) \, dx = \int_0^{\frac{1}{2}\pi} \log(\frac{1}{2} \sin 2x) \, dx$

$= \int_0^{\frac{1}{2}\pi} (-\log 2 + \log \sin 2x) \, dx$

$= -\frac{1}{2}\pi \log 2 + \int_0^{\pi} \frac{1}{2} \log \sin t \, dt$

$= -\frac{1}{2}\pi \log 2 + I,$

$I = -\frac{1}{2}\pi \log 2$

5

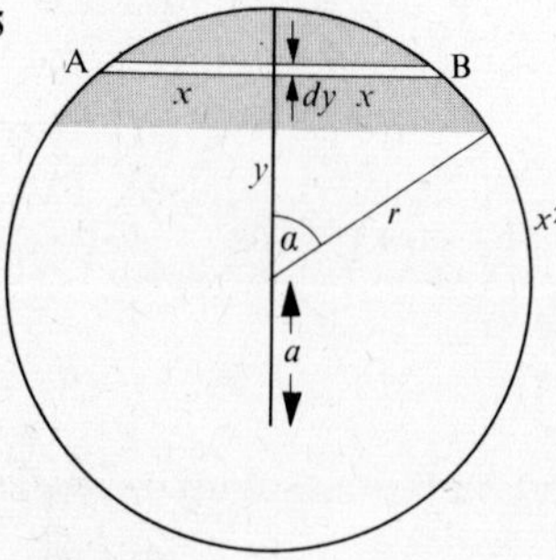

Figure 79

Volume of ring, $v = \int_{r\cos\alpha}^{r} 2\pi(a+y)2x \, dy$

$= 4\pi \int_{r\cos\alpha}^{r} (a+y)\sqrt{(r^2-y^2)} \, dy$

$= \frac{4}{3}\pi r^3 \sin^3\alpha + \pi r^2(2\alpha - \sin 2\alpha)a$

$= 4 \times \frac{4}{3}\pi r^3 \sin^3 \alpha.$

Thus $a = \dfrac{4r \sin^3\alpha}{2\alpha - \sin 2\alpha}$. 7.5

Surface area of ring is

$\int_{-\alpha}^{\alpha} 2\pi(a+y)r \, d\theta + 2\pi(a + r\cos\alpha)2r\sin\alpha$

$= 4\pi r a\alpha + 4\pi r^2 \sin\alpha + 4\pi a r \sin\alpha \cos\alpha + 4\pi r^2 \sin\alpha.$

Hence $\dfrac{\text{Surface area of ring}}{\text{Surface area of spheres}}$

$= \dfrac{4\pi\{ra(\alpha + \sin\alpha) + r^2(\sin\alpha\cos\alpha + \sin\alpha)\}}{4 \times 4\pi r^2 \sin^2\alpha}$

$= \dfrac{1}{4}\dfrac{\sin\alpha\cos\alpha + \sin\alpha}{\sin^2\alpha} + \dfrac{\sin\alpha(\alpha + \sin\alpha)}{2\alpha - \sin 2\alpha}$

(substituting for a using equation **7.5**)

$= \dfrac{1}{4}\cot\dfrac{1}{2}\alpha + \dfrac{\sin\alpha(\alpha+\sin\alpha)}{2\alpha - \sin 2\alpha}.$

6

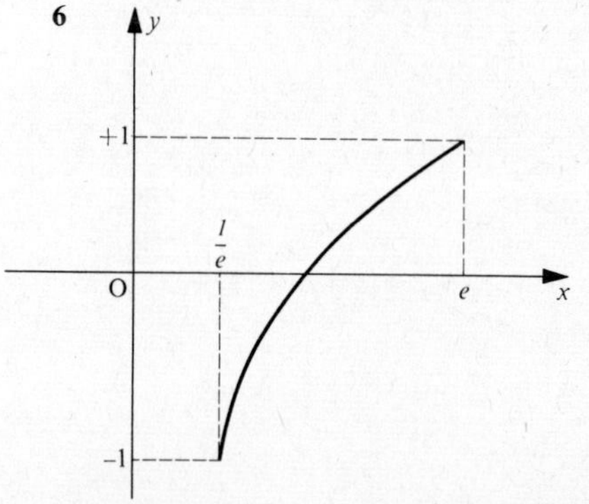

Figure 80

(a) Volume of wine, $v = \int \pi x^2 \, dy = \int_{e^{-1}}^{e} \pi x^2 \frac{1}{x} \, dx$

$= \frac{1}{2}\pi(e^2 - e^{-2}).$

(b) Centroid of full glass above x-axis,

$\bar{y} = \dfrac{1}{v}\int_{e^{-1}}^{e} y \, \pi x \, dx = \dfrac{\pi}{v}\int_{e^{-1}}^{e} x \log_e x \, dx,$

and on integration by parts:

$\bar{y} = \dfrac{e^4 + 3}{2(e^4 - 1)},$

hence answer is $\bar{y} + 1 = \dfrac{3e^4 + 1}{2(e^4 - 1)}.$

(c) Let (x', y') be point on glass surface when half full.

i.e. $\frac{1}{2}v = \frac{1}{4}\pi(e^2 - e^{-2}) = \int_{e^{-1}}^{x'} \pi x \, dx.$

Hence $x' = \dfrac{1}{e}\left[\dfrac{e^4+1}{2}\right]^{\frac{1}{2}}$

and $y' + 1 = \log_e x' + 1 = \dfrac{1}{2}\log_e \dfrac{e^4+1}{2}.$

7

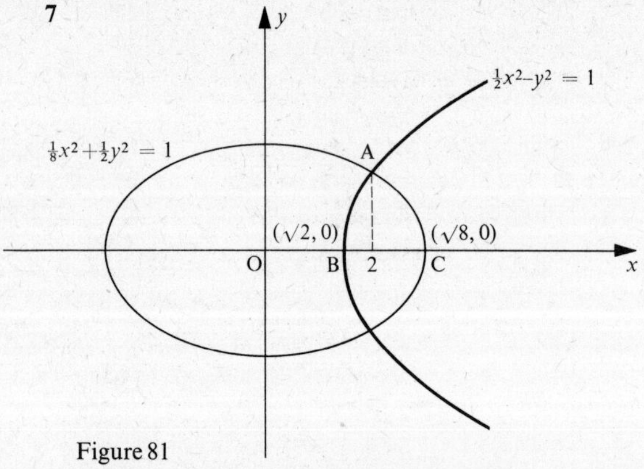

Figure 81

Volume generated by rotation of area ABC is

$$\int_{\sqrt2}^{2}\pi(\tfrac12x^2-1)\,dx+\int_{2}^{2\sqrt2}2\pi(1-\tfrac18x^2)\,dx=\tfrac23\pi(5\sqrt2-6).$$

8
$$\int_{-1}^{1}\frac{1}{(1-t^2)^{\frac12}}\frac{dt}{a-t}=\frac{1}{(a^2-1)^{\frac12}}\int_{-1}^{1}\frac{du}{(1-u^2)^{\frac12}}$$

$$=\frac{1}{(a^2-1)^{\frac12}}\left[\sin^{-1}u\right]_{-1}^{+1}=\frac{\pi}{(a^2-1)^{\frac12}}.$$

$$\int_{-1}^{1}(1-t^2)^{\frac12}\frac{dt}{a-t}=\int_{-1}^{1}\frac{1-t^2}{a-t}\frac{dt}{(1-t^2)^{\frac12}}$$

$$=\int_{-1}^{1}\frac{1}{(1-t^2)^{\frac12}}\left[t+a-\frac{a^2-1}{a-t}\right]dt$$

$$=-\left[(1-t^2)^{\frac12}\right]_{-1}^{1}+a\left[\sin^{-1}t\right]_{-1}^{1}$$

$$-(a^2-1)\frac{\pi}{(a^2-1)^{\frac12}}$$

$$=\pi\{a-(a^2-1)^{\frac12}\}.$$

Substitute $\cos\theta=t$ in above integral, then latter becomes

$$\int_{0}^{\pi}\frac{\sin^2\theta\,d\theta}{a-\cos\theta}$$

and putting given integral in similar form, i.e.

$$\int_{0}^{\pi}\frac{\sin^2\theta}{1-2x\cos\theta+x^2}\,d\theta=\frac{1}{2x}\int\frac{\sin^2\theta\,d\theta}{(1+x^2)/2x-\cos\theta}$$

$$=\frac{\pi}{2x}\left[\frac{1+x^2}{2x}-\left\{\left(\frac{1+x^2}{2x}\right)^2-1\right\}\right]=\frac12\pi.$$

9 The area is

$$\int_{0}^{\tan^{-1}\frac43}3\sin x\,dx+\int_{\tan^{-1}\frac43}^{\frac12\pi}4\cos x\,dx=2\text{ units.}$$

The volume of revolution is

$$\pi\int_{\tan^{-1}\frac43}^{\frac12\pi}(3\sin x)^2\,dx=\frac{9\pi}{2}\left\{\frac12\pi-\tan^{-1}\frac43+\frac{12}{25}\right\}$$

= 16·0 correct to two significant figures.

10

(i) Two integrations by parts are needed.

(ii) Use the substitution $u=\tan x$. The integral becomes

$$\int\frac{7+2u^2}{(9+4u^2)(1+u^2)}\,du.$$

Use partial fractions and show this to come to $x-\frac13\tan^{-1}(\frac23\tan x)+c$.

11 The given substitution transforms the first integral into

$$\int_{0}^{\frac14\pi}\cos2\theta\,d\theta=\tfrac12.$$

For the second integral, the same substitution yields

$$\int_{0}^{1}\frac{dx}{1+(1-x)^2}=\frac{\pi}{4}.$$

12

(i) Divide out and show the integral is $\frac12x^2-\frac12\log(1+x^2)+c$

(ii) The substitution $u=\sin x$ gives the result $2\sqrt{(\sin x)}+c$.

(iii) Substitute $a-x=\sin\theta$, and the integral becomes $-\frac12a^2\int(1+\cos2\theta)\,d\theta$, which equals

$$-\frac12\left[a^2\sin^{-1}\left(1-\frac{x}{a}\right)+(a-x)\sqrt{(2ax-x^2)}\right]+c.$$

In the last part it is probably preferable to use the substitution $y=1+\cos\theta$, and the integral becomes

$$\int\left[1-\frac{2}{y}\right]dy=1+\cos\theta-2\log(1+\cos\theta).$$

13

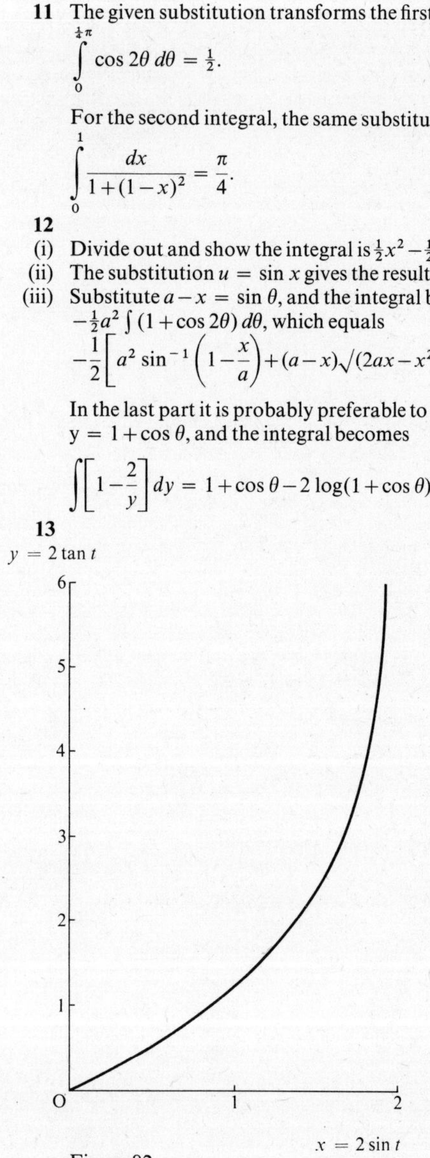

Figure 82

(i) Area $\displaystyle\int_0^{\sqrt{3}} y\,dx = \int_0^{\frac{1}{3}\pi} 2(\tan t)2\cos t\,dt = 4\int_0^{\frac{1}{3}\pi} \sin t\,dt = 2.$

(ii) Average ordinate is $\displaystyle\frac{1}{\sqrt{3}}\int_0^{\sqrt{3}} y\,dx = \frac{2}{\sqrt{3}} = \frac{2}{3}\sqrt{3}.$

14

(a)

(i) $\frac{8}{3}\log_e 2-\frac{7}{9}$

(ii) $\frac{38}{9}$

(b) Centroid at $(\frac{3}{8}, \frac{2}{5})$

16 Expressing $\displaystyle\frac{2x^3+5x^2+2x+4}{(x-1)^2(9x^2+4)} = \frac{A}{x-1}+\frac{B}{(x-1)^2}+\frac{Cx+D}{9x^2+4}$

and evaluating A, B, C, D we find $A = D = 0, B = 1, C = 2.$ Hence, integral is

$$\int\frac{1}{(x-1)^2}\,dx+\int\frac{2x}{9x^2+4}\,dx = \frac{-1}{x-1}+\frac{1}{9}\log(9x^2+4)+c.$$

Substitute $u = \sin\theta$,

then $$\int_{\sin^{-1}(3/4)}^{\sin^{-1}(3\sqrt{3}/4)} \frac{\cos\theta\,d\theta}{\sqrt{(5+4\cos^2\theta)}} = \int_{3/4}^{3\sqrt{3}/4}\frac{du}{\sqrt{(9-4u^2)}}$$

$$= \left[\frac{1}{2}\sin^{-1}\frac{2u}{3}\right]_{3/4}^{3\sqrt{3}/4} = \frac{\pi}{12}.$$

17 Area $\displaystyle= \int_{-\infty}^{\infty}\frac{dx}{2+x^2} = 2\int_0^{\infty}\frac{dx}{2+x^2}$

$$= 2\left[\frac{1}{\sqrt{2}}\tan^{-1}\frac{x}{\sqrt{2}}\right]_0^{\infty} = \frac{\pi}{\sqrt{2}}.$$

Volume of revolution $\displaystyle= \int_{\frac{1}{4}}^{\frac{1}{2}}\pi x^2\,dy$

$$= \pi\int_{\frac{1}{4}}^{\frac{1}{2}}\left[\frac{1}{y}-2\right]dy$$

$$= \pi\left[\log y-2y\right]_{\frac{1}{4}}^{\frac{1}{2}}$$

$$= \pi(\log 2-\tfrac{1}{2}).$$

18

(a) On integrating by parts,

$$\int\log\left[1+\frac{a^2}{x^2}\right]\times 1\,dx = x\log\left[1+\frac{a^2}{x^2}\right]-\int\frac{x}{1+a^2/x^2}\left[\frac{-2a^2}{x^3}\right]dx$$

$$= x\log\left[1+\frac{a^2}{x^2}\right]+2a^2\int\frac{dx}{x^2+a^2}$$

$$= x\log\left[1+\frac{a^2}{x^2}\right]+2a\tan^{-1}\frac{x}{a}+c.$$

(b) $$\int\sqrt{(x^2+x+1)}\,dx = \int\sqrt{\left\{\left(x+\frac{1}{2}\right)^2+\left(\frac{\sqrt{3}}{2}\right)^2\right\}}\,dx$$

$$= \frac{3}{4}\int\cosh^2\theta\,d\theta$$

$$\left(\text{on substituting } x+\frac{1}{2} = \frac{\sqrt{3}}{2}\sinh\theta\right);$$

$$= \frac{3}{8}\int(1+\cosh 2\theta)\,d\theta$$

$$= \frac{3}{8}\left[\theta+\frac{1}{2}\sinh 2\theta\right]+c$$

$$= \frac{3}{8}\sinh^{-1}\frac{2x+1}{\sqrt{3}}+ \;+\frac{3}{8}\,\frac{2x+1}{\sqrt{3}}\sqrt{\left[1+\left(\frac{2x+1}{\sqrt{3}}\right)^2\right]}+c$$

$$= \frac{3}{8}\sinh^{-1}\frac{2x+1}{\sqrt{3}}+\frac{2x+1}{4}\sqrt{(x^2+x+1)}+c.$$

J Mixed integration and differentiation: Answers and hints

1 $f(x) = +0{\cdot}101$ (maximum at $x = 5-\sqrt{6}$).

$f(x) = +9{\cdot}90$ (minimum at $x = 5+\sqrt{6}$).

Area $\displaystyle= \int_2^3 f(x)\,dx = \int_2^3\left[x+\frac{6}{x-5}\right]dx$

$$= \left[\tfrac{1}{2}x^2+6\log(x-5)\right]_2^3 = 0{\cdot}067.$$

2

(ii) Substitute $x = \operatorname{sech} y$ and integrate by parts,

i.e. $$\int\operatorname{sech}^{-1}x\,dx = \int y\frac{d}{dy}\operatorname{sech} y\,dy = y\operatorname{sech} y-\int\operatorname{sech} y\,dy.$$

To evaluate latter integral use

$$\operatorname{sech} y = \frac{2}{e^y+e^{-y}}$$

and substitution $u = e^y$.

3 Equation of tangent is $y\cos\theta+x\sin\theta = a\sin\theta\cos\theta$;

$AB = a.$

By symmetry $\bar{y} = 0$,

$$\bar{x} = \frac{\displaystyle\int_0^{\frac{1}{2}\pi}3a^4\sin^7\theta\cos^5\theta\,d\theta}{\displaystyle\int_0^{\frac{1}{2}\pi}3a^3\sin^7\theta\cos^2\theta\,d\theta} = \frac{21}{128}a.$$

4

(i) Now $\dfrac{dy}{dx} = \dfrac{dy}{du}\dfrac{du}{dx} = (u+2)\dfrac{dy}{du}$,

$$\frac{d^2y}{dx^2} = \frac{d}{dx}\left[(u+2)\frac{dy}{du}\right] = \frac{dy}{dx} + (u+2)\frac{d^2y}{du^2}\frac{du}{dx}$$

$$= \frac{dy}{dx} + (u+2)^2\frac{d^2y}{du^2}$$

$$= \frac{dy}{dx} + e^{2x}.y\,e^{-2x},$$

hence $\dfrac{d^2y}{dx^2} - \dfrac{dy}{dx} - y = 0.$

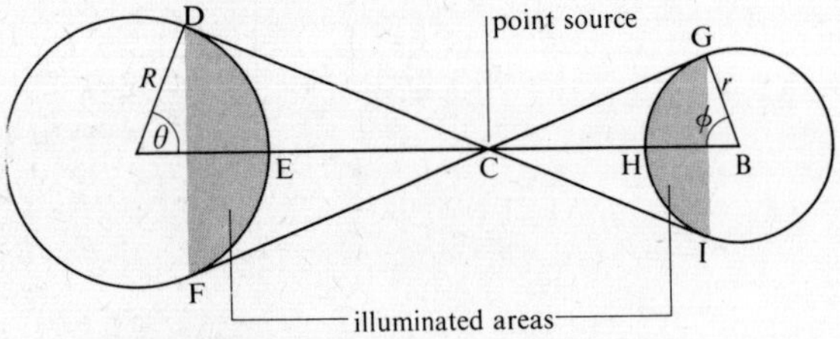

Figure 83

(ii) Area of spheres illuminated,

$$S = \int_0^{\theta} 2\pi R(\sin\theta)\,R\,d\theta + \int_0^{\phi} 2\pi r(\sin\phi)\,r\,d\phi$$

$$= 2\pi R^2(1-\cos\theta) + 2\pi r^2(1-\cos\phi),$$

where $\cos\theta = \dfrac{R}{\text{AC}} \equiv \dfrac{R}{x}$, $\quad \cos\phi = \dfrac{r}{\text{CB}} \equiv \dfrac{r}{\text{AB}-x}$.

$\dfrac{dS}{dx} = 0, \quad \dfrac{d^2S}{dx^2} < 0,$

when $\dfrac{\text{AC}}{\text{CB}} = \left(\dfrac{R}{r}\right)^{\frac{3}{2}}$ (i.e. maximum S condition).

5

(i) Let $y = \cos^{-1}\dfrac{a\cos x + b}{a + b\cos x}$

and show that $\dfrac{dy}{dx} = \dfrac{\sqrt{(a^2-b^2)}}{a+b\cos x}$.

Minimum dy/dx corresponds to $\cos x = 1$, maximum dy/dx occurs when $\cos x = -1$, hence result.

(ii) Expanding integral into partial fractions, we obtain

$$-\int_{-1}^{1}\frac{dx}{2-x} + \int_{-1}^{1}\frac{3}{(2-x)^2}\,dx - \int_{-1}^{1}\frac{x+1}{x^2+3}\,dx = -\log 3 + 2 - \frac{1}{\sqrt{3}}\frac{\pi}{3}.$$

6 On subtracting the curve equations we obtain $xy(y-x) = a^2(y-x)$, so curves intersect when $y = x$ or $xy = a^2$. On substituting $y = x$ into first curve equation we obtain $x^3 + a^2x - 2a^3 = 0$, which has one real root of $x = a$. Therefore, the curves have a single common point (a, a). Also $dy/dx = -1$ at (a, a) for both curves, hence curves must touch at this point.

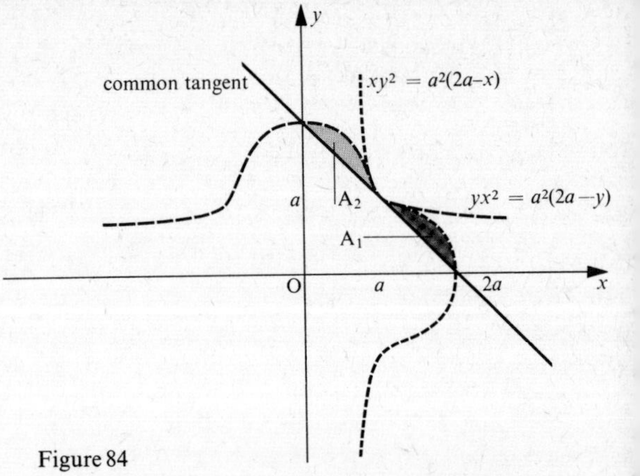

Figure 84

Area enclosed by first curve and common tangent is

$$A_1 = \int_a^{2a} a\sqrt{\left[\frac{2a-x}{x}\right]}\,dx - \tfrac{1}{2}a^2 \text{ (use substitution } x = 2a\sin^2\theta).$$

Area enclosed by second curve and common tangent is

$$A_2 = \int_0^a \frac{2a^3}{x^2+a^2}\,dx - (a^2 + \tfrac{1}{2}a^2) = 2a^2\left[\tan^{-1}\frac{x}{a}\right]_0^a - \frac{3}{2}a^2$$

$$= \tfrac{1}{2}(\pi-3)a^2 = A_1.$$

7

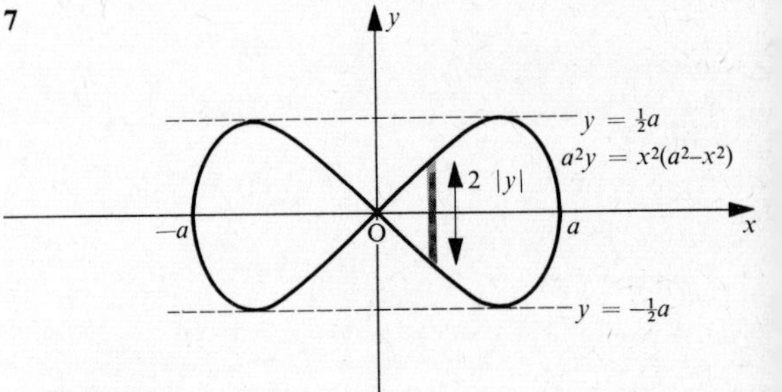

Figure 85

$\dfrac{dy}{dx} = 0$ when $x = \pm\dfrac{1}{\sqrt{2}}a$

and in which case $y = \pm\frac{1}{2}a$, hence the latter are tangents touching the curve at

$$\left(\pm\frac{1}{\sqrt{2}}a, +\frac{1}{2}a\right), \left(\pm\frac{1}{\sqrt{2}}a, \frac{1}{2}a\right) \text{ respectively.}$$

Area enclosed by one loop $= 2\displaystyle\int_0^a y\,dx = \tfrac{2}{3}a^2$.

8 At points of inflexion $d^2y/dx^2 = 0$. This condition gives $x = \pm m\pi$ $(m = 0, 1, 2, 3, \ldots)$ and $y = x$. Also d^2y/dx^2 changes sign as x increases through $\pm m\pi$. Thus points of inflexion occur at equally spaced intervals of π along $y = x$.

$$\frac{dy}{dx} = 0 \quad \text{when} \quad x = 2\pi, 4\pi, \ldots;$$
$$\frac{dy}{dx} = 4 \quad \text{when} \quad x = \pi, 3\pi, \ldots;$$

hence tangents at points of inflexion are parallel to $y = 0$ or $y = 4x$.

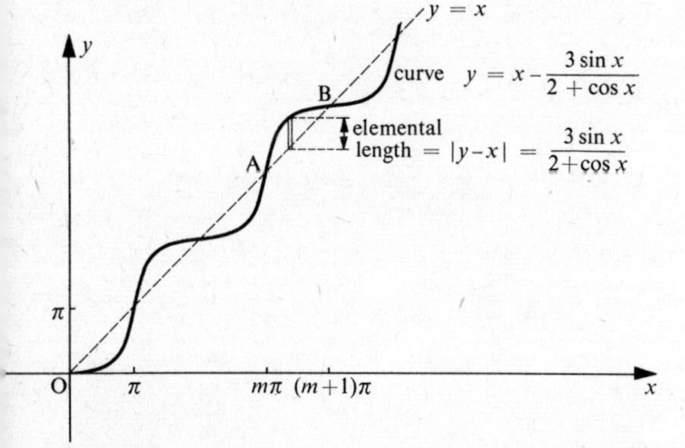

Figure 86

Area between curve and $y = x$ from A ($x = m\pi$) to B ($x = (m+1)\pi$) is

$$\int_{m\pi}^{(m+1)\pi} (y-x)\,dx = \int_{m\pi}^{(m+1)\pi} \left[\frac{-3\sin x}{2+\cos x}\right] dx = 3\log_e 3 = \log_e 27.$$

9 $$\frac{d}{dx}\{\log(y+z)\} = \frac{1}{y+z}\frac{1}{2}\left[\frac{1}{y}+\frac{1}{z}\right]2(px+q) = \frac{px+q}{yz}; \qquad \textbf{7.6}$$
$$\frac{d}{dx}\{\log(y-z)\} = \frac{-(px+q)}{yz}.$$

(i) Substitute $p = 1, q = 0, r = 1, s = 4$ in y and z and use equation **7.6**.

(ii) $$\int_0^1 \frac{dx}{\sqrt{\{(x+4)(x+9)\}}} = \frac{3+2\sqrt{2}}{5}$$

(put $p = 0, q = \frac{1}{2}, r = 4, s = 9$).

(iii) $$\int \frac{dx}{\sqrt{(x^2+a^2)}} = \log\{x+\sqrt{(x^2+a^2)}\}$$

(put $p = 1, q = 0, r = 0, s = a^2$).

10

(i) $$\log_e \cos x = -\frac{x^2}{2} - \frac{x^4}{12} - \ldots.$$

The evenness of the cosine function indicates that no odd powers of x are present.

(ii)

(a) $$\int_0^1 x\,e^{-\frac{1}{2}x^2}\,dx = \int_0^{\frac{1}{2}} e^{-y}\,dy = 1 - e^{-\frac{1}{2}}.$$

(b) $$\int_{\frac{1}{2}\pi}^{\pi} \cos^3 x\,dx = -\tfrac{2}{3}.$$

11

(a) $$\frac{d}{dx}e^{x^2} = 2x\,e^{x^2},$$

$$\int x\,e^{x^2}\,dx = \tfrac{1}{2}x\,e^{x^2},$$

$$\int_0^1 x^3 e^{x^2}\,dx = \int_0^1 x^2 . x\,e^{x^2}\,dx$$

$$= \left[x^2 \tfrac{1}{2} e^{x^2}\right]_0^1 - \int_0^1 2x\tfrac{1}{2}e^{x^2}\,dx$$

$$= \tfrac{1}{2}e - \left[\tfrac{1}{2}e^{x^2}\right]_0^1 = \tfrac{1}{2}.$$

(b) $$\int_0^{\frac{1}{2}} \sqrt{\frac{x}{1-x}}\,dx = \int_0^{\frac{1}{4}\pi} \frac{\sin\theta}{\cos\theta} 2\sin\theta\cos\theta\,d\theta = \int_0^{\frac{1}{4}\pi} 2\sin^2\theta\,d\theta$$

$$= \int_0^{\frac{1}{4}\pi} (1-\cos 2\theta)\,d\theta = \tfrac{1}{4}\pi - \tfrac{1}{2}$$

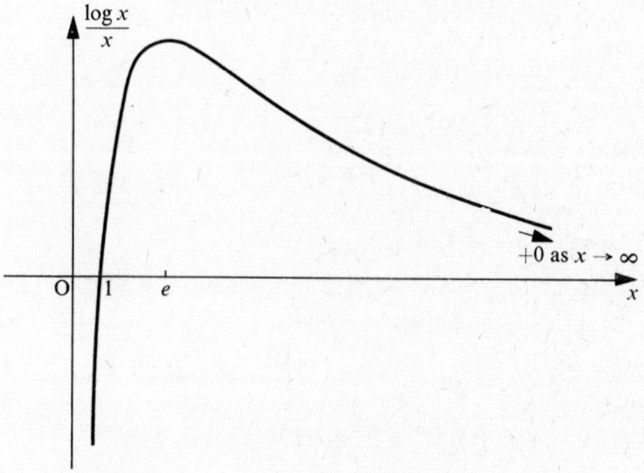

Figure 87

12 $$\frac{dy}{dx} = \frac{x(1/x) - \log_e x}{x^2} = 0 \quad \text{when} \quad x = e, y = \frac{1}{e},$$

i.e. turning point at $(e, 1/e)$. This point is a *maximum*, since dy/dx changes from positive to negative as x increases through e.

Area $\triangle$OQN $= \dfrac{1}{2}\sqrt{e}\dfrac{1}{2\sqrt{e}} = \dfrac{1}{4}$;

$$\int_1^e \frac{\log_e x}{x}\,dx = \int_0^1 \log_e x\,d(\log_e x) = \left[\tfrac{1}{2}(\log_e x)^2\right]_0^1 = \tfrac{1}{2}.$$

13

(a) $$\frac{d}{dx}\sinh x = \cosh x.$$

(b) Area $= \frac{1}{2}\cosh t \sinh t - \displaystyle\int_0^t \sinh^2 t\,dt = \frac{1}{2}t.$

Chapter Eight
Further Calculus Topics

A Further differentiation: Theory summary

1 *Power series expansions*

(a) *Maclaurin's theorem.*

$$f(x) = f(0)+f'(0)x+f''(0)\frac{x^2}{2!}+\ldots+f^{(r)}(0)\frac{x^r}{r!}+\ldots.$$

(b) *Taylor's theorem.*

$$f(x+a) = f(a)+f'(a)x+f''(a)\frac{x^2}{2!}+\ldots+f^{(r)}(a)\frac{x^r}{r!}+\ldots.$$

(c) *Finite series forms.*

$$f(x) = f(0)+f'(0)x+\ldots+f^{(n-1)}(0)\frac{x^{n-1}}{(n-1)!}+f^{(n)}(\theta x)\frac{x^n}{n!},$$

$$f(x+a) = f(a)+f'(a)x+\ldots+$$
$$+f^{(n-1)}(a)\frac{x^{n-1}}{(n-1)!}+f^{(n)}(a+\theta x)\frac{x^n}{n!},$$

where $0 < \theta < 1$.

2 *Leibnitz's theorem*
(To find nth derivative of a product)

$$\frac{d^n}{dx^n}(uv) = uv^{(n)}+nu^{(1)}v^{(n-1)}+\frac{n(n-1)}{2!}u^{(2)}v^{(n-2)}+\ldots+$$
$$+{}_nC_r u^{(r)}v^{(n-r)}+\ldots+nu^{(n-1)}v^{(1)}+u^{(n)}v.$$

3 *L'Hopital's rule.*
If $f(a) = g(a) = 0, f'(a) = g'(a) = 0, \ldots$
$f^{(n-1)}(a) = g^{(n-1)}(a) = 0$, but $g^{(n)}(a) \neq 0$ and $f^{(n)}(x), g^{(n)}(x)$ are continuous, then

$$\lim_{x\to a}\frac{f(x)}{g(x)} = \frac{f^{(n)}(a)}{g^{(n)}(a)}.$$

4 *Partial differentiation*

(a) If we have a function $f = f(x, y)$ of two variables x, y, then the partial derivatives with respect to x and y are written as

$$\frac{\partial f}{\partial x} \quad \text{and} \quad \frac{\partial f}{\partial y},$$

where for $\partial f/\partial x$ we differentiate f as a function of x regarding y as constant, and for $\partial f/\partial y$ we differentiate f with respect to y regarding x as constant.

(b) $$\frac{\partial^2 f}{\partial x^2} = \frac{\partial}{\partial x}\left[\frac{\partial f}{\partial x}\right], \quad \frac{\partial^2 f}{\partial y^2} = \frac{\partial}{\partial y}\left[\frac{\partial f}{\partial y}\right], \quad \frac{\partial}{\partial x}\left[\frac{\partial f}{\partial y}\right] = \frac{\partial^2 f}{\partial x\,\partial y}.$$

In most cases $$\frac{\partial^2 f}{\partial x\,\partial y} = \frac{\partial^2 f}{\partial y\,\partial x}.$$

(c) *Total differential.* If $f = f(x, y)$,

$$df = \frac{\partial f}{\partial x}dx+\frac{\partial f}{\partial y}dy.$$

(d) *Increment formula.* If $f = f(x, y)$ and small changes $\delta x, \delta y$ occur in x and y,

$$\delta f \approx \frac{\partial f}{\partial x}\delta x+\frac{\partial f}{\partial y}\delta x.$$

(e) If $f = f(x, y)$,

$$\frac{df}{dx} = \frac{\partial f}{\partial x}+\frac{\partial f}{\partial y}\frac{dy}{dx} \quad \text{where } y = y(x);$$

and if $x = x(t), y = y(t)$,

$$\frac{df}{dt} = \frac{\partial f}{\partial x}\frac{dx}{dt}+\frac{\partial f}{\partial y}\frac{dy}{dt}.$$

(f) If $f = f(u, v)$ and $u = u(x, y), v = v(x, y)$,

$$\frac{\partial f}{\partial x} = \frac{\partial f}{\partial u}\frac{\partial u}{\partial x}+\frac{\partial f}{\partial v}\frac{\partial v}{\partial x},$$

$$\frac{\partial f}{\partial y} = \frac{\partial f}{\partial u}\frac{\partial u}{\partial y}+\frac{\partial f}{\partial v}\frac{\partial v}{\partial y}.$$

B Further differentiation: Illustrative worked problems

1

(a) Evaluate

(i) $\displaystyle\lim_{x\to 0}\frac{e^x-1}{x}$.

(ii) $\displaystyle\lim_{x\to 1}\frac{\log_e x}{\sin 2\pi x}$.

(b) By using Maclaurin's theorem, show that the expansion of $y = \log_e(1+\sin x)$ in ascending powers of x starts with the terms

$$x-\tfrac{1}{2}x^2+\tfrac{1}{6}x^3-\tfrac{1}{12}x^4.$$

(c) Use Taylor's series to expand $\log_e(a+x)$ in ascending powers of x, ($a > 0$ and $|x| < a$).

Solution.

(a)

(i) $f(x) = e^x-1, \quad g(x) = x;$
$f'(x) = e^x, \quad g'(x) = 1.$

Using L'Hopital's rule,

$$\lim_{x\to 0}\frac{e^x-1}{x} = \frac{f'(0)}{g'(0)} = \left[\frac{e^x}{1}\right]_{x=0} = 1.$$

(ii) $f(x) = \log_e x, \qquad g(x) = \sin 2\pi x;$

$f'(x) = \frac{1}{x}, \qquad g'(x) = 2\pi \cos 2\pi x.$

$$\lim_{x\to 1} \frac{\log_e x}{\sin 2\pi x} = \frac{f'(1)}{g'(1)} = \left[\frac{1/x}{2\pi \cos 2\pi x}\right]_{x=1} = \frac{1}{2\pi}.$$

(b) $y = \log_e(1+\sin x), \qquad y(0) = \log_e 1 = 0;$

$y' = \frac{\cos x}{1+\sin x}, \qquad y'(0) = 1;$

$y'' = \frac{-1}{1+\sin x}, \qquad y''(0) = -1;$

$y^{(3)} = \frac{\cos x}{(1+\sin x)^2}, \qquad y^{(4)}(0) = 1;$

$$y^{(4)} = \frac{(1+\sin x)^2(-\sin x)-2(1+\sin x)\cos^2 x}{(1+\sin x)^4},$$

$y^{(4)}(0) = -2.$

$$y = y(0)+y'(0)x+y''(0)\frac{x^2}{2!}+y^{(3)}(0)\frac{x^3}{3!}+y^{(4)}(0)\frac{x^4}{4!}+\ldots$$

$$= 0+x-\frac{1}{2!}x^2+\frac{1}{3!}x^3-\frac{2}{4!}x^4+\ldots$$

$$= x-\tfrac{1}{2}x^2+\tfrac{1}{6}x^3-\tfrac{1}{12}x^4+\ldots.$$

(c) Let $f(x) = \log_e x$,

then $f'(x) = \frac{1}{x}, \qquad f''(x) = \frac{-1}{x^2},$

$$f^{(r)}(x) = (-1)^{r-1}\frac{1}{r}\frac{1}{x^r},$$

and thus

$$\log_e(a+x) = f(a+x) = f(a)+f'(a)x+f''(a)\frac{x^2}{2!}+\ldots+$$

$$+f^{(r)}(a)\frac{x^r}{r!}+\ldots$$

$$= \log_e a+\frac{1}{a}x-\frac{1}{2a^2}x^2+\frac{1}{3a^3}x^3+\ldots+$$

$$+(-1)^{r-1}\frac{1}{ra^r}x^r+\ldots.$$

2 If $y = \cosh p\theta$ and $x = \sin\theta$, where p is a real constant, prove that

$$(1-x^2)\frac{d^2y}{dx^2}-x\frac{dy}{dx}-p^2y = 0.$$

By differentiating this equation n times with respect to x prove that, when $x = 0$,

$$\frac{d^{n+2}y}{dx^{n+2}} = (n^2+p^2)\frac{d^ny}{dx^n} \quad (n = 0, 1, 2, \ldots).$$

Hence obtain the values of

$$\frac{d^{2n}y}{dx^{2n}} \text{ and } \frac{d^{2n+1}y}{dx^{2n+1}} \text{ when } \theta = 0.$$

[L, J 1966, FM V, Q 6]

Solution. If $y = \cosh p\theta$,

then $\frac{dy}{d\theta} = p\sinh p\theta;$

and if $x = \sin\theta$,

$$\frac{dx}{d\theta} = \cos\theta \quad\text{and}\quad \frac{d\theta}{dx} = \frac{1}{\cos\theta}.$$

Now $\frac{dy}{dx} = \frac{dy}{d\theta}\frac{d\theta}{dx} = \frac{p\sinh\theta}{\cos\theta},$

$$\frac{d^2y}{dx^2} = \frac{d}{d\theta}\left[\frac{p\sinh\theta}{\cos\theta}\right]\frac{d\theta}{dx}$$

$$= \frac{p}{\cos^3\theta}(p\cosh p\theta\cos\theta+\sinh p\theta\sin\theta).$$

Thus $(1-x^2)\frac{d^2y}{dx^2} = \cos^2\theta\frac{d^2y}{dx^2}$

$$= \sin\theta\frac{p\sinh\theta}{\cos\theta}+p^2\cosh p\theta$$

$$\equiv x\frac{dy}{dx}+p^2y,$$

hence $(1-x^2)\frac{d^2y}{dx^2}-x\frac{dy}{dx}-p^2y = 0.$

On using Leibnitz's theorem,

$$\frac{d^n}{dx^n}\left[(1-x^2)\frac{d^2y}{dx^2}\right] = (1-x^2)y^{(n-2)}-2nxy^{(n+1)}-$$

$$-2\frac{n(n-1)}{2!}y^{(n)},$$

$$\frac{d^n}{dx^n}\left[x\frac{dy}{dx}\right] = xy^{(n+1)}+ny^{(n)},$$

$$\frac{d^n}{dx^n}\left[(1-x^2)\frac{d^2y}{dx^2}-x\frac{dy}{dx}-p^2y\right]_{x=0} = y^{(n+2)}-n^2y^{(n)}-p^2y^{(n)}$$

$$= 0,$$

i.e. $y^{(n+2)} = (n^2+p^2)y^{(n)}$ when $x = 0$ (or $\theta = 0$).

Now when $\theta = 0$, $y = 1$, $y' = 0$, $y'' = p^2$ and, on applying the recurrence relation just found,

$y^{(3)} = (1^2+p^2)y' = 0, \quad y^{(4)} = (2^2+p^2)y'' = (2^2+p^2)p^2$, etc.

Clearly all odd-order derivatives are zero, thus $y^{(2n+1)} = 0$. Also, if $2n-2$ replaces n in the recurrence relation,

$$y^{(2n)} = \{(2n-2)^2+p^2\}\ldots(4^2+p^2)(2^2+p^2)p^2.$$

3 Given that $z = f(u, v)$, where $u = x^2-y^2$ and $v = 2xy$, prove that

$$\frac{\partial z}{\partial x} = 2x\frac{\partial z}{\partial u}+2y\frac{\partial z}{\partial v} \quad\text{and}\quad \frac{\partial z}{\partial y} = -2y\frac{\partial z}{\partial u}+2x\frac{\partial z}{\partial v}.$$

Show that the equation

$$y\frac{\partial z}{\partial x}+x\frac{\partial z}{\partial y} = (x^2+y^2)z$$

can be written as $\partial f/\partial v = \tfrac{1}{2}f$.

Show that the solution of the equation is

$$z = e^{xy}F(x^2-y^2),$$

where $F(u)$ is an arbitrary function of u.

[O & C, S 1967, M & HM V S, Q 12]

Solution. Now $\dfrac{\partial z}{\partial x} = \dfrac{\partial z}{\partial u}\dfrac{\partial u}{\partial x} + \dfrac{\partial z}{\partial v}\dfrac{\partial v}{\partial x}$,

but $\dfrac{\partial u}{\partial x} = \dfrac{\partial}{\partial x}(x^2-y^2) = 2x$ and $\dfrac{\partial v}{\partial x} = \dfrac{\partial}{\partial x}(2xy) = 2y$,

therefore $\dfrac{\partial z}{\partial x} = 2x\dfrac{\partial z}{\partial u} + 2y\dfrac{\partial z}{\partial v}$. **8.1**

Similarly, $\dfrac{\partial z}{\partial y} = \dfrac{\partial z}{\partial u}\dfrac{\partial u}{\partial y} + \dfrac{\partial z}{\partial v}\dfrac{\partial v}{\partial y} = -2y\dfrac{\partial z}{\partial u} + 2x\dfrac{\partial z}{\partial v}$. **8.2**

Using equations **8.1** and **8.2**, the given equation can be written as

$$2y\left[x\frac{\partial z}{\partial u} + y\frac{\partial z}{\partial v}\right] + 2x\left[-y\frac{\partial z}{\partial u} + x\frac{\partial z}{\partial v}\right] = (x^2+y^2)z,$$

hence $2(x^2+y^2)\dfrac{\partial z}{\partial v} = (x^2+y^2)z$,

i.e. $\dfrac{\partial z}{\partial v} = \dfrac{1}{2}z$ or $\dfrac{\partial f}{\partial v} = \dfrac{1}{2}f$. **8.3**

On integrating equation **8.3** we obtain

$\log_e z = \frac{1}{2}v + C(u)$

and letting the arbitrary function $C(u) = \log_e F(u)$,

$\log_e \dfrac{z}{F(u)} = \dfrac{1}{2}v$,

i.e. $z = F(u)\,e^{\frac{1}{2}v} = e^{xy}F(x^2-y^2)$.

C Further differentiation: Problems

Answers and hints will be found on pp. 113–14.

1 If $y = e^x \sin x$, prove, by induction or otherwise, that

$$\frac{d^n y}{dx^n} = 2^{\frac{1}{2}n}e^x \sin(x + \tfrac{1}{4}n\pi).$$

Hence, or otherwise, expand $e^x \sin x$ in powers of x as far as the term in x^3. Express in their simplest forms the coefficients of x^{4r} and x^{4r+1} in the expansion. [C, S 1967, PM III, Q 8]

2 If $y = (\sec x + \tan x)^p$, prove that

$$\cos x\frac{dy}{dx} = py.$$

By successive differentiations of this equation, obtain the values, when $x = 0$, of

$$y, \frac{dy}{dx}, \frac{d^2y}{dx^2}, \frac{d^3y}{dx^3}.$$

Hence obtain the first four terms in the expansion of $(\sec x + \tan x)^{\frac{1}{2}}$ in powers of x. [C, S 1966, PM III, Q 8]

3

(i) Evaluate $\displaystyle\lim_{x\to 0}\frac{\cos\log(1+x) - \cos x}{x^2(e^x-1)}$.

(ii) If r, n are positive integers and $r \leqslant n$, show that

$$D^r(x+1)^n = \frac{n!}{(n-r)!}(x+1)^{n-r},$$

where D denotes d/dx.

Using Leibnitz's theorem, or otherwise, show that when $x = 0$

$$D^n(x^2-1)^n = n!\sum_{r=0}^{n}(-1)^r({}_nC_r)^2 = (-1)^{\frac{1}{2}n}\left[\frac{n!}{(\frac{1}{2}n)!}\right]^2,$$

if n is even, and obtain the result when n is odd. [L, S 1966, FM V, Q 3]

4

(i) If $V = e^{x+y}\cos(x-y)$ and $u = \partial V/\partial x$, $v = \partial V/\partial y$, prove that

$$u + v = 2V,$$

$$\frac{\partial u}{\partial x} + \frac{\partial v}{\partial y} = 0.$$

(ii) Given that $x = \frac{1}{2}(t - t^{-1})$, $y = \frac{1}{2}(t^4 - t^{-4})$, where t is positive, prove that

(a) $\sqrt{(1+x^2)}\dfrac{dy}{dx} = 2(t^4 + t^{-4})$,

(b) $(1+x^2)\dfrac{d^2y}{dx^2} + x\dfrac{dy}{dx} - 16y = 0$.

[O & C, S 1967, M (Sc & Stat) III S, Q 5]

5 If ϕ is a function of u and v, and

$$u = x^3 - 3xy^2, \qquad v = 3x^2y - y^3,$$

prove that

$$\frac{\partial \phi}{\partial x} = 3(x^2-y^2)\frac{\partial \phi}{\partial u} + 6xy\frac{\partial \phi}{\partial v},$$

$$\frac{\partial \phi}{\partial y} = -6xy\frac{\partial \phi}{\partial u} + 3(x^2-y^2)\frac{\partial \phi}{\partial v}$$

and that $\dfrac{\partial^2\phi}{\partial x^2} + \dfrac{\partial^2\phi}{\partial y^2} = 9(x^2+y^2)^2\left[\dfrac{\partial^2\phi}{\partial u^2} + \dfrac{\partial^2\phi}{\partial v^2}\right]$.

[O & C, S 1966, M & HM V S, Q 11]

D Further integration: Theory summary

1 *Integration by successive reduction*

(a) *Based on the integration by parts formula.*

$$\int u\frac{dv}{dx}\,dx = uv - \int v\frac{du}{dx}\,dx.$$

Example. If $I_n = \displaystyle\int x^n e^{ax}\,dx$,

$$I_n = \frac{1}{a}x^n e^{ax} - \frac{n}{a}I_{n-1}.$$

(b) *Reduction formulae for trignometrical integrals.*

$$\int_0^{\frac{1}{2}\pi} \cos^m\theta\, d\theta = \frac{m-1}{m}\int_0^{\frac{1}{2}\pi} \cos^{m-2}\theta\, d\theta \quad (m \geqslant 2).$$

$$\int_0^{\frac{1}{2}\pi} \sin^n\theta\, d\theta = \frac{n-1}{n}\int_0^{\frac{1}{2}\pi} \sin^{n-2}\theta\, d\theta \quad (n \geqslant 2).$$

$$\int_0^{\frac{1}{2}\pi} \cos^m\theta \sin^n\theta\, d\theta = \frac{m-1}{m+n}\int_0^{\frac{1}{2}\pi} \cos^{m-2}\theta \sin^n\theta\, d\theta \quad (m \geqslant 2, n \geqslant 0),$$

$$= \frac{n-1}{m+n}\int_0^{\frac{1}{2}\pi} \cos^m\theta \sin^{n-2}\theta\, d\theta \quad (m \geqslant 0, n \geqslant 2),$$

$$= \frac{(m-1)(m-3)\ldots(n-1)(n-3)\ldots}{(m+n)(m+n-2)(m+n-4)\ldots} \times \ldots \times$$

$$\times \begin{cases} \frac{1}{2}\pi & \text{(if } m, n \text{ both even)} \\ 1 & \text{(otherwise).} \end{cases}$$

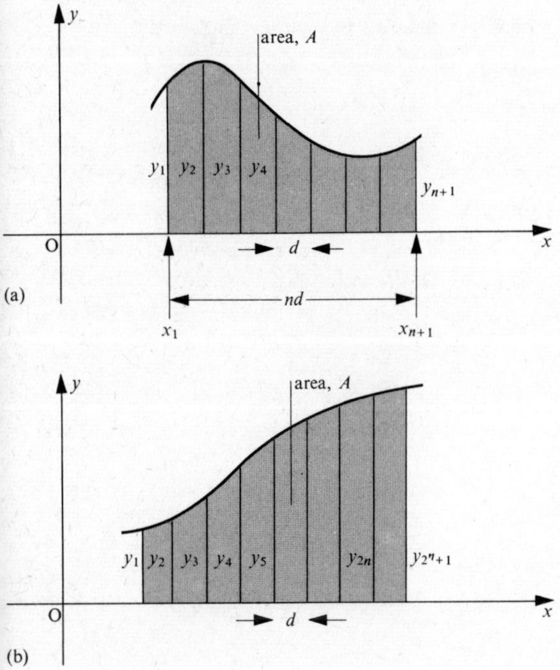

Figure 88

2 *Approximate integration*

(a) *Trapezium method.* Divide the required area into any number, say n, strips by $n-1$ equidistant ordinates (as shown in Figure 88a), then the area A is given approximately by

$$A \approx \tfrac{1}{2}d(y_1+2y_2+2y_3+\ldots+2y_n+y_{n+1}),$$

where $d = (x_{n+1}-x_1)/n$ is the width of the strips.

(b) *Simpson's rule.* Divide the required area into an even number, say $2n$, strips by $2n-1$ equidistant ordinates (see Figure 88b), then

$$A \approx \tfrac{1}{3}d\{y_1+y_{2n+1}+2(y_3+y_5+\ldots+y_{2n-1})+ +4(y_2+y_4+\ldots+y_{2n})\},$$

where d = width of strip.

3 *Inequalities associated with integrals*

(a) If F_{max} and F_{min} are the greatest and least values of $f(x)$ in the range $a \leqslant x \leqslant b$, then

$$F_{\text{min}}(b-a) \leqslant \int_a^b f(x)\, dx \leqslant F_{\text{max}}(b-a).$$

(b) If $0 < f(x) < g(x)$ for $a \leqslant x \leqslant b$, then

$$0 < \int_a^b f(x)\, dx < \int_a^b g(x)\, dx.$$

E Further integration: Illustrative worked problems

1 If $I_n = \displaystyle\int \frac{x^n\, dx}{\sqrt{(x^2+2x+2)}} \quad (n \geqslant 0),$

obtain a reduction formula in the form

$$aI_n + bI_{n-1} + cI_{n-2} = x^{n-1}\sqrt{(x^2+2x+2)} + \text{a constant},$$

finding a, b and c in terms of n.
Hence prove that

$$\int_0^1 \frac{x^2\, dx}{\sqrt{(x^2+2x+2)}} = \frac{1}{2}\left[3\sqrt{2}-2\sqrt{5}+\log\frac{2+\sqrt{5}}{1+\sqrt{2}}\right].$$

[L, S 1968, FM V, Q 8]

Solution. Now $\displaystyle I_n = \int \frac{x^{n-1}(x+1)-x^{n-1}}{\sqrt{(x^2+2x+2)}}\, dx$

$$= \int \frac{x^{n-1}(x+1)}{\sqrt{(x^2+2x+2)}}\, dx - I_{n-1}, \qquad \textbf{8.4}$$

but, on integrating by parts,

$$\int \frac{x^{n-1}(x+1)}{\sqrt{(x^2+2x+2)}}\, dx$$

$$= x^{n-1}\sqrt{(x^2+2x+2)}-(n-1)\int x^{n-2}\sqrt{(x^2+2x+2)}\, dx$$

$$= x^{n-1}\sqrt{(x^2+2x+2)}-(n-1)\int \frac{x^{n-2}(x^2+2x+2)}{\sqrt{(x^2+2x+2)}}\, dx$$

$$= x^{n-1}\sqrt{(x^2+2x+2)}-(n-1)(I_n+2I_{n-1}+2I_{n-2}).$$

Hence, on substituting in equation **8.4**, rearranging and including a constant of integration,

$$nI_n+(2n-1)I_{n-1}+2(n-1)I_{n-2} = x^{n-1}\sqrt{(x^2+2x+2)}+ \\ +\text{a constant}.$$

Thus $a = n,\quad b = 2n-1,\quad c = 2(n-1).$

Using the reduction formula with $n = 2, n = 1$ and changing to definite integrals, we obtain

$$2I_2+3I_1+2I_0 = [x\sqrt{(x^2+2x+2)}]_0^1 = \sqrt{5},$$
$$I_1+I_0 = [\sqrt{(x^2+2x+2)}]_0^1 = \sqrt{5}-\sqrt{2}.$$

Hence $I_2 = \int_0^1 \frac{x^2\,dx}{\sqrt{(x^2+2x+2)}} = \frac{1}{2}(3\sqrt{2}-2\sqrt{5}+I_0),$

where $I_0 = \int_0^1 \frac{dx}{\sqrt{(x^2+2x+2)}} = \int_0^1 \frac{dx}{\sqrt{[(x+1)^2+1]}}$

$$= \Big[\log[x+1+\sqrt{\{(x+1)^2+1\}}]\Big]_0^1 = \log\frac{2+\sqrt{5}}{1+\sqrt{2}}.$$

Thus $\int_0^1 \frac{x^2\,dx}{\sqrt{(x^2+2x+2)}} = \frac{1}{2}\left[3\sqrt{2}-2\sqrt{5}+\log\frac{2+\sqrt{5}}{1+\sqrt{2}}\right].$

2 A gas expands in a cylinder of constant cross-section with a movable piston closing one end. If the law of expansion is pv^γ = constant, and the initial and final values of the pressure and volume are $p_1, v_1; p_2, v_2$ respectively, show that work of amount $(p_1v_1-p_2v_2)/(\gamma-1)$ is done during the expansion. Readings of volume and pressure taken during the expansion are:

v	2	3	4	5	6	7	8	9	10	11	12
p	500	305	212	159	127	104	88	76	67	59	53

Use Simpson's rule to find the work done and deduce the value of γ for the gas. [L, S 1967, FM VI, Q 8]

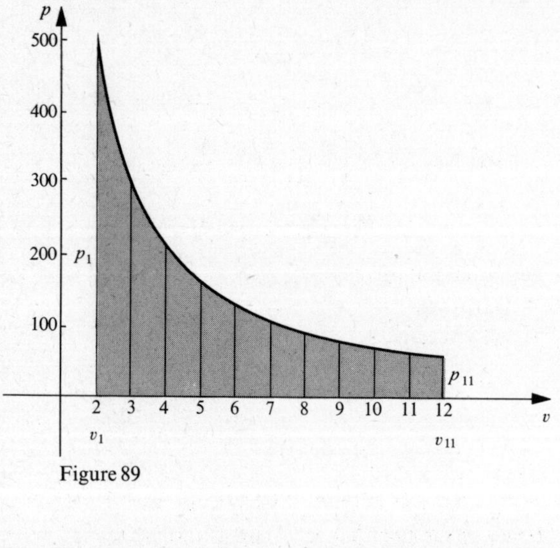

Figure 89

Solution. Now $p_1v_1^\gamma = pv^\gamma = p_2v_2^\gamma = c,$ a constant. Work done in expansion, W, is given by

$$W = \int_{v_1}^{v_2} p\,dv = \int_{v_1}^{v_2} \frac{c}{v^\gamma}\,dv$$
$$= \left[\frac{cv^{1-\gamma}}{1-\gamma}\right]_{v_1}^{v_2}$$
$$= \frac{1}{\gamma-1}(cv_1^{1-\gamma}-cv_2^{1-\gamma})$$
$$= \frac{1}{\gamma-1}(p_1v_1-p_2v_2).$$

Also work done for expansion from $v_1 = 2$ to $v_{11} = 12$ equals area under the curve of Figure 89 between ordinates $p_1 = 500$ and $p_{11} = 53$. Hence, using Simpson's rule,

$$W = \frac{1}{\gamma-1}(p_1v_1-p_2v_2) = \frac{1}{\gamma-1}(1000-636) = \frac{1}{\gamma-1}364$$
$$\approx \tfrac{1}{3}\times 1\times\{500+53+2(212+217+88+67)+ \\ +4(305+159+104+76+59)\}$$
$$\approx 1451.$$

Hence $\gamma \approx 1+\frac{364}{1451} = 1{\cdot}251.$

3 Show by graphical considerations that for $r > 0$,

$$\frac{1}{(r+1)^3} < \int_r^{r+1} \frac{dt}{t^3} < \frac{1}{r^3},$$

and deduce that

$$\int_n^\infty \frac{dt}{t^3} < \sum_n^\infty \frac{1}{r^3} < \int_n^\infty \frac{dt}{t^3}+\frac{1}{n^3},$$

where n is a positive integer. Hence show that

$$\sum_1^\infty \frac{1}{r^3} = \sum_1^{n-1}\frac{1}{r^3}+\frac{n+1}{2n^3}$$

to an error of less than $\pm\frac{1}{2n^3}.$

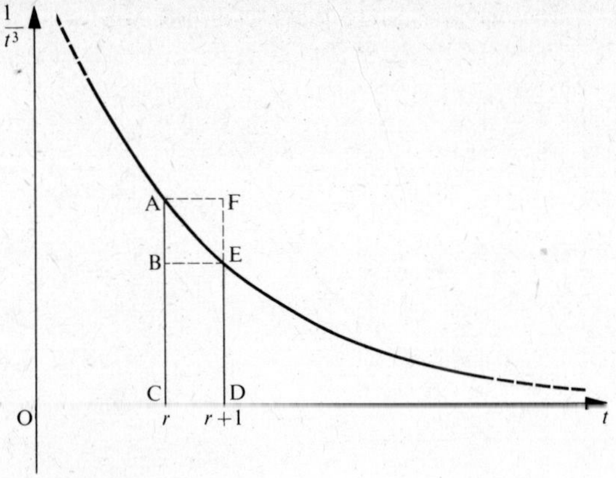

Figure 90

Solution. In Figure 90

$$AC = \frac{1}{r^3}, \quad ED = \frac{1}{(r+1)^3}, \quad CD = 1;$$

$$\text{area BCDE} = \frac{1}{(r+1)^3}, \quad \text{area ACDF} = \frac{1}{r^3};$$

$$\text{also} \quad \text{area ACDE} = \int_r^{r+1} \frac{dt}{t^3}.$$

Hence since area BCDE < area ACDE < ACDF,

$$\frac{1}{(r+1)^3} < \int_r^{r+1} \frac{dt}{t^3} < \frac{1}{r^3}.$$

It also follows, by summing the above result for $r = 1$ to $r = \infty$,

$$\text{that} \quad \sum_n^\infty \frac{1}{(r+1)^3} < \int_n^\infty \frac{dt}{t^3} < \sum_n^\infty \frac{1}{r^3}, \qquad \textbf{8.5}$$

$$\text{but} \quad \sum_n^\infty \frac{1}{(r+1)^3} = \sum_n^\infty \frac{1}{r^3} - \frac{1}{n^3}. \qquad \textbf{8.6}$$

Thus, using relations **8.5** and **8.6**,

$$\sum_n^\infty \frac{1}{r^3} < \int_n^\infty \frac{dt}{t^3} + \frac{1}{n^3}, \quad \sum_n^\infty \frac{1}{r^3} > \int_n^\infty \frac{dt}{t^3}.$$

$$\text{Hence} \quad \int_n^\infty \frac{dt}{t^3} < \sum_n^\infty \frac{1}{r^3} < \int_n^\infty \frac{dt}{t^3} + \frac{1}{n^3}. \qquad \textbf{8.7}$$

Evaluating the integrals in relation **8.7**, we have

$$\frac{1}{2n^2} < \sum_n^\infty \frac{1}{r^3} < \frac{1}{2n^2} + \frac{1}{n^3}.$$

Hence $\sum_n^\infty \frac{1}{r^3}$

lies between $\frac{1}{2n^2}$ and $\frac{1}{2n^2} + \frac{1}{n^3}$,

and so, by taking their mean value,

$$\sum_n^\infty \frac{1}{r^3} = \frac{n+1}{2n^3} \pm \theta \frac{1}{2n^3} \quad (0 < \theta < 1),$$

$$\sum_1^\infty \frac{1}{r^3} = \sum_1^{n-1} \frac{1}{r^3} + \sum_n^\infty \frac{1}{r^3}$$

$$= \sum_1^{n-1} \frac{1}{r^3} + \frac{n+1}{2n^3} \quad \text{with an error less than } \pm \frac{1}{2n^3}.$$

F Further integration: Problems

Answers and hints will be found on pp. 114–15.

1 If $I_n = \int_0^\pi \frac{\cos n\theta \, d\theta}{5 + 4\cos\theta}$,

prove that, if n is an integer greater than unity,

$I_n + I_{n-2} = -\frac{5}{2} I_{n-1}$.

Prove that $I_0 = \frac{1}{3}\pi$ and find the values of I_1 and I_3.

[L, J 1966, FM V, Q 7]

2 If $I_m = \int_1^e (\log x)^m \, dx$,

where m is a positive integer, prove that

$I_m = e - m I_{m-1}$.

Hence, or otherwise, prove that

$I_m = e\{1 - m + m(m-1) - \ldots + (-1)^m m!\} + (-1)^{m+1} m!$

[O, S 1967, PM II, Q 11]

3 If $u_n = \int (\sec x + \tan x)^n \, dx$,

prove that, if $n \neq 1$,

$(n-1)(u_n + u_{n-2}) = 2(\sec x + \tan x)^{n-1} + c$.

Find $\int_{-\frac{1}{6}\pi}^{\frac{1}{6}\pi} (\sec x + \tan x)^{-1} \, dx$,

and hence find $\int_{-\frac{1}{6}\pi}^{\frac{1}{6}\pi} (\sec x + \tan x)^{-3} \, dx$.

[O & C, S 1967, P S V, Q 6]

4

(i) The area enclosed by the part of the cycloid $x = a(\theta + \sin\theta)$, $y = a(1 - \cos\theta)$, for values of θ from 0 to 2π, and the x-axis, is revolved through four right angles about the y-axis. Find the volume generated.

(ii) Evaluate $\int_0^{\frac{1}{2}} \frac{dx}{\sqrt{(1-x^2)}}$.

Hence, using five ordinates, find, with the help of the tables provided and by using Simpson's rule, the value of π to three significant figures. [L, J 1967, FM V, Q 6]

5 Given that $I = \int_0^1 \frac{\{t(t-1)\}^4}{t^2+1}\,dt,$

prove that $I = \frac{22}{7} - \pi.$

Prove also that, if $0 \leqslant t \leqslant 1,$

$t(1-t) \leqslant \frac{1}{4}.$

Use these results to show that $\frac{22}{7}$ differs from π by less than $\pi/2^{10}$. [O, S 1966, P III S, Q 6]

6 Prove that, if $n > 1$,

$$\int_0^{\frac{1}{2}\pi} \sin^{n-1}x \cos x(\cos x - \sin x)^2\,dx > 0.$$

By expanding the integrand, prove that

$$\int_0^{\frac{1}{2}\pi} \sin^n x \cos^2 x\,dx < \frac{1}{2n}.$$

By writing the last integral as

$$\int_0^{\frac{1}{2}\pi} \cos x \frac{d}{dx}\left[\frac{\sin^{n+1}x}{n+1}\right] dx$$

and integrating by parts, prove that

$$\int_0^{\frac{1}{2}\pi} \sin^{n+2}x\,dx < \frac{n+1}{2n}.$$

[O & C, S 1967, P II, Q 8]

7 If K is a constant, and if $f(x) < K$ for $a \leqslant x \leqslant b$, prove that

$$\int_a^b f(x)\,dx < K(b-a).$$

Deduce that if $0 \leqslant t \leqslant \delta < 1$, where δ is a constant,

$$\lim_{n\to\infty} \int_0^t \frac{x^n\,dx}{1+x} = 0.$$

If n is a positive integer, prove that

$$\int_0^t \frac{dx}{1+x} = t - \frac{t^2}{2} + \frac{t^3}{3} - \ldots + (-1)^{n-1}\frac{t^n}{n} + (-1)^n \int_0^t \frac{x^n\,dx}{1+x},$$

and obtain an expression for $\log(1+t)$ as an infinite series of powers of t. [W, S 1969, P S, Q 10]

G Further applications of calculus: Theory summary

1 *Length of an arc of a curve*

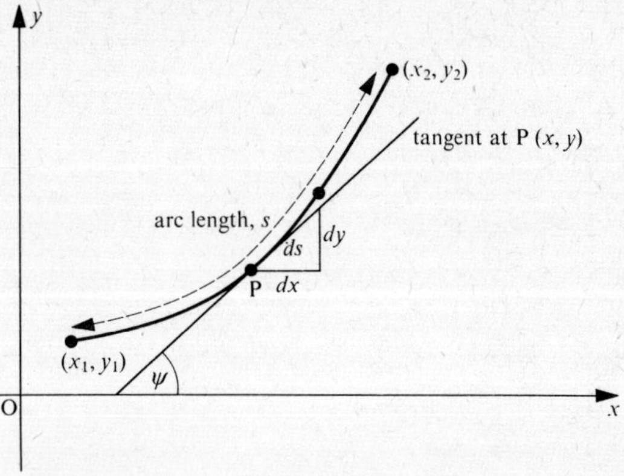

Figure 91

(a) *In Cartesian coordinates.* Length PQ $= \delta s$.

Elemental length, $ds = \lim_{P\to Q} \delta s = \sec\psi\,dx \quad \left(\psi = \tan^{-1}\frac{dy}{dx}\right)$

$$= \sqrt{\{(dx)^2 + (dy)^2\}}$$

$$= \sqrt{\left[1 + \left(\frac{dy}{dx}\right)^2\right]}\,dx$$

$$= \sqrt{\left[\left(\frac{dx}{dy}\right)^2 + 1\right]}\,dy.$$

Total arc length, $s = \int_{x_1}^{x_2} \sqrt{\left[1 + \left(\frac{dy}{dx}\right)^2\right]}\,dx$

$$= \int_{y_1}^{y_2} \sqrt{\left[\left(\frac{dx}{dy}\right)^2 + 1\right]}\,dy.$$

(b) *In parametric form.* If $x = x(t)$, $y = y(t)$,

$$\left[\frac{ds}{dt}\right]^2 = \left[\frac{dx}{dt}\right]^2 + \left[\frac{dy}{dt}\right]^2,$$

$$s = \int_{t_1}^{t_2} \sqrt{\left[\left(\frac{dx}{dt}\right)^2 + \left(\frac{dy}{dt}\right)^2\right]}\,dt.$$

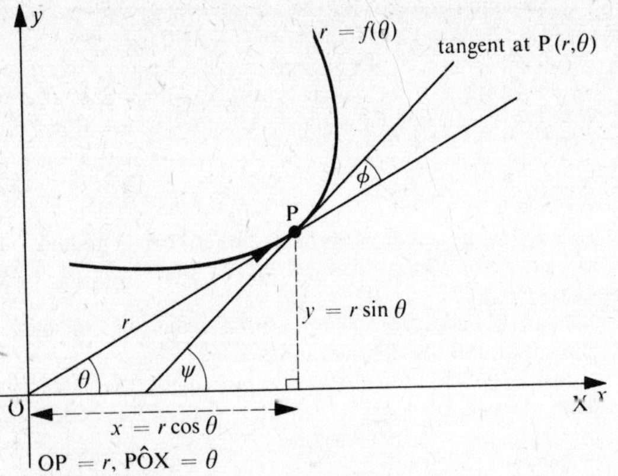

Figure 92

2 *Polar coordinates*

(a) Point $P(x, y)$ may be alternatively defined in polar coordinates (r, θ), where

$$x = r\cos\theta, \qquad y = r\sin\theta;$$

$$r = \sqrt{(x^2+y^2)}, \qquad \theta = \tan^{-1}\frac{y}{x}.$$

(b) *Arc length.* $(ds)^2 = (dr)^2 + (r\,d\theta)^2$.

$$ds = \sqrt{\left[1+r^2\left(\frac{d\theta}{dr}\right)^2\right]}\,dr, \qquad s = \int_{r_1}^{r_2}\sqrt{\left[1+r^2\left(\frac{d\theta}{dr}\right)^2\right]}\,dr,$$

or $$ds = \sqrt{\left[\left(\frac{dr}{d\theta}\right)^2+r^2\right]}\,d\theta, \qquad s = \int_{\theta_1}^{\theta_2}\sqrt{\left[\left(\frac{dr}{d\theta}\right)^2+r^2\right]}\,d\theta.$$

(c) $$\tan\phi = r\frac{d\theta}{dr}, \qquad \sin\phi = r\frac{d\theta}{ds}, \qquad \cos\phi = \frac{dr}{ds}.$$

(d) Area of sector enclosed by an arc $r = f(\theta)$ and the radial lines $\theta = \alpha, \theta = \beta$ is

$$\int_{\alpha}^{\beta}\tfrac{1}{2}r^2\,d\theta.$$

3 *Radius of curvature*

(a) *Definition.* Radius of curvature, $$\rho = \frac{ds}{d\psi};$$

curvature, $$\frac{1}{\rho} = \frac{d\psi}{ds}.$$

(b) *In Cartesian form.*

$$\rho = \frac{\{1+(dy/dx)^2\}^{\frac{3}{2}}}{d^2y/dx^2}.$$

(c) *In parametric form.*

$$\rho = \frac{(\dot{x}^2+\dot{y}^2)^{\frac{3}{2}}}{\dot{x}\ddot{y}-\dot{y}\ddot{x}} \quad \left(\dot{x} = \frac{dx}{dt}, \ddot{x} = \frac{d^2x}{dt^2}, \ldots\right).$$

(d) *Newton's formula.* The radius of curvature for curves of the form

$$y = ax+bx^2+cx^3\,g(x)$$

at the origin (0, 0) is $\rho = \lim_{x\to 0}\dfrac{x^2}{2y}$.

4 *Pappus–Guldin theorems*

(a) The surface area A obtained by rotating an arc of a plane curve about an axis in its plane, which does not intersect the arc, is

$$A = 2\pi s\bar{y},$$

where s is the length of the arc and $\bar{y}$ is the distance of the centroid of the arc from the axis.

(b) The volume v obtained by rotating a plane area about an axis in its plane, which does not intersect the area, is

$$v = 2\pi S\bar{y},$$

where S is the area rotated and $\bar{y}$ is the distance of the centroid of the area from the axis.

H Further applications of calculus: Illustrative worked problems

1 The arc AB of the curve $r = a(1-\cos\theta)$ joins the points where $\theta = 0$ and $\theta = \pi$. Find:

(a) the length of the arc AB and the distance of its centroid from the initial line,

(b) the surface area of the solid formed by rotating the arc AB completely about the initial line. [L, J 1968, FM VI, Q 6]

Solution. (a) Element of arc length,

$$ds = \sqrt{(dr^2+r^2\,d\theta^2)} = \sqrt{\left[\left(\frac{dr}{d\theta}\right)^2+r^2\right]}\,d\theta$$
$$= a\sqrt{[\sin^2\theta+(1-\cos\theta)^2]}\,d\theta$$
$$= a\sqrt{2}\sqrt{(1-\cos\theta)}\,d\theta$$
$$= 2a\sin\tfrac{1}{2}\theta\,d\theta.$$

$$\text{Arc length AB} = \int_0^{\pi}2a\sin\tfrac{1}{2}\theta\,d\theta$$
$$= \left[-4a\cos\tfrac{1}{2}\theta\right]_0^{\pi} = 4a.$$

$$\text{Centroid: } \bar{y} = \frac{1}{4a}\int_0^{\pi}y\,ds$$
$$= \frac{1}{4a}\int_0^{\pi}r(\sin\theta)2a\sin\tfrac{1}{2}\theta\,d\theta$$
$$= \tfrac{1}{2}a\int_0^{\pi}(1-\cos\theta)\sin\theta\sin\tfrac{1}{2}\theta\,d\theta$$
$$= \tfrac{1}{2}a\int_0^{\pi}\sin\theta\sin\tfrac{1}{2}\theta\,d\theta - \tfrac{1}{2}a\int_0^{\pi}\cos\theta\sin\theta\sin\tfrac{1}{2}\theta\,d\theta$$
$$= \tfrac{1}{4}a\int_0^{\pi}[\cos\tfrac{1}{2}\theta-\cos\tfrac{3}{2}\theta]\,d\theta - \tfrac{1}{4}a\int_0^{\pi}\sin 2\theta\sin\tfrac{1}{2}\theta\,d\theta$$

$$= \tfrac{1}{4}a\left[2\sin\tfrac{1}{2}\theta - \tfrac{2}{3}\sin\tfrac{3}{2}\theta\right]_0^\pi -$$

$$-\tfrac{1}{8}a\int_0^\pi (\cos\tfrac{3}{2}\theta - \cos\tfrac{5}{2}\theta)\,d\theta$$

$$= \tfrac{1}{4}a\times\tfrac{8}{3} - \tfrac{1}{8}a\left[\tfrac{2}{3}\sin\tfrac{3}{2}\theta - \tfrac{2}{5}\sin\tfrac{5}{2}\theta\right]_0^\pi$$
$$= \tfrac{2}{3}a - \tfrac{1}{8}a(-\tfrac{16}{15}) = \tfrac{4}{5}a.$$

(b) Applying the Pappus–Guldin theorem we have

Surface area $= 2\pi(\text{length of arc AB})\bar{y} = 2\pi\times 4a\times\tfrac{4}{5}a$
$= \tfrac{32}{5}\pi a^2.$

2 A spiral curve is given by the parametric equation $(a > 2b,\ t \geqslant 0)$

$x = a\sin t + be^{-t}(\cos t - \sin t),$
$y = -a\cos t + be^{-t}(\cos t + \sin t).$

Show that $\psi = t$ and find the radius of curvature at the typical point. Show that for large values of t the radius of curvature is very nearly equal to a and x^2+y^2 is very nearly equal to a^2.
[O & C, S 1967, M & HM III, Q 3]

Solution. Now $\tan\psi = \dfrac{dy}{dx}$ by definition,

but $\dfrac{dx}{dt} = a\cos t + be^{-t}(-\sin t - \cos t) - be^{-t}(\cos t - \sin t)$
$= (a - 2b\,e^{-t})\cos t,$

and $\dfrac{dy}{dt} = a\sin t + be^{-t}(-\sin t + \cos t) - be^{-t}(\cos t + \sin t)$
$= (a - 2be^{-t})\sin t.$

$\dfrac{dy}{dx} = \dfrac{dy/dt}{dx/dt} = \tan t = \tan\psi,$

hence $\psi = t.$

The radius of curvature at any point is

$\rho = \dfrac{ds}{d\psi} = \dfrac{ds}{dt}$ (since $\psi = t$),

but $\left(\dfrac{ds}{dt}\right)^2 = \left(\dfrac{dx}{dt}\right)^2 + \left(\dfrac{dy}{dt}\right)^2$
$= (\cos^2 t + \sin^2 t)(a - 2be^{-t})^2$
$= (a - 2be^{-t})^2$

and therefore $\rho = a - 2be^{-t}.$

Also, when t is very large $e^{-t} \to 0$ and $\rho \to a$. Further, since the exponential term tends to zero then $e^{-t}(\cos t \pm \sin t)$ must tend to zero. Hence

$x \to a\sin t,\quad y \to -a\cos t$

and therefore $x^2 + y^2 \to a^2\sin^2 t + a^2\cos^2 t = a^2.$

I Further applications of calculus: Problems

Answers and hints will be found on pp. 115–18.

1 A circle of radius a rolls on the outside of the circumference of a fixed circle of radius $4a$. Obtain the coordinates of a point P on the circumference of the rolling circle in the form

$x = a(5\cos\theta - \cos 5\theta),$
$y = a(5\sin\theta - \sin 5\theta),$

referred to the centre of the fixed circle as origin and the diameter of the fixed circle through the initial point of contact as x-axis.

Find the curvature at a point on the locus of P, and hence show that the locus has no points of inflexion.

Prove that the total length of this locus as θ increases from 0 to 2π is $40a$.
[L, J 1968, FM V, Q 6]

2 A tangent is drawn to the circle $x^2+y^2 = a^2$ and P is the foot of the perpendicular drawn to the tangent from the point $C(\tfrac{1}{2}a, 0)$. Taking C as pole and the initial line along the positive x-axis, find the polar equation of the locus of P.

The locus of P is rotated through two right angles about the x-axis to form a solid of revolution. Prove that the surface area of the solid is given by

$$\tfrac{1}{2}\pi a^2\int_0^\pi (5 - 4\cos\theta)^{\frac{1}{2}}(2 - \cos\theta)\sin\theta\,d\theta$$

and evaluate this integral. [L, J 1966, FM V, Q 5]

3 A curve is given in terms of a parameter θ by the equations $x = a(\cos\theta + \log\tan\tfrac{1}{2}\theta),\ y = a\sin\theta$, where $0 \leqslant \theta \leqslant \pi$. A sketch of the curve is given in [Figure 93].

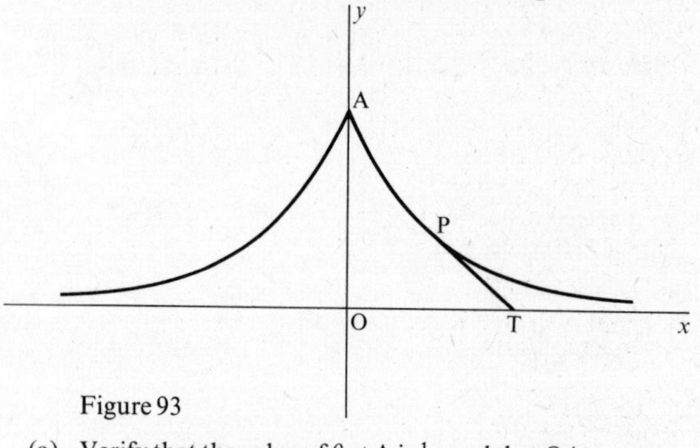

Figure 93

(a) Verify that the value of θ at A is $\tfrac{1}{2}\pi$ and that OA $= a$.

(b) Prove that the tangent at the point P, where $\theta = \beta$, is inclined at an angle β to the x-axis.

(c) Prove that PT $= a$ whatever the value of β.

(d) Prove that the length of the arc AP is $a\log\operatorname{cosec}\beta$.

(e) Find the area of the curved surface generated by the rotation of the arc AP through four right angles about the line OX.
[O, S 1966, P I, Q 9]

4 Prove that the total length of the curve given parametrically by the equations $x = a\cos^3 t,\ y = a\sin^3 t$ is $6a$.

Prove also that it encloses an area of $\tfrac{3}{8}\pi a^2$.
[C, S 1966, P III, Q 10]

5

(i) Sketch the curve whose equation in polar coordinates is $r = a(1+\cos\theta)$. Find the areas of the two portions into which the straight line $\theta = \frac{1}{2}\pi$ divides the upper half of the figure.

(ii) Find the volume of the solid generated by rotating the curve $r^2 = a^2\cos\theta (r > 0)$ about the line $\theta = 0$.

[O & C, S 1967, M for Sc III S, Q 2]

6 If $xy = c^2$ prove that

$$\frac{dy}{dx} = -\frac{y}{x} \quad \text{and} \quad \frac{d^2y}{dx^2} = \frac{2y}{x^2}.$$

The normal at the point $P(ct, c/t)$ on the rectangular hyperbola $xy = c^2$ meets the curve again at Q. Prove that the length of PQ is twice the radius of curvature of the curve at P.

[AEB, S 1968, P & A II, Q 4]

7

(a) Find the rectangular Cartesian coordinate equation of the curve whose polar coordinate equation is

$$r = \frac{a}{1+\cos\theta},$$

where $a > 0$. Sketch the curve.

(b) Sketch the curve

$$r^2 = a^2\cos 2\theta,$$

and find the area enclosed by it.

The curve is circumscribed by a rectangle with two sides parallel to the line $\theta = 0$. Find the area of the rectangle.

[JMB, S 1966, FM I, Q 3]

8 Find the radius of curvature of the parabola $y^2 = 4ax$ at the point $P(at^2, 2at)$, and show that the corresponding centre of curvature is the point $Q(2a+3at^2, -2at^3)$.

Find the length of the path described by Q when P moves on the parabola from the vertex to the point $(4a, 4a)$.

[JMB, S 1966, FM II, Q 3]

9 Find the length of the curve

$$x = t-\sin t, \qquad y = 1-\cos t$$

between the points $t = 0$ and $t = 2\pi$. Find also the surface area generated when this portion of the curve is rotated through one complete revolution about the x-axis.

[W, S 1969, P II, Q 9]

10 A curve is defined parametrically by the equations

$$x = e^\theta\cos\theta, \qquad y = e^\theta\sin\theta.$$

Show that the tangent at any point on the curve makes a constant angle with the line joining that point to the origin.

Find the length of the curve between the points $\theta = \alpha$ and $\theta = \beta$.

[W, S 1968, P II, Q 4]

J Further differentiation: Answers and hints

1 $y(0) = 0, \quad y'(0) = 2^{\frac{1}{2}}\sin\frac{1}{4}\pi = 1,$

$y''(0) = 2\sin\frac{1}{2}\pi = 2,$

$y^{(3)}(0) = 2^{\frac{3}{2}}\sin\frac{3}{4}\pi = 2,$

hence, using Maclaurin's theorem,

$$y = e^x\sin x = x+x^2+\tfrac{1}{3}x^3+\ldots.$$

$y^{(4r)}(0) = 2^{2r}\sin r\pi = 0,$

coefficient of x^{4r} is 0.

$y^{(4r+1)}(0) = 2^{2r+\frac{1}{2}}\sin(r+\frac{1}{4}\pi) = 2^{2r}\cos r\pi = (-1)^r 2^{2r},$

coefficient of x^{4r+1} is $\dfrac{(-1)^r 2^{2r}}{(4r+1)!}$.

2 $1, p, p^2, p^3+p.$

$$(\sec x+\tan x)^{\frac{1}{2}} = 1+\tfrac{1}{2}x+\tfrac{1}{8}x^2+\tfrac{5}{48}x^3+\ldots.$$

3

(i) $$\lim_{x\to 0}\frac{\cos\log(1+x)-\cos x}{x^2(e^x-1)}$$

$$= \lim_{x\to 0}\frac{\cos(x-\frac{1}{2}x^2+\frac{1}{3}x^3-\ldots)-(1-x^2/2!+x^4/4!-\ldots)}{x^2(1+x+x^2/2!+x^3/3!+\ldots-1)}$$

$$= \lim_{x\to 0}\frac{\{1-\frac{1}{2}(x-\frac{1}{2}x^2+\frac{1}{3}x^3-\ldots)\}-(1-\frac{1}{2}x^2+\ldots)}{x^3+\ldots}$$

$$= \lim_{x\to 0}\frac{\frac{1}{2}x^3+\ldots}{x^3+\ldots} = \frac{1}{2}.$$

$$\left(\text{Using L'Hopital's rule, limit } = \frac{f^{(3)}(0)}{g^{(3)}(0)} = \frac{3}{6} = \frac{1}{2}.\right)$$

(ii) If result is assumed valid for $D^r(x+1)^n$, then

$$D^{r+1}(x+1)^n = \frac{n!}{\{n-(r+1)\}!}(x+1)^{n-r-1},$$

but $\quad D^1(x+1)^n = n(x+1)^{n-1} = \dfrac{n!}{(n-1)!}(x+1)^{n-1},$

hence by induction $\quad D^r(x+1)^n = \dfrac{n!}{(n-r)!}(x+1)^{n-r}.$

Let $\quad y = (x^2-1)^n = (x-1)^n(x+1)^n,$

then, using Leibnitz' theorem,

$$y^{(n)} = (x-1)^nD^n(x+1)^n + {}_nC_1 D^1(x-1)^nD^{n-1}(x+1)^n+\ldots+$$
$$+{}_nC_rD^r(x-1)^nD^{n-r}(x+1)^n+\ldots+D^n(x-1)^n(x+1)^n;$$

and, when $x = 0$,

$$y^{(n)} = \sum_0^n {}_nC_r\frac{n!}{(n-r)!}(-1)^{n-r}\frac{n!}{r!}$$

$$= n!\sum_0^n ({}_nC_r)^2(-1)^r \quad \text{if } n \text{ is even.}$$

Now $\quad (x^2-1)^n = x^{2n}-nx^{2(n-1)}+{}_nC_2x^{2(n-2)}+\ldots+(-1)^n.$

Thus if n is odd $D^n(x^2-1)^n$ will give terms involving powers of x,

hence when $x = 0, \quad D^n(x^2-1)^n = 0.$

When n is even, middle term in $(x^2-1)^n$ is $(-1)^{\frac{1}{2}n}\{n!/(\frac{1}{2}n)!\}x^n$, hence the nth derivative of this term is $(-1)^{\frac{1}{2}n}\{n!/(\frac{1}{2}n)!\}^2$, which is also the value of $D^n(x^2-1)^n$ when $x = 0$.

4

(i) $u = e^y \dfrac{\partial}{\partial x}\{e^x \cos(x-y)\} = e^{x+y}\{\cos(x-y)-\sin(x-y)\}$,

$v = e^x \dfrac{\partial}{\partial y}\{e^y \cos(x-y)\} = e^{x+y}\{\cos(x-y)+\sin(x-y)\}$,

hence $u+v = 2e^{x+y}\cos(x-y) = 2V$.

Also
$$\frac{\partial u}{\partial x} = e^y \frac{\partial}{\partial x}[e^x\{\cos(x-y)-\sin(x-y)\}] = -2e^{x+y}\sin(x-y)$$

and
$$\frac{\partial v}{\partial y} = e^x \frac{\partial}{\partial y}[e^y\{\cos(x-y)+\sin(x+y)\}] = +2e^{x+y}\sin(x-y),$$

therefore $\dfrac{\partial u}{\partial x}+\dfrac{\partial v}{\partial y} = 0$.

5
$$\frac{\partial \phi}{\partial x} = \frac{\partial \phi}{\partial u}\frac{\partial u}{\partial x}+\frac{\partial \phi}{\partial v}\frac{\partial v}{\partial x} = 3(x^2-y^2)\frac{\partial \phi}{\partial u}+6xy\frac{\partial \phi}{\partial v},$$
$$\frac{\partial \phi}{\partial y} = \frac{\partial \phi}{\partial u}\frac{\partial u}{\partial y}+\frac{\partial \phi}{\partial v}\frac{\partial v}{\partial y} = -6xy\frac{\partial \phi}{\partial u}+3(x^2-y^2)\frac{\partial \phi}{\partial v}.$$

Now
$$\frac{\partial^2 \phi}{\partial x^2} = 6x\frac{\partial \phi}{\partial u}+ +3(x^2-y^2)\left[\frac{\partial^2 \phi}{\partial u^2}3(x^2-y^2)+\frac{\partial^2 \phi}{\partial u\,\partial v}6xy\right]+ +6y\frac{\partial \phi}{\partial v}+6xy\left[\frac{\partial^2 \phi}{\partial u\,\partial v}3(x^2-y^2)+\frac{\partial^2 \phi}{\partial v^2}6xy\right],$$

and
$$\frac{\partial^2 \phi}{\partial y^2} = -6x\frac{\partial \phi}{\partial u}-6xy\left[\frac{\partial^2 \phi}{\partial u^2}(-6xy)+ +\frac{\partial^2 \phi}{\partial u\,\partial v}3(x^2-y^2)\right]-6y\frac{\partial \phi}{\partial v}+ +3(x^2-y^2)\left[\frac{\partial^2 \phi}{\partial u\,\partial v}(-6xy)+\frac{\partial^2 \phi}{\partial v^2}3(x^2-y^2)\right];$$

hence
$$\frac{\partial^2 \phi}{\partial x^2}+\frac{\partial^2 \phi}{\partial y^2} = 9(x^2+y^2)^2\left(\frac{\partial^2 \phi}{\partial u^2}+\frac{\partial^2 \phi}{\partial v^2}\right).$$

K Further integration: Answers and hints

1
$$I_n+I_{n-2} = \int_0^\pi \frac{\cos n\theta+\cos(n-2)\theta}{5+4\cos\theta}\,d\theta$$
$$= \int_0^\pi \frac{2\cos(n-1)\theta\cos\theta}{5+4\cos\theta}\,d\theta$$
$$= 2\int_0^\pi \cos(n-1)\theta\left[\frac{1}{4}-\frac{5}{4}\frac{1}{5+4\cos\theta}\right]d\theta$$
$$= \frac{1}{2}\left[\frac{\sin(n-1)\theta}{n-1}\right]_0^\pi-\frac{5}{2}I_{n-1} = -\frac{5}{2}I_{n-1}.$$

On substituting $t = \tan\frac{1}{2}\theta$,
$$I_0 = \int_0^\pi \frac{d\theta}{5+4\cos\theta} = \int_0^\infty \frac{2\,dt}{3^2+t^2}$$
$$= \left[\tfrac{2}{3}\tan^{-1}\tfrac{1}{3}t\right]_0^\infty = \tfrac{1}{3}\pi;$$
$$I_1 = \int_0^\pi \frac{\cos\theta}{5+4\cos\theta}\,d\theta = \int_0^\pi\left[\frac{1}{4}-\frac{5}{4}\frac{1}{5+4\cos\theta}\right]d\theta$$
$$= \tfrac{1}{4}\pi-\tfrac{5}{4}I_0 = -\tfrac{1}{6}\pi.$$

Using recurrence relation,
$$I_3 = -\tfrac{5}{2}(-\tfrac{5}{2}I_1-I_0)-I_1 = -\tfrac{1}{24}\pi.$$

2
$$I_m = \int_1^e 1.(\log x)^m\,dx = \left[x(\log x)^m\right]_1^e-\int xm(\log x)^{m-1}\frac{1}{x}\,dx$$
$$= e-mI_{m-1}. \qquad \textbf{8.8}$$

Second part may be proved by repeated application of formula **8.8** until $m = 2$.

3
$$u_n+u_{n-2} = \int(\sec x+\tan x)^{n-2}(\sec^2 x+2\sec x\tan x+ +\tan^2 x+1)\,dx$$
$$= \int 2(\sec x+\tan x)^{n-2}\sec x(\sec x+\tan x)\,dx$$
$$= 2\int y^{n-2}\,dy$$
$$= 2\frac{y^{n-1}}{n-1}+c' \quad \text{(by putting } y = \sec x+\tan x\text{)};$$

i.e. if $n \neq 1$
$$(n-1)(u_n+u_{n-2}) = 2(\sec x+\tan x)^{n-1}+c.$$
$$v_{-1} \equiv \int_{-\frac{1}{6}\pi}^{\frac{1}{6}\pi}(\sec x+\tan x)^{-1}\,dx = \int_{-\frac{1}{6}\pi}^{\frac{1}{6}\pi}\frac{\cos x}{1+\sin x}\,dx$$
$$= \left[\log_e(1+\sin x)\right]_{-\frac{1}{6}\pi}^{\frac{1}{6}\pi} = \log_e 3.$$

Using recurrence relation with $n = -1$, we find
$$-2(v_{-1}+v_{-3}) = 2[\sec x+\tan x]_{-\frac{1}{6}\pi}^{\frac{1}{6}\pi}.$$

Hence
$$v_{-3} \equiv \int_{-\frac{1}{6}\pi}^{\frac{1}{6}\pi}(\sec x+\tan x)^{-3}\,dx = 2\tfrac{2}{3}-\log 3$$

4

(i) Volume generated $= 2\pi\displaystyle\int xy\,dx$
$$= 2\pi\int_0^{2\pi} xy\frac{dx}{d\theta}\,d\theta$$
$$= 2\pi a^3\int_0^{2\pi}(\theta+\sin\theta)\sin^2\theta\,d\theta$$
$$= 2\pi^3 a^3.$$

$$\int_0^{\frac{1}{2}} \frac{dx}{\sqrt{(1-x^2)}} = [\sin^{-1} x]_0^{\frac{1}{2}}$$

$$= \tfrac{1}{6}\pi.$$

$$\approx \tfrac{1}{3} \times 0{\cdot}125\{1+1{\cdot}155+2\times 1{\cdot}033+ +4(1{\cdot}008+1{\cdot}079)\}$$

$$= 0{\cdot}524.$$

Hence $\pi \approx 3{\cdot}14$.

$$I = \int_0^1 \left[t^6 - 4t^5 + 5t^4 - 4t^2 + 4 - \frac{4}{t^2+1}\right] dt = \frac{22}{7} - \pi.$$

If $y = t(1-t)$, $dy/dt = 0$ and $d^2y/dt^2 < 0$ when $t = \frac{1}{2}$, hence $y_{\max} = \frac{1}{4}$ and therefore

$t(1-t) \leqslant \frac{1}{4}$.

Also $\quad I = \dfrac{22}{7} - \pi < \displaystyle\int_0^1 \frac{(\frac{1}{4})^4\, dt}{t^2+1} = \frac{\pi}{2^{10}},$

i.e. $\dfrac{22}{7} - \pi < \dfrac{\pi}{2^{10}}.$

6 Over range $0 \leqslant x \leqslant \frac{1}{2}\pi$ all product terms in integrand are positive, therefore

Integral > 0.

$$\int_0^{\frac{1}{2}\pi} \sin^{n-1} x \cos x (1 - 2\sin x \cos x)\, dx$$

$$= \int_0^{\frac{1}{2}\pi} \sin^{n-1} x \cos x\, dx - 2\int_0^{\frac{1}{2}\pi} \sin^n x \cos^2 x\, dx$$

$$= \left[\frac{\sin^n x}{n}\right]_0^{\frac{1}{2}\pi} - 2\int_0^{\frac{1}{2}\pi} \sin^n x \cos^2 x\, dx > 0.$$

Hence $\dfrac{1}{n} - 2\displaystyle\int_0^{\frac{1}{2}\pi} \sin^n x \cos^2 x\, dx > 0$

or $\displaystyle\int_0^{\frac{1}{2}\pi} \sin^n x \cos^2 x\, dx < \frac{1}{2n}.$

$$\int_0^{\frac{1}{2}\pi} \cos x \frac{d}{dx}\left[\frac{\sin^{n+1} x}{n+1}\right] dx$$

$$= \left[\cos x \frac{\sin^{n-1} x}{n+1}\right]_0^{\frac{1}{2}\pi} + \int_0^{\frac{1}{2}\pi} \sin x \frac{\sin^{n+1} x}{n+1}\, dx < \frac{1}{2n}.$$

$$\int_0^{\frac{1}{2}\pi} \sin^{n+2} x\, dx < \frac{n+1}{2n}.$$

7 As $f(x) < K$, $\displaystyle\int_a^b f(x)\, dx < \int_a^b K\, dx = K[x]_a^b = K(b-a).$

For $0 \leqslant t \leqslant \delta < 1$,

$$0 < \frac{x^n}{1+x} < t^n \leqslant \delta^n$$

for $n > 0$ over range $0 \leqslant x \leqslant t$.

Therefore $\quad 0 < \displaystyle\int_0^t \frac{x^n\, dx}{1+x} < \delta^n \to 0$ as $n \to \infty$, since $\delta < 1$.

Now $\dfrac{1}{1+x} = 1 - x + x^2 - x^3 + \ldots + (-1)^{n-1} x^{n-1} + \dfrac{(-1)^n x^n}{1+x}.$

Thus $\displaystyle\int_0^t \frac{dx}{1+x} = t - \frac{t^2}{2} + \frac{t^3}{3} - \ldots (-1)^{n-1}\frac{t^n}{n} + (-1)^n \int_0^t \frac{x^n\, dx}{1+x}.$

Also $\displaystyle\int_0^t \frac{dx}{1+x} = [\log(1+x)]_0^t = \log(1+t),$

hence, taking the limit as $n \to \infty$ in the previous result gives

$$\log(1+t) = t - \frac{t^2}{2} + \frac{t^3}{3} - \ldots + 0.$$

L Further applications of calculus: Answers and hints

1 From Figure 94, P has coordinates
$x = 5a\cos\theta - a\cos(2\pi - 5\theta) = a(5\cos\theta - \cos 5\theta)$,
$y = 5a\sin\theta + a\sin(2\pi - 5\theta) = a(5\sin\theta - \sin 5\theta)$.

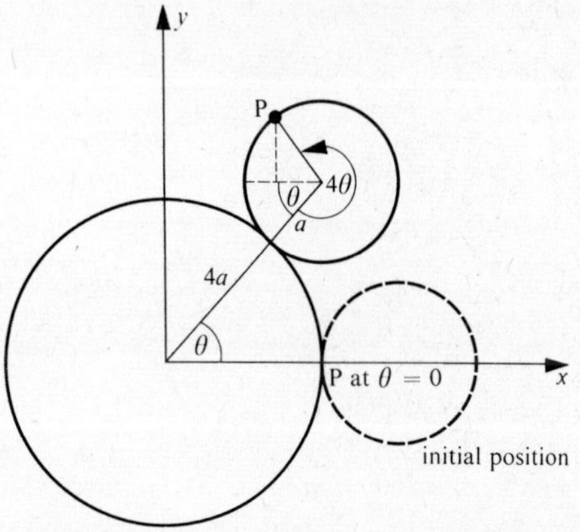

Figure 94

Now $\dfrac{dy}{dx} = \dfrac{dy/d\theta}{dx/d\theta} = \dfrac{5a(\cos\theta - \cos 5\theta)}{5a(\sin 5\theta - \sin\theta)}$

$$= \frac{\sin 3\theta \sin 2\theta}{\cos 3\theta \sin 2\theta}$$

$$= \tan 3\theta \quad (\text{note } \psi = 3\theta),$$

also $\dfrac{d^2y}{dx^2} = 3\sec^2 3\theta \dfrac{d\theta}{dx}$

$$= \frac{3\sec^3 3\theta}{10a \sin 2\theta},$$

hence the curvature at P,

$$\frac{1}{\rho} = \frac{d^2y/dx^2}{\{1+(dy/dx)^2\}^{\frac{3}{2}}} = \frac{3}{10a\sin 2\theta}.$$

There are no points of inflexion since $\rho \propto \sin 2\theta$ can never be infinite.

As $\rho = ds/d\psi$, required arc length,

$$s = 4\int_0^{\frac{1}{6}\pi} \tfrac{1}{3} \times 10a \sin 2\theta \, d(3\theta)$$

$$= 40a\left[-\tfrac{1}{2}\cos 2\theta\right]_0^{\frac{1}{6}\pi} = 40a.$$

2 From Figure 95 it is seen that $\triangle$s OAB, CPB are similar, and hence

$$r = a\frac{\text{CB}}{\text{OB}} = a(1 - \tfrac{1}{2}\cos\theta), \text{ the locus of } P(r, \theta).$$

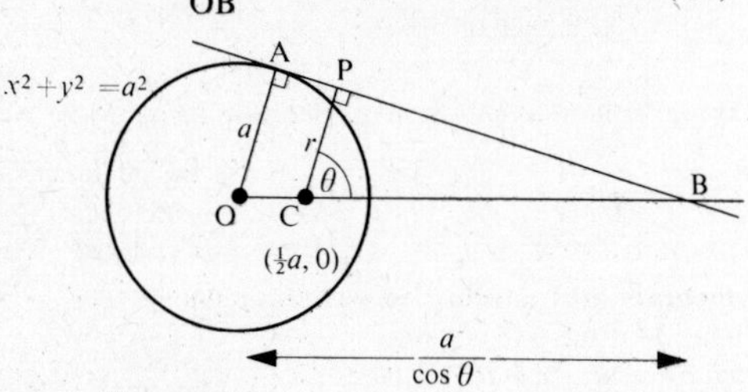

Figure 95

Surface area of solid of revolution is

$$S = \int 2\pi r \sin 0 \, ds$$

$$= 2\pi \int_0^{\pi} r\sin\theta \left[\left(\frac{dr}{d\theta}\right)^2 + r^2\right]^{\frac{1}{2}} d\theta$$

$$= 2\pi a^2 \int_0^{\pi} (1-\tfrac{1}{2}\cos\theta)\sin\theta\left[\tfrac{1}{4}\sin^2\theta + (1-\tfrac{1}{2}\cos\theta)^2\right]^{\frac{1}{2}} d\theta$$

$$= \tfrac{1}{2}\pi a^2 \int_0^{\pi} (5-4\cos\theta)^{\frac{1}{2}}(2-\cos\theta)\sin\theta \, d\theta.$$

To evaluate the latter, use substitution $u^2 = 5 - \cos\theta$, then

$$S = \tfrac{1}{16}\pi a^2 \int_1^3 (3u^2 + u^4)\, du = \tfrac{93}{20}\pi a^2.$$

3

(a) At $x = 0$, $\cos\theta + \log\tan\frac{1}{2}\theta = 0$,

which is satisfied by $\theta = \frac{1}{2}\pi$, as $\cos\frac{1}{2}\pi = 0$, $\log\tan\frac{1}{4}\pi = \log 1 = 0$.

(b) Show that $\dfrac{dy}{dx} = \dfrac{dy/d\theta}{dx/d\theta} = \tan\theta$,

result then follows.

(c) $\text{PT} = y \operatorname{cosec}\beta$

$$= (a\sin\beta)\operatorname{cosec}\beta = a.$$

(d) $\text{AP} = \displaystyle\int \sqrt{\left[1 + \left(\frac{dx}{dy}\right)^2\right]}\, dy = \int_\beta^{\frac{1}{2}\pi} a\cot\theta\, d\theta$

$$= a\left[\log\sin\theta\right]_\beta^{\frac{1}{2}\pi} = a\log\operatorname{cosec}\beta.$$

(e) $\text{Area} = \displaystyle\int 2\pi y\, ds = \int_\beta^{\frac{1}{2}\pi} 2\pi a^2 \sin\theta\cot\theta\, d\theta = 2\pi a^2(1-\sin\beta).$

4

$$\frac{ds}{dt} = \sqrt{\left[\left(\frac{dx}{dt}\right)^2 + \left(\frac{dy}{dt}\right)^2\right]} = \pm 3a\sin t\cos t.$$

$$\text{Total arc length} = 4\int_0^{\frac{1}{2}\pi} 3a\sin t\cos t\, dt = 12a\left[\tfrac{1}{2}\sin^2 t\right]_0^{\frac{1}{2}\pi} = 6a.$$

$$\text{Area enclosed} = 4\int_0^{\frac{1}{2}\pi} -y\frac{dx}{dt}\, dt = 12a^2\int_0^{\frac{1}{2}\pi}\cos^2 t\sin^4 t\, dt$$

$$= \tfrac{3}{8}\pi a^2.$$

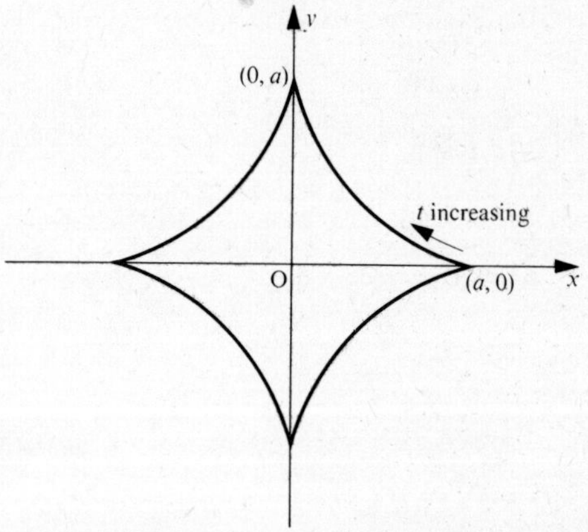

Figure 96 Curve $x = a\cos^3 t$, $y = a\sin^3 t$; an astroid

5

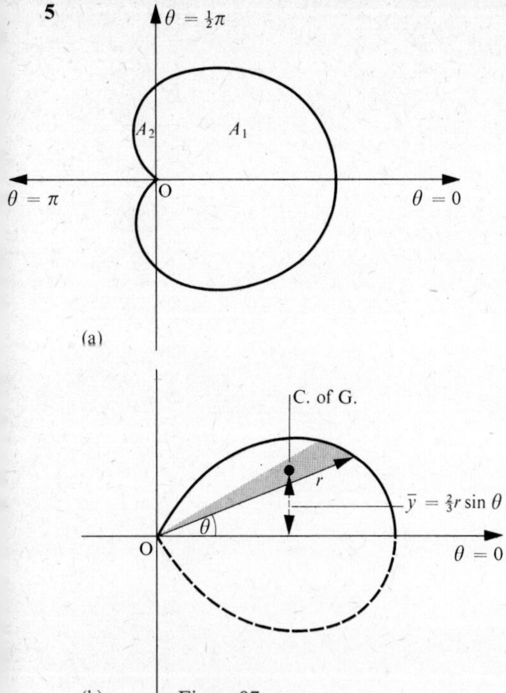

(b) Figure 97

(i) $A_1 = \int_0^{\frac{1}{2}\pi} \frac{1}{2}r^2\, d\theta = \frac{1}{2}a^2 \int_0^{\frac{1}{2}\pi} (1+\cos\theta)^2\, d\theta = \frac{1}{2}a^2(\frac{3}{4}\pi+2).$

$A_2 = \int_{\frac{1}{2}\pi}^{\pi} \frac{1}{2}r^2\, d\theta = \frac{1}{2}a^2(\frac{3}{4}\pi-2).$

(ii) Applying Pappus's theorem to sector element,
$dv = \frac{1}{2}r^2\, d\theta \times 2\pi \times \frac{2}{3}r \sin\theta.$

$$\text{Total volume} = \int_0^{\frac{1}{2}\pi} \frac{2}{3}\pi r^3 \sin\theta\, d\theta$$

$$= \frac{2}{3}\pi a^3 \int_0^{\frac{1}{2}\pi} \cos^{\frac{3}{2}}\theta \sin\theta\, d\theta$$

$$= \frac{2}{3}\pi a^3 \left[\frac{2}{5}\cos^{\frac{5}{2}}\theta\right]_0^{\frac{1}{2}\pi} = \frac{4}{15}\pi a^3.$$

6 Radius of curvature at P is

$$\frac{(1+y^2/x^2)^{\frac{3}{2}}}{2y/x} = \frac{c(1+t^4)^{\frac{3}{2}}}{2t^3}.$$

Equation of normal at P is

$$y - \frac{c}{t} = \left(\frac{x}{y}\right)_t (x-ct)$$

and point of intersection with
$y = c^2/x$ gives Q as $(-c/t^3, -ct^3)$.

Hence $\text{PQ} = \sqrt{\left[\left(ct+\frac{c}{t^3}\right)^2 + \left(\frac{c}{t}+ct^3\right)^2\right]}$

$$= c(1+t^4)^{\frac{3}{2}}\frac{1}{t^3},$$

$\text{PQ} = 2\rho.$

7

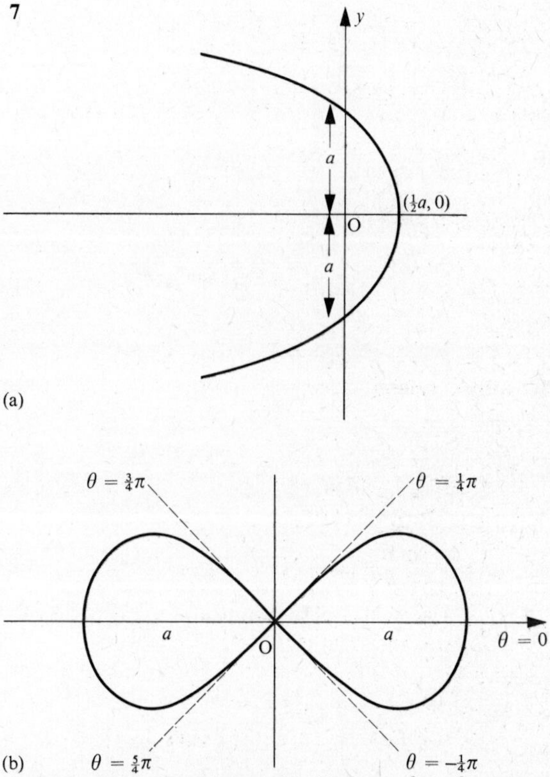

Figure 98 (a) Curve for part (a). (b) Curve $r^2 = a^2 \cos 2\theta$

(a) $y^2 = a(a-2x)$, a parabola.

(b) Area enclosed is a^2, area of rectangle is $a^2\sqrt{2}$.

8 Radius of curvature $= 2a(t^2+1)^{\frac{3}{2}}$.
Length of path of Q $= 2a(5\sqrt{5}-1)$.

9 Length of curve $= \displaystyle\int_0^{2\pi} \sqrt{(\dot{x}^2+\dot{y}^2)}\, dt$

$$= \int_0^{2\pi} \sqrt{\{(1-\cos t)^2+\sin^2 t\}}\, dt$$

$$= \int_0^{2\pi} \sqrt{(2-2\cos t)}\, dt$$

$$= \sqrt{2}\int_0^{2\pi} \sqrt{(2\sin^2 \tfrac{1}{2}t)}\, dt$$

$$= 4\left[-\cos \tfrac{1}{2}t\right]_0^{2\pi} = 8.$$

Surface area of revolution $= \displaystyle\int_0^{2\pi} 2\pi y\, ds$

$$= 2\pi \int_0^{2\pi} (1-\cos t)\, 2\sin \tfrac{1}{2}t\, dt$$

$$= 8\pi \int_0^{2\pi} \sin^3 \tfrac{1}{2}t\, dt$$

$$= \tfrac{64}{3}\pi.$$

10 Slope of tangent,

$$\frac{dy}{dx} = \frac{\dot{y}}{\dot{x}}$$

$$= \frac{e^{\theta}(\sin\theta+\cos\theta)}{e^{\theta}(\cos\theta-\sin\theta)}$$

$$= \frac{1+\tan\theta}{1-\tan\theta} = \tan(\theta+\tfrac{1}{4}\pi).$$

Slope of line joining point and origin is $y/x = \tan\theta$, hence angle between tangent and line is $\frac{1}{4}\pi$, a constant.

Length of curve $= \displaystyle\int_{\alpha}^{\beta} \sqrt{(\dot{x}^2+\dot{y}^2)}\, d\theta$

$$= \int_{\alpha}^{\beta} \sqrt{(2e^{2\theta})}\, d\theta$$

$$= \sqrt{2}(e^{\beta}-e^{\alpha}).$$

Chapter Nine
Differential Equations

A Formation and solution of differential equations: Theory summary

1 *Formation*

(a) *First order*. An equation $F(x, y, c) = 0$ is given, involving an arbitrary constant c. On differentiating this equation once, it is possible to eliminate c yielding the differential equation

$$f\left(x, y, \frac{dy}{dx}\right) = 0.$$

For example, if $y = cx^2$,

then $$\frac{dy}{dx} = 2cx$$

and so $$\frac{dy}{dx} - \frac{2y}{x} = 0,$$

a result independent of c.

(b) *Second order*. An equation $F(x, y, a, b) = 0$ involving two arbitrary constants a, b is given. Here, two differentiations are necessary in order to form a differential equation which does not contain a, b.

For example, if $y = ae^x + bx$,

then $$\frac{dy}{dx} = ae^x + b, \quad \frac{d^2y}{dx^2} = ae^x$$

and so

$$y = \frac{d^2y}{dx^2} + x\left[\frac{dy}{dx} - \frac{d^2y}{dx^2}\right], \quad \text{or} \quad (x-1)\frac{d^2y}{dx^2} - x\frac{dy}{dx} + y = 0,$$

which is of the form $f(x, y, dy/dx, d^2y/dx^2) = 0$.

2 *Mathematical expression of given problems as differential equations*

Examples:

(a) In physics, Newton's law of cooling states that 'the rate of cooling of a body is proportional to the difference between its temperature and that of the surroundings.'

In terms of a differential equation, this is

$$\frac{d\theta}{dt} = k(\theta - \theta_0),$$

where θ is the temperature at time t and θ_0 is the external temperature.

(b) The law of motion for 'a particle of mass m which moves along the x-axis under resistance proportional to the velocity v while accelerated by a constant force F' is

$$m\frac{dv}{dt} = F - kv, \quad \text{or} \quad m\frac{d^2x}{dt^2} + k\frac{dx}{dt} = F,$$

as $v = dx/dt$.

3 *Solution of first-order differential equations*

The integration of a first-order equation will introduce one arbitrary constant. In an actual problem this constant may be determined from the initial condition supplied.

(a) If the differential equation has the form

$$\frac{dy}{dx} + f(x)g(y) = 0,$$

the solution is effected by 'separating the variables' and integrating, with the result that

$$\int \frac{dy}{g(y)} + \int f(x)\,dx = c,$$

where c is constant.

(b) *The 'linear' type*. $\dfrac{dy}{dx} + P(x)y = Q(x)$,

where P and Q are functions of x only, can be solved by multiplying through by an 'integrating factor', $R(x) = e^{\int P(x)\,dx}$, enabling the solution to be obtained from

$$\frac{d}{dx}(yR) = QR,$$

which gives $y = \dfrac{1}{R}\left(\displaystyle\int QR\,dx + c\right)$.

(c) *Other types*. Substitutions are used to transform the given equation to type (a) or (b) above. One important example is the 'homogeneous equation',

$$\frac{dy}{dx} = f\left(\frac{y}{x}\right).$$

The substitution $y = vx$ is made giving

$$x\frac{dv}{dx} + v = f(v).$$

The variables can now be separated as in (a) above, and integration yields

$$\int \frac{dv}{f(v) - v} = x + c,$$

v is then replaced by y/x for the final solution.

4 *Families of solutions*

Consider only differential equations of the first order,

$$\frac{dy}{dx} = f(x, y).$$

This differential equation gives rise to a set of curves in the xy plane, and each curve has a gradient at a given point (x, y) on it given by the above equation.

Another family may have a differential equation

$$\frac{dy}{dx} = F(x, y)$$

and the curves of this set may intersect those of the previous set. If the curves of the first set intersect those of the second set at right angles, then

$$\left(\frac{dy}{dx}\right)_1\left(\frac{dy}{dx}\right)_2 = -1,$$

so that $f(x, y)\,F(x, y) = -1$.

Hence if one family is given by

$$\frac{dy}{dx} = f(x, y)$$

the orthogonal family will be given by

$$\frac{dy}{dx} = -\frac{1}{f(x, y)}.$$

5 *Solution of second-order differential equations*
These are equations of the general form

$$\frac{d^2y}{dx^2} = F\left(x, y, \frac{dy}{dx}\right).$$

The solution will include two arbitrary constants.

(a) *Some simple cases.*

(i) $\dfrac{d^2y}{dx^2} = f(x)$: two direct integrations yield the solution.

(ii) $\dfrac{d^2y}{dx^2} = f(y)$: here make the substitution $p = dy/dx$ so that

$$p\frac{dp}{dy} = f(y).$$

On integrating, $\frac{1}{2}p^2 = \int f(y)\,dy + c \equiv g(y)$, say,

and so $p \equiv \dfrac{dy}{dx} = \sqrt{\{2g(y)\}}$,

and thus the solution is

$$x + A = \int \frac{dy}{\sqrt{\{2g(y)\}}}.$$

Note: In the special but important case when $f(y) = m^2y$, $g(y)$ will take the form $\frac{1}{2}m^2(y^2+a^2)$,

thus $x + A = \dfrac{1}{m}\displaystyle\int \frac{dy}{\sqrt{(y^2+a^2)}} = \frac{1}{m}\sinh^{-1}\frac{y}{a}$

$y = a\sinh(mx+b)$.
This may also be expressed in the form

$$y = A\sinh mx + B\cosh mx$$

or $y = Ce^{mx} + De^{-mx}$.

This special case may be obtained more easily as shown by the methods in (b) below.

(iii) $\dfrac{d^2y}{dx^2} = f\left(\dfrac{dy}{dx}\right)$.

The substitution $p = dy/dx$, as in (ii), gives

$$\frac{dp}{dx} \equiv p\frac{dp}{dy} = f(p)$$

and, on separating the variables, y or x can be solved in terms of p. A second integration completes the solution.

(b) Next we take the second-order equation.

$$\frac{d^2y}{dx^2} + b\frac{dy}{dx} + cy = f(x),$$

where $f(x)$ is a function consisting of simple polynomials, sine and cosine functions, and exponential and hyperbolic sine and cosine functions.

The complementary function (c.f.). This forms part of the general solution of the above equation. It is the solution of

$$\frac{d^2y}{dx^2} + b\frac{dy}{dx} + cy = 0.$$

The trial substitution $y = e^{mx}$ yields the 'auxiliary equation' $m^2 + bm + c = 0$. Let the roots be m_1 and m_2.

(i) If $m_1 \neq m_2$, then the differential equation has the solution $y = A_1e^{m_1x} + A_2e^{m_2x}$ (A_1 and A_2 arbitrary).
If m_1 and m_2 are complex $m_1, m_2 = p \pm iq$, say, then the solution may be written as

$$y = e^{px}(B_1\cos qx + B_2\sin qx)$$

or $y = Ce^{px}\sin(qx+\phi)$.

(ii) If $m_1 = m_2 = m$, so both roots are real and equal, then $y = e^{mx}(C_1x + C_2)$
N.B. In the above solutions, a, b, A, B, C, ϕ are arbitrary constants.

The particular integral (p.i.). This is the remaining part of the general solution. It is a particular solution of

$$\frac{d^2y}{dx^2} + b\frac{dy}{dx} + cy = f(x)$$

and its form depends on $f(x)$ and the form taken by the c.f. The general solution to the differential equation will be the sum of the c.f., y_c and the p.i., y_p,

i.e. $y = y_c + y_p$.

For example, solve

$$\frac{d^2y}{dx^2} - y = x^2 + e^{2x} + e^{-x}.$$

For the c.f., solve $m^2 - 1 = 0$,
whence $m = \pm 1$,
so that $y_c = A_1e^x + A_2e^{-x}$.

For the p.i., we try one of the form

$$y_p = px^2 + qx + r + ae^{2x} + bxe^{-x}.$$

The choice of the term xe^{-x} is made, since a term e^{-x} occurs both in the expression y_c and in the function $f(x)$.
p, q, r, a, b are to be determined by substitution back into the differential equation.

Since $\dfrac{d^2y_p}{dx^2} = 2p + 4ae^{2x} + b(x-2)e^{-x}$,

we must satisfy the identity

$$-px^2 - qx + (2p-r) + 3ae^{2x} - 2be^{-x} = x^2 + e^{2x} + e^{-x}.$$

Thus we must have $-p = 1$, $-q = 0$, $2p - r = 0$, $3a = 1$ and $-2b = 1$,

hence $y_P = -x^2 - 2 + \frac{1}{3}e^{2x} - \frac{1}{2}xe^{-x}$.

Finally, $y = y_c + y_p$ can be found.

(c) *Other types of second-order equation.* Consider, for example, the second-order homogeneous equation

$$x^2\frac{d^2y}{dx^2}+bx\frac{dy}{dx}+cy = f(x).$$

It is convenient here to introduce the substitution $x = e^t$ and use t as the independent variable:

$$\frac{dy}{dt} = \frac{dy}{dx}\frac{dx}{dt} \equiv x\frac{dy}{dx},$$

$$\frac{d^2y}{dt^2} = \frac{dx}{dt}\frac{dy}{dx}+x\frac{d^2y}{dx^2}\frac{dx}{dt} = x\frac{dy}{dx}+x^2\frac{d^2y}{dx^2},$$

hence $x^2\dfrac{d^2y}{dx^2} = \dfrac{d^2y}{dt^2}-\dfrac{dy}{dt}$

and so the differential equation becomes

$$\frac{d^2y}{dt^2}+(b-1)\frac{dy}{dt}+cy = f(e^t)$$

This is solved by the above methods, for y in terms of t, and the solution may be subsequently expressed in terms of x.

B Differential equations: Illustrative worked problems

1 The motion of a particle is represented by the differential equation

$$\frac{d^2x}{dt^2}+k\frac{dx}{dt}+64x = 0,$$

x being the distance of the particle from a fixed point at time t. When $t = 0, x = 0$ and $dx/dt = u$. Find the position of the particle when $t = 1$ if (a) $k = 20$, (b) $k = 16$, (c) $k = 14$.

[L, J 1968, FM VI, Q 8]

Solution. The differential equation is

$$\frac{d^2x}{dt^2}+k\frac{dx}{dt}+64x = 0,$$

x denoting distance, and t time.

(a) $k = 20$. By substituting the trial solution $x = e^{mt}$ into the differential equation, form the auxiliary equation

$$m^2+20m+64 = 0.$$

Solving, $m = -4$ and -16, and so the general solution for x is

$$x = Ae^{-4t}+Be^{-16t}.$$

When $t = 0, x = 0$, so $A+B = 0$ and x has the form

$$x = A(e^{-4t}-e^{-16t}).$$

This makes $\dfrac{dx}{dt} = A(-4e^{-4t}+16e^{-16t})$,

and this equals u when $t = 0$. Thus $u = 12A$,

whence $A = \dfrac{u}{12}$ and $x = \dfrac{u}{12}(e^{-4t}-e^{-16t})$.

When $t = 1$, $x = \dfrac{u}{12}(e^{-4}-e^{-16})$.

(b) $k = 16$. The auxiliary equation is $m^2+16m+64 = 0$ and $m = -8$, repeated.

$$x = (C+Dt)e^{-8t}$$

is the general solution for x. When $t = 0, x = 0$, so $C = 0$ and $x = Dte^{-8t}$, with the constant D still to be determined.

$$\frac{dx}{dt} = D(1-8t)e^{-8t}$$

and this equals u at $t = 0$. Hence $D = u$, so that $x = ute^{-8t}$. When $t = 1, x = ue^{-8}$.

(c) $k = 14$. In this case the auxiliary equation is $m^2+14m+64 = 0$, that is $(m+7)^2+15 = 0$, which has roots $m = -7\pm i\sqrt{15}$, giving the general solution,

$$x = e^{-7t}(E\cos\sqrt{15}t+F\sin\sqrt{15}t).$$

Since $x = 0$ when $t = 0$, $E = 0$ and so x takes the form

$$x = Fe^{-7t}\sin\sqrt{15}t.$$

Hence $\dfrac{dx}{dt} = F(-7\sin\sqrt{15}t+\sqrt{15}\cos\sqrt{15}t)e^{-7t}$.

This is u when $t = 0$, therefore $u = F\sqrt{15}$, so $F = u/\sqrt{15}$, making

$$x = \frac{u}{\sqrt{15}}e^{-7t}\sin\sqrt{15}t.$$

When $t = 1$, $x = \dfrac{ue^{-7}}{\sqrt{15}}\sin\sqrt{15}$.

2

(i) Find the general solution of the equation

$$x\frac{dy}{dx}-y = x.$$

(ii) Solve the equation

$$2(1-xy)\frac{dy}{dx} = y^2,$$

given that $y = 1$ when $x = 0$.

(iii) Use the substitution $z = xy$ to solve the equation

$$x\frac{d^2y}{dx^2}+2\frac{dy}{dx}+xy = 0$$

and find y, given that $y = 1$, $dy/dx = 0$ when $x = \frac{1}{2}\pi$.

[L, S 1967, FM V, Q 7]

Solution.

(i) First convert the differential equation to standard form:

$$\frac{dy}{dx}-\frac{1}{x}y = 1.$$

The integrating factor is

$$e^{\int dx/x} = e^{-\log x} = \frac{1}{x}.$$

Multiplying the differential equation (in standard form) by this integrating factor, we obtain

$$\frac{d}{dx}\left(\frac{1}{x}y\right) = \frac{1}{x},$$

which, on integration, gives

$$y = x(\log x+c),$$

where c is an arbitrary constant.

(ii) Rearranging the terms in the equation it may be seen that

$$2\frac{dy}{dx} = 2xy\frac{dy}{dx}+y^2 \equiv \frac{d}{dx}(xy^2)$$

Integrating, $2y = xy^2+c.$

Since $y = 1$ when $x = 0, c = 2$

and hence $2(y-1) = xy^2,$

i.e. $x = 2\left[\frac{1}{y}-\frac{1}{y^2}\right].$

(iii) The substitution $z = xy$ gives

$$\frac{dz}{dx} = x\frac{dy}{dx}+y, \qquad \frac{d^2z}{dx^2} = x\frac{d^2y}{dx^2}+2\frac{dy}{dx},$$

so the given differential equation is transformed to

$$\frac{d^2z}{dx^2}+z = 0,$$

whose general solution is

$$z = a\cos x+b\sin x$$

i.e. $xy = a\cos x+b\sin x$ or $y = \frac{1}{x}(a\cos x+b\sin x).$

When $x = \frac{1}{2}\pi$, it is given that $y = 1$, therefore $b = \frac{1}{2}\pi$.
Differentiate the expression (xy):

$$x\frac{dy}{dx}+y = -a\sin x+\tfrac{1}{2}\pi\cos x.$$

When $x = \frac{1}{2}\pi$, $dy/dx = 0$ and $y = 1$, thus the last result gives $1 = -a.$

Hence, finally, $xy = \frac{1}{2}\pi\sin x-\cos x$

or $$y = \frac{1}{x}\left(\frac{\pi}{2}\sin x-\cos x\right).$$

3

(i) By means of the substitution $y = x^{-4}z$ reduce the differential equation

$$\frac{dy}{dx}+\frac{4}{x}y = \frac{1}{x^2}\sin x$$

to one in which the variables are separable and solve the equation with the condition that $y = 0$ when $x = \frac{1}{2}\pi$.

(ii) Find the solution of the differential equation

$$\frac{d^2x}{dt^2}+9x = 27t+25\cos 2t+15\sin 2t,$$

with the conditions that when $t = 0, x = dx/dt = 0.$

[O & C, S 1967, M for S, Stats III (S), Q 7]

Solution.

(i) The substitution $y = x^{-4}z$ gives $\frac{dy}{dx} = -4x^{-5}z+x^{-4}\frac{dz}{dx}$,

and the differential equation becomes

$$\left[-4x^{-5}z+x^{-4}\frac{dz}{dx}\right]+4x^{-5}z = \frac{1}{x^2}\sin x.$$

i.e. $\frac{1}{x^4}\frac{dz}{dx} = \frac{1}{x^2}\sin x$

and therefore $dz = x^2\sin x\,dx.$

On integrating by parts,

$$z = -x^2\cos x+\int(\cos x)2x\,dx$$

$$= -x^2\cos x+2\left(x\sin x-\int(\sin x)1\,dx\right)$$

$$= -x^2\cos x+2x\sin x+2\cos x+c.$$

Since $y = 0$ and consequently $z = 0$ for $x = \frac{1}{2}\pi$, we find $c = -\pi$ and hence, finally,

$$y = x^{-4}(-x^2\cos x+2x\sin x+2\cos x-\pi).$$

(ii) Substituting $x = Ae^{mt}$ into the differential equation

$$\frac{d^2x}{dt^2}+9x = 0$$

gives the auxiliary equation $m^2+9 = 0$ and hence $m = \pm 3i$.
Thus the complementary function for the given equation is

$$x = x_c = a\cos 3t+b\sin 3t.$$

The particular integral has the form

$$x = x_p = 3t+A\cos 2t+B\sin 2t,$$

and to find A and B we must satisfy

$$(-4A\cos 2t-4B\sin 2t)+9(A\cos 2t+B\sin 2t) = 25\cos 2t+15\sin 2t.$$

Hence, comparing coefficients of $\cos 2t$, $\sin 2t$, we have

$$5A = 25, \qquad 5B = 15,$$

i.e. $A = 5, \qquad B = 3.$

The complete general solution is

$$x = x_c+x_p = a\cos 3t+b\sin 3t+3t+5\cos 2t+3\sin 2t.$$

In the problem, $x = 0$ when $t = 0$, whence $a = 0$; thus

$$\frac{dx}{dt} = 3b\cos 3t+3-10\sin 2t+6\cos 2t.$$

Since $dx/dt = 0$ when $t = 0$,

$$0 = 3b+3+6,$$

i.e. $b = -3.$

Thus finally $x = 3(t-\sin 3t+\sin 2t)+5\cos 2t.$

4 If a conic has foci at the points $(c, 0)$ and $(-c, 0)$, prove that it satisfies the equation

$$\left[x\frac{dy}{dx}-y\right]\left[x+y\frac{dy}{dx}\right] = c^2\frac{dy}{dx}.$$

If two such conics intersect, prove that they intersect at right angles. [O & C, S 1966, M & HM V (PS), Q 6]

Solution.
The equation of the conic is of the form

$$\frac{x^2}{a^2}+\frac{y^2}{a^2(1-e^2)} = 1.$$

It is given essentially that $ae = c$ ($e \equiv$ eccentricity), so

$$\frac{x^2}{1}+\frac{y^2}{1-e^2} = \frac{c^2}{e^2} \qquad \textbf{9.1}$$

and differentiating equation **9.1** we obtain:

$$2x+\frac{2y\,dy/dx}{1-e^2} = 0. \qquad \mathbf{9.2}$$

On eliminating e^2 from equations **9.1** and **9.2**,

$$x^2+\frac{y^2}{(-y/x)\,dy/dx} = \frac{c^2}{1+(y/x)\,dy/dx},$$

$$\left[1+\frac{y}{x}\frac{dy}{dx}\right]\left[x^2\frac{dy}{dx}-xy\right] = c^2\frac{dy}{dx}$$

or $$\left[x\frac{dy}{dx}-y\right]\left[x+y\frac{dy}{dx}\right] = c^2\frac{dy}{dx}.$$

Let respective slopes (i.e. dy/dx) of two intersecting conics be p_1 and p_2.

Then $c^2 = \frac{1}{p_1}(xp_1-y)(x+yp_1) = \frac{1}{p_2}(xp_2-y)(x+yp_2)$,

i.e. $p_2(xp_1-y)(x+yp_1) = p_1(xp_2-y)(x+yp_2)$

and, on simplifying,

$$xy(p_1-p_2)(1+p_1p_2) = 0.$$

Thus $p_1p_2 = -1$ or $\left[\frac{dy}{dx}\right]_1\left[\frac{dy}{dx}\right]_2 = -1,$

proving that the tangents and therefore the conics intersect at right angles.

C Differential equations: Problems

Answers and hints will be found on pp. 125–7.

1

(i) Find the general solutions of the differential equations

(a) $(x+2)\frac{dy}{dx}+1+y-x^2-x^2y = 0,$

(b) $\frac{dy}{dx} = \frac{\sec x-y}{\sin x}.$

(ii) A body takes 20 minutes to cool down from 80 °C to 50 °C. If the room temperature is constant at 10 °C and the rate of cooling of the body is proportional to the difference between its temperature and that of the room, find the time taken by the body to cool down from 50 °C to 30 °C.

[L, J 1968, FM VI, Q 7]

2 A particle falls from rest under gravity, the resistance of the air being kv^2 per unit mass, where v is the speed of the particle and k is a constant. Show that v has a limiting value c and find the speed acquired in falling a distance x in terms of c and x.

If the particle is projected vertically upwards with the speed c, show that it will attain a height $(c^2/2g)\log 2$.

[L, S 1967, FM V, Q 8]

3

(i) Use the substitution $x+y = u$ to transform the differential equation

$$(x+2y+2)\,dx+x(x+y)\,dy = 0$$

into another containing variables x and u only.

Hence obtain the solution in terms of x and y.

(ii) Solve the differential equation

$$x(1+x^2)\frac{dy}{dx}+(1-x^2)y = x^2(1+x^2)^2\log x,$$

given that, when $x = 2$, $dy/dx = 0$.

[L, J 1967, FM VI, Q 8]

4

(i) The differential equation

$$\frac{d^2y}{dx^2}+a\frac{dy}{dx}+by = 0,$$

where a and b are constants, is satisfied by

$$y = e^{-2x}(2\cos 3x+3\sin 3x)$$

for all values of x. Find a and b.

Show that the general solution of the differential equation is $y = e^{-2x}(A\cos 3x+B\sin 3x)$.

(ii) By substituting $y = p_0+p_1x+p_2x^2$, where p_0, p_1 and p_2 are constants, in

$$\frac{d^2y}{dx^2}+5\frac{dy}{dx}+4y = 4x^2,$$

find a solution of the differential equation.

Let this solution be $f(x)$. By putting $y = f(x)+v$, where v is a function of x, find the general solution of the differential equation.

[L, S 1966, F M VI, Q 7]

5

(i) By substituting $y = vx$, where v is a function of x, solve the differential equation

$$(x^2-y^2)+5xy\frac{dy}{dx} = 0.$$

(ii) If $x = e^t$, prove that

$$x^2\frac{d^2y}{dx^2}+x\frac{dy}{dx} = \frac{d^2y}{dt^2}.$$

Find the general solution of

$$x^2\frac{d^2y}{dx^2}+3x\frac{dy}{dx}+ky = 0$$

in terms of real functions of x in each of the cases

(a) $k < 1$, (b) $k = 1$, (c) $k > 1$.

[L, J 1966, FM VI, Q 8]

6

A chemical X is decomposing into two chemicals Y and Z. The total mass of X, Y and Z remains constant throughout the decomposition and, when the mass of X remaining is x g, the rates of information of Y and Z are qx and rx g/s respectively, where q and r are constants. Initially the mass of X present is A g and the masses of Y and Z present are zero. At time T s from this instant the mass of X present is a g. Prove that

$$q+r = \frac{1}{T}\log_e\frac{A}{a},$$

and that the mass of X left at t s after the start is

$$A\left[\frac{a}{A}\right]^{t/T}\text{ g.}$$

[O, S 1967, M II, Q 10]

7

(i) Find the general solution of the differential equation

$$x\frac{dy}{dx}-y = x^2e^{-x}.$$

(ii) Find the general solution of the differential equation

$$\frac{d^2y}{dx^2}-4y = 3e^{-x}.$$

[O, S 1967, FM 1 & P II, Q 12]

8

(i) Solve the equation

$$\frac{d^2y}{dx^2}+4\frac{dy}{dx} = 8,$$

given that, when $x = 0, y = dy/dx = 0$.

(ii) The tangent at a variable point P on a curve meets the axis of x at T, the normal at P meets it at N. If PN is constant, prove that the curve is a circle.

If, on the other hand, PT is a constant k, show that the equation of the curve can be expressed in the parametric form

$$y = k \sin \psi, \qquad x = k(\log \tan \tfrac{1}{2}\psi + \cos \psi).$$

[C, S 1967, P III, Q 11]

9

(i) Solve the differential equation

$$\frac{dy}{dx} = e^{x+2y}.$$

(ii) Find the general solution of the differential equation

$$\frac{d^2y}{dx^2}+y = x+2 \sin x$$

and the particular solution if $y = dy/dx = 1$ when $x = 0$.

[C, S 1967, P S, Q 11]

10

Verify that $y = \cos x$ satisfies the equation

$$\frac{d^2y}{dx^2}+y = 0.$$

If the general solution of the equation is written $z \cos x$, show that

$$\frac{d^2z}{dx^2}\cos x-2\frac{dz}{dx}\sin x = 0.$$

By integrating this equation prove that its general solution is $z = A \tan x+B$ where A and B are arbitrary constants. Hence deduce that the general solution of the original equation is $y = A \sin x+B \cos x$.

[O & C, S 1967, M & HM II, Q 10]

11

A family of parabolas is formed by shifting the parabola $y^2 = 4ax$ through any distance parallel to the y-axis. Write down the equation of the family (in terms of x, y and a parameter λ), and find the differential equation (not containing λ) satisfied by the family.

By replacing dy/dx by $-dx/dy$ and solving the resulting equation, show that a curve which cuts the parabolas at right angles at one (at least) of its points of intersection with them has an equation of the form

$$9a(y-c)^2 = 4x^3.$$

Find the (unique) value of c for the particular member of this system of curves which cuts the parabola $y^2 = 4ax$ at right angles at the point $(a, 2a)$.

[O & C, S 1967, M & HM V, Q 8]

12

(i) By Newton's law of cooling the rate of change at time t of the surface temperature T of a sphere in an atmosphere of constant temperature T_0 is proportional to the difference of these temperatures. Form a differential equation for T and show that, if $T = 3T_0$ when $t = 0$ and $T = 2T_0$ when $t = 1$,

$$T = T_0(1+2^{1-t}).$$

(ii) It is found that when quantities of two liquids A and B are boiling in the same container the ratio of the rates of vaporization of the two liquids being vaporized at any instant is a fixed multiple of the ratio of the amounts remaining. If x and y are the respective amounts of A and B remaining at any instant show that $b^kx = ay^k$, where k is a constant and a and b are the initial values of x and y.

[O & C, S 1967, M for Sc III, Q 6]

13

(i) If $y = e^{-x}v(x)$, show that

$$\frac{dy}{dx} = e^{-x}\left[\frac{dv}{dx}-v\right],$$

and find an expression for d^2y/dx^2.

The function y is known to satisfy the differential equation

$$\frac{d^2y}{dx^2}+2\frac{dy}{dx}+5y = e^{-x}.$$

Use your results to show that the function v satisfies the differential equation

$$\frac{d^2v}{dx^2}+4v = 1.$$

Solve this equation for v, and hence obtain the general solution for y.

(ii) Solve the differential equation

$$\frac{dy}{dx}+\frac{x^2+\sin^2 y}{1+x^2} = 1.$$

[O & C, S 1966, M for Sc I, Q 11]

14 The rate of installing washing machines in a country is proportional to the number of households in the country which do not possess washing machines; the rate of scrapping of installed washing machines is proportional to the total number W remaining installed. No household has more than one washing machine installed. If H is the number of households in the country, prove that

$$\frac{dW}{dt} = kH-(k+k_1)W,$$

where k, k_1 are positive constants.

The number of households is growing at a steady rate, so that $H = a+bt$, where a, b are positive. Solve for W in terms of a, b, t, k, k_1 and an arbitrary constant c by finding constants α, β and μ such that

$$^{\mu}W = \alpha+\beta t+ce^{\mu t}$$

satisfies the differential equation. Find the proportion of households ultimately possessing washing machines.

[O & C, S 1966, M for Sc II, Q 10]

15 The differential equation of a family of curves is given by

$$\frac{dy}{dx}+2y = e^x.$$

Prove that the equation of the curve of the family passing through the point (0, 0) is $3y = e^x - e^{-2x}$. The area bounded by this curve, the x-axis and the ordinate $x = \log_e 2$ is rotated through four right angles about the x-axis. Calculate the volume of the solid of revolution generated.

[AEB, S 1968, P & A I, Q 5]

16

(a) Find the solution of the equation

$$\frac{dy}{dx}+2xy = 2x^3$$

for which $y = 1$ when $x = 0$.

(b) Use the substitution $y = z-x$ to find the general solution of the equation

$$\frac{dy}{dx} = \cos(x+y).$$

[JMB, S 1967, P II, Q 8]

17

(a) Use the substitution $y = vx$ to transform the equation

$$x^2\frac{dy}{dx}+y^2-(\alpha+\beta+1)xy+\alpha\beta x^2 = 0,$$

where α and β are different constants, into an equation involving v, dv/dx and x. Hence solve the equation.

(b) A particle moves along the x-axis with velocity v; show that its acceleration is $v\,dv/dx$. If the acceleration is $-k(v^3+\lambda^2 v)$, where k and λ are positive constants, and if the particle is projected from the origin with velocity $u(> 0)$, show that v becomes zero for a value of x not exceeding $\pi/(2k\lambda)$, however great the value of u.

[JMB, S 1967, P S, Q 8]

18

(a) Solve the equation

$$\frac{dy}{dx}+2y\tan x = \sin x$$

with the condition that $y = 0$ when $x = 0$.

(b) A variable x, which is a function of t, satisfies the differential equation

$$\frac{dx}{dt} = kx(a-x),$$

where k and a are positive constants. Given that $x = \frac{1}{5}a$ at $t = 0$ and $x = \frac{2}{3}a$ at $t = 3$, find the value of ka. Find also the value of x when $t = 5$.

[JMB, S 1966, P II, Q 9]

19 The equation

$$\frac{x^2}{a^2+\lambda}+\frac{y^2}{b^2+\lambda} = 1$$

represents a family of curves as λ varies. Show that through any point P of the plane there pass at most two curves of the family.

Prove that, for such a curve,

$$\left[\frac{x}{y}+\frac{dy}{dx}\right]\left[\frac{x}{y}-\frac{dx}{dy}\right] = \frac{a^2-b^2}{y^2},$$

and deduce that the two curves through P intersect there at right angles.

[W, S 1969, P S, Q 4]

D Differential equations: Answers and hints

1

(i)

(a) The differential equation can be written in the form

$$\frac{1}{1+y}\frac{dy}{dx}+\frac{1-x^2}{x+2} = 0,$$

integration of which gives

$$\log(1+y)-3\log(x+2)-\tfrac{1}{2}x^2+2x = \text{constant}.$$

(b) The differential equation is a linear form,

$$\frac{dy}{dx}+(\operatorname{cosec} x)y = \frac{1}{\sin x\cos x}.$$

The integrating factor is $\operatorname{cosec} x-\cot x$, and the equation will integrate to give

$$y(\operatorname{cosec} x-\cot x) = \log(\sec x+\tan x)-\tfrac{1}{2}\tan\tfrac{1}{2}x+\text{constant}.$$

(ii) The law of cooling is $-d\theta/dt = k(\theta-10)$, integrating to $\theta-10 = ae^{-kt}$. The conditions $t = 0, \theta = 80°$; $t = 20$, $\theta = 50°$ show that $a = 70, k = \frac{1}{20}\log\frac{7}{4}$. Then when $\theta = 30°$,

$$t = 20\frac{\log_e\frac{7}{2}}{\log_e\frac{7}{4}}.$$

2 For the downward motion,

$$v\frac{dv}{dx} = g-kv^2.$$

Integration yields

$$v^2 = \frac{g}{k}(1-e^{-2kx}),$$

after using the condition that $v = 0$ initially. Show from this that $v \to \sqrt{(g/k)} = c$, and hence show that $v = c(1-e^{-2gx/c^2})^{\frac{1}{2}}$. In the upward motion under the differential equation

$$v\frac{dv}{dx} = -(g+kv^2),$$

obtain

$$\log\left(\frac{g+kv^2}{g+kc^2}\right) = -2kx,$$

on using that $v = c$ initially, and deduce that

$$-\frac{2g}{c^2}x = \log\frac{1}{2}\left(1+\frac{v^2}{c^2}\right).$$

Put $v = 0$ to show that the maximum height is

$$\frac{c^2}{2g}\log 2.$$

3

(i) $y = u-x$, $dy = du-dx$ and the given equation is transformed into

$$\frac{du}{dx} = \frac{(u+1)(x-2)}{ux}.$$

Integrate by separating the variables and show that

$$y = \log\frac{x+y+1}{x^2}+c.$$

(ii) Write the differential equation in standard linear form, and show that it then has an integrating factor of $x/(x^2+1)$. Show that the general solution is

$$y = (x^2+1)\left[\frac{x^2}{3}\log x - \frac{x^2}{9} + \frac{c}{x}\right].$$

Use the condition that when $x = 2$, $dy/dx = 0$ to find that, here, $c = 8(\frac{1}{9} - 2\log 2)$.

4

(i) Substitute the expression for y into the differential equation. From the identity obtained, find two simultaneous equations for a, b and hence show that $a = 4$ and $b = 13$. The roots of the auxiliary equation become $m = -2 \pm 3i$, and this gives the general solution as quoted.

(ii) Show that $p_0 = \frac{21}{8}$, $p_1 = -\frac{5}{2}$ and $p_2 = 1$. Then using that $f'' + 5f' + 4f = 4x^2$, show that the substitution $y = f(x) + v$ transforms the differential equation to $v'' + 5v' + 4v = 0$. Hence show that $v = ae^{-x} + be^{-4x}$. The general solution is then the sum of $f(x)$ and v,

i.e. $y = ae^{-x} + be^{-4x} + \frac{21}{8} - \frac{5}{2}x + x^2$.

5

(i) Show that

$$x\frac{dv}{dx} = -\frac{4v^2+1}{5v},$$

and hence that $x^2(1+4v^2)^{\frac{5}{4}}$ = constant or $Ax^2 = (x^2+4y^2)^5$.

(ii) The substitution gives the differential equation

$$\frac{d^2y}{dt^2} + 2\frac{dy}{dt} + ky = 0.$$

The auxiliary quadratic equation has roots $m = -1 \pm \sqrt{(1-k)}$.

If $k < 1$, $y = \frac{1}{x}(ax^n + bx^{-n})$ $(n = +\sqrt{(1-k)})$.

If $k = 1$, $y = \frac{1}{x}(c + d\log x)$.

If $k > 1$, $y = \frac{1}{x}\{E\cos(n\log x) + F\sin(n\log x)\}$

$(n = +\sqrt{(k-1)})$.

6 As Rate of change of X + rate of formation of Y and Z = 0,

we obtain $\frac{dx}{dt} + (q+r)x = 0$.

On integrating by separation of variables and using the fact that at $t = 0$, $x = A$, we find given expressions.

7

(i) This linear equation has solution $y = x(c - e^{-x})$.

(ii) The general solution of the second order equation is $y = ae^{2x} + be^{-2x} - e^{-x}$.

8

(i) Show first that $y' + 4y = 8x$, and hence that $y = 2x - \frac{1}{2} + \frac{1}{2}e^{-4x}$.

(ii) If P has coordinates (x, y), show that when PN $= a$,

$$a^2 = y^2\left[1 + \left(\frac{dy}{dx}\right)^2\right].$$

Solve for $y\,dy/dx$ and integrate to show that $-\sqrt{(a^2-y^2)} = x + c$, giving the equation of a circle. Given PT $= k$ instead, show that

$$y\frac{dx}{dy} = \sqrt{(k^2-y^2)}.$$

Make the substitution $y = k\sin\psi$ and show that

$$x = k\int\frac{\cos^2\psi}{\sin\psi}\,d\psi$$

and hence obtain the given answer for x.

9

(i) The solution is $e^x = -\frac{1}{2}e^{-2y} + c$.

(ii) The general solution is the sum of the c.f. $y_c = a\cos x + b\sin x$ and the p.i. $y_p = x - x\cos x$. The particular solution required will be found to have $a = 1$ and $b = 1$.

10 The first two parts are straightforward. In the last part, show, by putting $dz/dx = p$, that integration yields $p\cos^2 x = A$. A second integration gives z and hence y.

11 The family of parabolas has equation $(y-\lambda)^2 = 4ax$, and the differential equation of this family is $x(dy/dx)^2 = a$. The orthogonal family is characterized by $a(dy/dx)^2 = x$. The value of c for which $9a(y-c)^2 = 4x^3$ cuts $y^2 = 4ax$ at right angles at $(a, 2a)$ is $c = \frac{8}{3}a$.

12

(i) The differential equation is $dT/dt = -k(T - T_0)$ and is solved by separating the variables.

(ii) The ratio of the rates of vaporization of the liquids is $-dx/dt : -dy/dt \equiv dx/dy = kx/y$, given. The solution is completed by separating the variables and integrating.

13

(i) $y'' = e^{-x}(v'' - 2v' + v)$ The general solution of $v'' + 4v = 1$ is $v = a\cos 2x + b\sin 2x + \frac{1}{4}$, whence that of y is $y = e^{-x}v$.

(ii) Simplify the differential equation to

$$\frac{dy}{dx} = \frac{\cos^2 y}{1+x^2}$$

Integration gives $x = \tan(c + \tan y)$.

14 The proof is straightforward. Substitute the expressions for H and W into the differential equation. From the identity obtained, show that

$$\mu = -(k+k_1), \qquad \beta = \frac{k}{k+k_1}b, \qquad \alpha = \frac{k\{a(k+k_1)-b\}}{(k+k_1)^2}.$$

The proportion of households ultimately possessing washing machines is

$$\lim_{t\to\infty}\frac{W}{H} = \lim_{t\to\infty}\frac{\alpha + \beta t + ce^{\mu t}}{a+bt} = \frac{\beta}{b} = \frac{k}{k+k_1}.$$

15 The integrating factor is e^{2x},

thus $\frac{d}{dx}(ye^{2x}) = e^x e^{2x}$.

On integrating, $ye^{2x} = \frac{1}{3}e^{3x} + c$,

but as curve passes through $(0, 0)$ $c = -\frac{1}{3}$, hence

$ye^{2x} = \frac{1}{3}(e^{3x} - 1)$ or $3y = e^x - e^{-2x}$.

$$\text{Volume generated} = \int_0^{\log 2} \pi y^2\, dx$$

$$= \int_0^{\log 2} \tfrac{1}{9}\pi(e^x - e^{-2x})^2\, dx$$

$$= \tfrac{1}{9}\pi\left[\tfrac{1}{2}e^{2x} + 2e^{-x} - \tfrac{1}{4}e^{-4x}\right]_0^{\log 2}$$
$$= \tfrac{47}{576}\pi.$$

16

(a) $y = x^2 - 1 + 2e^{-x^2}$.

(b) General solution is $\tan\frac{1}{2}(x+y) = x + \text{constant}$.

17

(a) Transformed equation is

$$x\frac{dv}{dx} + (v-\alpha)(v-\beta) = 0.$$

Solution is $x^{\alpha-\beta}\left[\dfrac{y-\alpha x}{y-\beta x}\right] = \text{a constant}.$

(b) Result follows from general solution:

$$\tan^{-1}\frac{v}{\lambda} = \tan^{-1}\frac{u}{\lambda} - \lambda kx.$$

18

(a) $y = \cos x(1-\cos x)$.

(b) $ka = \log_e 2$; $x = -\frac{8}{9}a$ when $t = 5$.

19 Take point P as (h, k). Then values of λ for curves through P must satisfy

$$(b^2+\lambda)h^2 + (a^2+\lambda)k^2 = (a^2+\lambda)(b^2+\lambda),$$

that is, a quadratic in λ which at the most can have only two real roots. Hence, at the most, only two curves can pass through P.

Let $a' = (a^2+\lambda)^{-1}$, $b' = (b^2+\lambda)^{-1}$ then equation of curve becomes $a'x^2 + b'y^2 = 1$. On differentiation we find:

$$\frac{dy}{dx} = -\frac{a'x}{b'y},$$

hence $$\frac{x}{y} + \frac{dy}{dx} = \frac{x}{y}\left[1 - \frac{a'}{b'}\right]$$

and $$\frac{x}{y} - \frac{dx}{dy} = \frac{x}{y}\left(1 + \frac{b'y^2}{a'x^2}\right) = \frac{x}{y}\frac{1}{a'x^2} = \frac{1}{a'xy},$$

thus $$\left[\frac{x}{y} + \frac{dy}{dx}\right]\left[\frac{x}{y} - \frac{dx}{dy}\right] = \frac{1}{a'y^2}\left[1 - \frac{a'}{b'}\right] = \frac{a^2-b^2}{y^2}.$$

Let slopes of two curves intersecting at P be y_1', y_2', then

$$\frac{a^2-b^2}{k^2} = \left[\frac{h}{k} + y_1'\right]\left[\frac{h}{k} - \frac{1}{y_1'}\right] = \left[\frac{h}{k} + y_2'\right]\left[\frac{h}{k} - \frac{1}{y_2'}\right],$$

from which we find $(1 + y_1'y_2')(y_1' - y_2') = 0$. Hence $y_1'y_2' = -1$, showing that the two curves intersect at right angles.

Index

Angles between planes and lines, 78
Apollonius' circle, 57
Approximate integration, 107
Arc length, 110
Area of sector in polar coordinates, 111
Area of triangle, 26
Area under curve, 90
Argument of a complex number, 29

Binomial series, 9

Centroid (centre of gravity), 91
Ceva's theorem, 71
Circles, 57
Circumcircle, 64
Cofactors of a determinant, 37
Combinations, 45
Complex numbers, 29
Compound angles, 26
Cone, 79
Conics, 58–60
Consistency of equations, 40
Convergence of series, 47
Cosine rule, 26
Cramer's rule, 40
Cross-ratio, 70
Cubic equations, 10
Curve sketching, 85

D'Alembert's ratio test, 47
De Moivre's theorem, 30
Derivatives, rules, tables, 84–5
Desargue's theorem (perspective triangles), 72
Descarte's rule of signs, 11
Determinants, 39
Differential equations, 119–20
Differential, total, 104
Differentiation, 84–5; of vectors, 78

Equations, algebraic, 10–11; trigonometric, 26
Euler line, 64
Euler relations, 30
Excircle, 64

Families of solutions of a differential equation, 199
Feuerbach's theorem, 64

General conic, 70
Guldin's theorems, 111

Harmonic ranges, 70
Homogeneous linear equations, 40
Hyperbolic functions, 30

Incircle, 64
Increments, 84, 104
Indices, 9
Induction, 10
Inequalities, 11; associated with integrals, 107
Inflexion points, 85
Integration rules, tables, 89–90
Inverse hyperbolic functions, 31
Inverse trigonometric functions, 26
Inversion, 72

Leibnitz's alternating series test, 47; theorem for nth derivative, 104
L'Hopital's rule, 104
Linear differential equations, 119
Linear equations, 40
Line-pair, 70
Logarithms, 9

Maclaurin-Cauchy integral test, 47
Maclaurin's series, 26, 104
Matrices, 39–40
Maxima, 85
Mean value, 91
Menelaus' theorem, 72
Minima, 85
Minor of a determinant, 39
Modulus of a complex number, 29
Multiple angles, 26

Newton's approximation, 11
Nine-point circle, 64
Non-homogenous linear equations, 41
Normals, 85

Pappus–Guldin theorems, 111
Partial differentiation, 104
Partial fractions, 9
Pedal triangle, 64
Pencils of conics, 70; harmonic pencils, 70
Permutations, 44
Perspective triangles, 72
Polar coordinates, 111
Pole and polar, 71
Polynomials, 11
Power series, 104
Probability, 45

Quadratic equations, 10
Quadrilateral and quadrangle, complete, 71

Radians, 25
Radius of curvature, 111
Rank of a matrix, 40
Rates of change, 85
Recurrence relations in sequences, 10
Reduction formulae, 106–7
Remainder theorem, 10
Root mean square (r.m.s.) value, 91
Roots, quadratic, 10; cubic, 11; complex, 30
Rotation of axes, 70

Scalar product of vectors, 77
Separation of variables, 119
Series, logarithmic and exponential, 9; finite, 10; for hyperbolic functions, 30
Simpson's rule, 108
Sine rule, 26
Sphere, 79
Straight lines, 55–7
Substitutions in integrals, 90

Tangents, 85
Taylor's series, 104
Tetrahedron, 79
Translation of axes, 70
Trapezium rule, 107
Trigonometrical ratios, 25

Vector, equation of straight line and plane, 77
Vectors, algebra of, 76–7
Volumes of revolution, 91